Renaturing Urban Rivers

Opportunities and Challenges of Involving Riverine Communities

Editor

Carlos Smaniotto Costa

Universidade Lusófona, Department of Architecture and Urban Planning and Interdisciplinary Research Centre for Architecture, Urbanism and Design (University of Lisbon), Lisboa, Portugal.

CRC Press
Taylor & Francis Group
Boca Raton London New York

CRC Press is an imprint of the
Taylor & Francis Group, an **informa** business
A SCIENCE PUBLISHERS BOOK

Cover credit: Images reproduced by kind permission of Emma Veronese, Edson Lourenço, Stefano Tornieri, Anouk Bélanger, Maxim Bonin, Luciana Bragança and Carlos Smaniotto Costa.

First edition published 2027
by CRC Press
2385 NW Executive Center Drive, Suite 320, Boca Raton FL 33431

and by CRC Press
4 Park Square, Milton Park, Abingdon, Oxon, OX14 4RN

CRC Press is an imprint of Taylor & Francis Group, LLC

Library of Congress Cataloging-in-Publication Data (applied for)

ISBN: 978-1-032-68869-5 (hbk)
ISBN: 978-1-032-71174-4 (pbk)
ISBN: 978-1-032-71173-7 (ebk)

DOI: 10.1201/9781032711737

Typeset in Times New Roman
by Prime Publishing Services

Preface

Let us travel throughout the world, from Europe to the Americas, Africa, and Asia, accompanying different efforts to integrate rivers into a healthier and regenerative urban life. The book *Renaturing Urban Rivers: Opportunities and Challenges of Involving Riverine Communities* brings together thirteen examples of community engagement in river restoration, rivers that under urbanisation pressures call for rethinking the way ecosystems are being treated. Each of these cases, by exploiting, discussing, and analysing the relationship between the rivers and the riverine communities, reveals unique perspectives.

Restoring and renaturing rivers is a multi-front task. It involves restoring green and blue infrastructure to regenerate urban environments, to mitigate climate change, improve water quality, enhance biodiversity, and create public open spaces. Collaborative processes and interaction patterns show that river restoration for a long-term success and to ensure benefits align with social needs require extensive community involvement. This is exactly where the book *Renaturing Urban Rivers* sets its objectives: to better understand the opportunities and challenges of involving riverine communities in restoration efforts.

This book focuses on modest reclamation efforts currently underway worldwide, modest because they are examples of place-based use of creative tools and ways to engage with different population groups across their communities. These cases are not ostentatious but of an attention-grabbing nature for their impact, knowledge advancement, and engagement results. This implies efforts to prevent the forced displacement of communities whose livelihoods and culture are tied to rivers and riverine ecosystems. They are a response to a call to bring experiences to the table in river restoration that, besides increasing the understanding of the problems, also provide valuable insights on the role of the communities in processes. These cases contribute thus to the nexus between urban development, rivers, and the engagement of riverine communities. Putting these cases in the same light aims to buttress the urban and environmental debate by providing a platform for sharing experiences, knowledge, and systematic identification of qualitative characteristics to provide a better understanding of causal relationships between these three nexuses.

By analysing urban design, ecological improvements, and social benefits, *Renaturing Urban Rivers* seeks to highlight the importance of bringing more nature back to an increasingly urbanised world and of improving the ability to mitigate hazards and reduce risks - all with the engagement of the riverine communities. Rivers and renaturing are used as umbrella concepts that encompass different kinds of watercourses within an urban environment and the efforts to make these watercourses transit from degraded, polluted, and sometimes invisible into active or vibrant ones, where nature becomes a place in the urban environment. The cases show that restoring and renaturing rivers is complex and far-reaching. Riverine communities build on strong ties to river, on the one side, but on the other side, they view rivers as worthless consumer goods.

Renaturing Urban Rivers is an insightful book. It presents and reviews advantages, benefits, difficulties and challenges faced by river restoration along with providing insights on environmental and social impact. The lack of space for rivers and the constant competition for urban land make it harder to apply the concept of river restoration and renaturing in their whole extent. Providing spaces for rivers means enabling natural processes to allow rivers room to breathe, restore their natural functions, integrate with their surroundings and create spaces for recreation and human-nature interaction. River restoration encompasses a comprehensive and synergistic array of actions and issues (i.e., legal, administrative, spatial) towards facilitating a closer (re)connection between the river and its *territory*. In urban environments this connection is particularly affected by the ties with the city and riverine communities. Restoring rivers' natural processes and, consequently, their functional attributes (geomorphology, freshwater provision for humans and ecosystems, offering habitats and supporting biodiversity, etc.) is a never-ending task. Different stakeholders have a say in the processes of land take and river restoration, as witnessed by the cases, and among them the academic disciplines play a crucial role in shaping the knowledge production and dissemination. With the objective of creating dynamic habitats (for humans and non-humans) and contributing to healthy, interconnected river-city systems, the cases show how riverine communities play a vital role in the process, be it through the indigenous and local knowledge, participation in decision-making, or communities' ability to foster stewardship, issues that are crucial for sustainable restoration efforts. Revitalising and renaturing rivers offers a variety of ecological and social and economic benefits, which would be irreplaceably lost.

Renaturing rivers is a multifaceted concept involving different issues. The chapters offer diverse insights and discuss how they contribute to, and impact urban development practices. In Chapter 1, Eliana do Pilar Rocha and Carlos Smaniotto reflect on the social, cultural and structural role of a river that flows through a low-income neighbourhood in the city of Curitiba (Brazil). Their research provides an overview of the river restoration and its integration as the backbone of a linear park into the urban fabric. Aleksandra Djukić, Jelena Marić, Nikola Mitrović and Emilija Jović, in their chapter, provide an analysis of riverfront megaprojects. Focusing on the development in the riverfronts of Belgrade, they reveal how such projects too often lack transparency and public inclusion and prioritise profit over community needs. Ana Vera et al. explore how stakeholders contribute to urban river recovery. Their chapter addresses a stakeholder mapping approach to examine and comprehend stakeholder engagement within the case studies provided by the RUN Project, a collaborative effort across eight Ibero-American countries. Stefano Tornieri and Agatino Rizzo introduce an inventory of landscape elements as an alternative to resource extraction fuelled in Arctic villages along the Torne River in Sweden. They explore the relationships between fishing techniques, seasonal practices, and the built environment, demonstrating how these elements collectively shape the community's identity and heritage. Anouk Bélanger, Maxim Bonin and Xavier Massicotte discuss the informal use of the shores of the St Laurent River in Montreal. A long process of industrial presence and privatisation of waterfront land have restricted the public access to riverbanks. The emerging water sport subculture and lifestyles are raising awareness and interest in urban blue spaces. Piotr Sikorski and Kinga Kimic examine ecosystem services provided by the Vistula River in Warsaw. Using three spatial levels (from city and district down to a site scale), their analysis focuses on greenspaces and their natural and social values and how these are perceived by the inhabitants. This analysis is the base for a discussion on nature-based solutions to improve ecosystem services in the city. Sagarika Nayak, Nazish Abid, Somi Sareen and Sameen Fatma delve into a complex system of use, pollution and shaping culture around the Ganga River in Kanpur, India. Their chapter emphasises the importance of sustainable urban planning and nature-based solutions to heal the river without losing its cultural essence. It also highlights that the real power in river regeneration lies in the hands of local communities. Marluci Menezes, Juliana Valentim Harayashiki and Silvia Mikami Pina study how walking can be a sensitive spatial practice to activate memory and foster the reconnection of communities with water landscapes. Considering cases in Lisbon, Yokohama, and Santos, they discuss how walking along waterbodies can be used to recover memories and stimulate the generation of urban narratives, creating thus water memoryscapes. Joaquín Cochero et al. explore the amalgamation of digital advancements and citizen science initiatives to assist in urban river recovery. They discuss how different projects enable participatory data collection, collaborative analysis, and policy influence by integrating local communities, digital technologies, and scientific methods in river monitoring, ecological change recording, and environmental awareness raising. Nagayamma Aragão, Luciana Souza Bragança and Alana Costa de Oliveira take the challenge to address the complex issue of land tenure regulation along the basin of the Arrudas River (Belo Horizonte). Their research shows how difficult it is to work in an area with socio-environmental conflicts, particularly when top-down approaches fostered mistrust and disregarded local social and environmental sensitivities. Edson Kiluanje Paulino Lourenço and Marluci Menezes address the uses by the population of the Cambamba River in Luanda. A river that faces pronounced pollution problems, with large amounts of garbage and open sewers. By analysing socio-spatial representations, they approach the river to show how it is being socially used. Bahanur Nasya et al., by using the *New European Bauhaus Impact Model*, assess stakeholder dynamics and local sustainability practices across the Savio River in Cesena. This analysis is meant to support Cesena's goal to become a climate-neutral city, ensuring long-term ecological and socio-economic resilience. Backed by the premise that anthropocentrism must be overcome and on the hypothesis of multispecies, Luciana Souza Bragança and Marcela Silviano Brandão discuss the game *Inventa(rio) Cidades* and the results in raising awareness for multispecies approach in urban rivers restoration. Understood as a dialogical tool and a playful approach to riverine communities, the authors analyse the shaping of environmental consciousness, while working on territories with conflicts. According to them, broadening the dialogue between stakeholders through games allow users to experiment and learn in a safe and motivating environment.

In the production of this book, several people have helped greatly, my warm thanks go to them all. Firstly, all the authors who trusted their work, experiences, and insights to be shared through this platform. Secondly, the authors who helped with the peer review process, ensuring high standards in scientific communication. Thirdly, the editors of Taylor & Francis for the trust placed in sharing experiences of engaging communities with their rivers. Organised under the umbrella of the Cyted Network RUN (Renaturalised Urban Rivers), the book *Renaturing Urban Rivers* and its chapters are situated at the crossroads of two global issues: watercourses and cities, which should not be understood as exclusive but integrative. The addressed insights serve as a valuable resource for those seeking deeper understanding of river ecosystem restoration and also towards providing valuable spaces for recreation and human–nature interaction. Renaturing urban rivers thus means learning to live with rivers and consequently adopting better urban development practices and establishing creative partnerships with riverine communities.

Carlos Smaniotto Costa
Lisbon, July 2025

Contents

1

Rivers as Administrative Boundaries and an Approach to Reading a Fluvial Landscape

The Case of the Cajuru Linear Park

Eliana do Pilar Rocha[1,*] and *Carlos Smaniotto Costa*[2]

1. Introduction

Urban growth, from a global perspective, has a significant impact on the environment; however, an unsustainable path is provoking a harmful imbalance between the natural and urban ecosystems. Different cities worldwide are making efforts to bring nature back to their urban landscapes, also as a means to guarantee economic prosperity coupled with ecological sustainability. Hermida et al. (2021) describe sustainability as a paradigm that should ensure the quality of life of current and future generations through citizen participatory processes. It is a synergistic balancing that integrates economic development with environmental, social and cultural equity. In the authors' terms, a city is a complex system with different patterns and layers that arise from the people's interactions with the environment. The functionality and balance of these socio-environmental processes must be based on both social and environmental well-being.

[1] Centro Universitário Claretiano, Geography Course Coordination, Urban Socio-Environmental Studies Group, Curitiba, Brazil.
[2] Universidade Lusófona, Department of Architecture and Urban Planning, and Interdisciplinary Research Center for Architecture, Urbanism and Design (CIAUD), Lisbon, Portugal. smaniotto. Eamil: costa@ulusofona.pt, 0000-0002-1896-4663
* Corresponding author: elianarocha@claretiano.edu.br, 000-0002-4397-7159

DOI: 10.1201/9781032711737-1

In recent decades, different social and environmental projects have been on the agenda in Brazil—both in election campaigns and in local action plans. This chapter focuses on the ontological and analytical underpinnings of a river for community development. It considers the morphological and socio-cultural features of a linear park along a river in the city of Curitiba; to make the argument that communities need to be engaged in the regeneration process of rivers that cross their environment. Empirically, the research is grounded in analysing the context of the River Atuba, whose south segment forms the backbone of the Linear Park of Cajuru (port. Parque Municipal Linear), henceforth PLM Cajuru. The overarching objective is to address, first, a socio-environmental analysis towards creating a deeper understanding of the community's use of the park and, second, to understand the role of management in sustaining the value of the park. This role is challenged by a typical situation, as the river segment in the PLM Cajuru forms the border between two municipalities, Curitiba and Pinhais. Politicians and decision-makers found very often rivers to be a convenient way to divide territories. This division results in a series of deficits in the park management, and this, in turn, affects the use of the park by the community. Backed by the preliminary results of the socio-environmental analysis, this chapter discusses the social use patterns of the local communities, the groups that assume the role of park rangers. It aims at building arguments about the river acting also as a maintenance border and the need to approach the park, the river and its banks as one management unit.

2. Rivers as Municipal Boundaries

Rivers, as other natural geographic features, have long been used to divide lands (Popelka and Smith 2020). On the reasons why waterways are used as political borders there is little research available - and even less on the municipal or local borders (Popelka and Smith 2020). According to the authors, rivers account for 23% of international boundaries. They also form 17% of state and provincial borders worldwide, and 12% of all county-level local borders. One issue that may facilitate using geographic features as borders, is the fact that borders are often not seen outside a map. Geographic features make a border line easily identifiable and thus more visible; and makes it inexpensive to demarcate (Querol 2019). As the course of the river can change overtime, those rivers that have been straightened and canalised offer a more constant border line. Following the thalweg has been largely used to mark the borders (Querol 2019; Popelka and Smith 2020).

One important aspect of using a river to define administrative or political borders, is that numerous domestic stakeholders share jurisdiction in water resource management (Popelka and Smith 2020). These stakeholders, according to the authors, ought not to be ignored when crafting water policies and instituting whole-basin management regimes. Curitiba's municipal area is also defined by four rivers: Passaúna, Barigui, Atuba and Iguaçu. Together, they represent 94.5% of Curitiba's boundaries with neighbouring municipalities (IAP 2009). Relevant for this study, is the River Atuba, as it is the backbone of the PLM Cajuru forming the border between Curitiba and neighbouring city of Pinhais for a length of 2,100 m, as depicted by Figure 1.

Figure 1: The extension and narrowness of the PML Cajuru between the municipalities of Curitiba and Pinhais. Source: Google Earth, adapted by the authors.

3. Curitiba's Urban Development and Rivers

Curitiba is the capital city of the Paraná State, and the core city in a vast metropolitan area that encompasses 29 municipalities. According to the 2022 census (IBGE 2024), it is the 8th most populated metropolitan area in Brazil with 3,560,000, concentrating 31% of the entire population of the Paraná State (AMEP n.d.). Despite having been awarded nationally and internationally the titles Model City and Ecological Capital since the 1970s, "Curitiba as the most cities in developing countries, still faces several environmental, social and mobility problems" (Rocha 2019, p. 203). Nowadays Curitiba, as almost its municipality area is built-up, faces rapid urban growth towards its borders. The urbanised area expanded dramatically due to the spread of auto-construction and social housing developments in the outskirts of the city. Thus, the metropolitan development is occurring in "hot spots" in the periphery of Curitiba, far from the central areas of the other cities, which, to date, have rarely been considered an integral part of the metropolitan area. This is the case of the city of Pinhais, whose main urban area expanded along the border to Curitiba. Pinhais until 1992 belonged to the municipality of Piraquara, whose core urban area is around 30km far from Curitiba. Pinhais, with 133,490 inhabitants (IBGE 2020) is the 3rd largest city in the metropolitan area and 10th largest city in Paraná. As a young municipality, and in particular due to its location near Curitiba, Pinhais became a leading economic area in Paraná. It is the 10th largest economic centre and an important service and trade hub in the region (AMEP n.d.).

Curitiba metropolitan area experienced an almost continuous physical growth, and is rapidly becoming a single, integrated mega-urban region. To better govern such an entity, the Metropolitan Area of Curitiba was established in 1973. Without a political and administrative power, this further exacerbated a multi-jurisdictional political division that covers, on the flip side, a single urban entity. This explains also the complexity for the management of environmental resources, in particular the cases of the watercourses, whose problems do not stop at political borders.

Six river basins drain the area of Curitiba, all running from the north southwards to the River Iguaçu, the main and best-known river. All six rivers, among them River Atuba, rise, flow through densely populated areas. The catchment area of the River Atuba is considered the second most urbanised in the Metropolitan Region. Due to population growth and economic drivers, the land use along the rivers faces increasing environmental challenges, such as water pollution, degradation of water quality including a loss in biodiversity, and water-related risks such as flooding and soil erosion. Another issue is the anthropogenic transformation of rivers, almost all six rivers are straightened, canalised and partly covered by roads, for example, the River Ivo. Despite its importance, the River Ivo, a tributary of the Belém River, known for playing a fundamental role in Curitiba's development as the first source of water supply, runs now hidden under the city centre.

As in most Brazilian cities, rivers and waterbodies in Curitiba show visible problems of degradation and socio-environmental vulnerability. Illegal garbage and raw sewage dumping into water bodies is a widespread problem across Brazil (IAP 2009), the River Atuba is not an exception. Being part of the River Iguaçu basin and flowing throughout different urban areas, it shows in its entire length anthropogenic modifications; it is straightened and canalised and wetlands along the banks were drained. This results in changes in the entire water cycle, including water availability and recharge, that get disturbed in course of land take for urban development. Flowing in the northeast side of the city of Curitiba, the Atuba River catchment area (port: bacia) encompasses other municipalities in the Curitiba Metropolitan Region, such as Pinhais, Colombo and Almirante Tamandaré (Figure 2). Its basin covers an area of 128.6 km^2, and it is fed by another 30 km^2 by the Bacacheri River, its main tributary. Since the Atuba River flows through the lowlands of the River Iguaçu, with its flat and marshy landscape pervaded by small hills, the little slope in the surrounding areas of the River Atuba resulted in the formation of an extensive floodplain. Such morphological pattern fostered settlements resulting in a very urbanised catchment, and a large stretch of it flowing through urban areas.

According to Richa and Schafer (2014), to facilitate data management and monitoring of the basin, the SUDERHSA (the state agency for water management)

Figure 2: An overview of the location of the Rio Atuba and its catchment area in the Metropolitan Area of Curitiba (RMC). Source: Richa and Schafer, 2014, p.02.

divides the basin into three sub-basins, as depicted in Figure 3. For the present research, the efforts concentrate in catchment area 3, in the southern part of the basin.

According to Martins (2014), greenspaces are legally protected in Curitiba by the *Sistema Municipal de Unidades de Conservação* (port. the municipal system of protected areas) through the Municipal Constitution Law (1990) that regulates the municipality's environmental concerns. In its Chapter VI, Art. 190, paragraph II, it states that the municipality has to establish conservation priority areas and place them under protection and municipal management. These areas are dedicated to achieving the long-term conservation of nature and environmental resources, and should encompass the protection of water resources, natural ecosystems, areas of interest for flora and fauna, and other natural and cultural. In the case of water resources these are usually socio-environmentally degraded areas, and thus subject to environmental impacts of macro- and micro- climatic factors, such as flooding and erosion.

One of the typologies of green infrastructure are linear parks, which are primarily implemented in areas that require environmental recovery. Assuming that they can help in the mitigation of environmental impacts, as they can alleviate erosion processes and the effects of high temperatures, in addition to serving as a recreational space for riverine communities (Duarte 2022). The linear parks are created to foster the protection of valley bottoms, to discourage land takes and occupations, and also as a measure for environmental recovery and qualification of the urban landscape. Parks, in addition to offering spaces for leisure, sports and cultural practices, contribute to enhancing the urban drainage and create links among the greenspace network (Soares 2014).

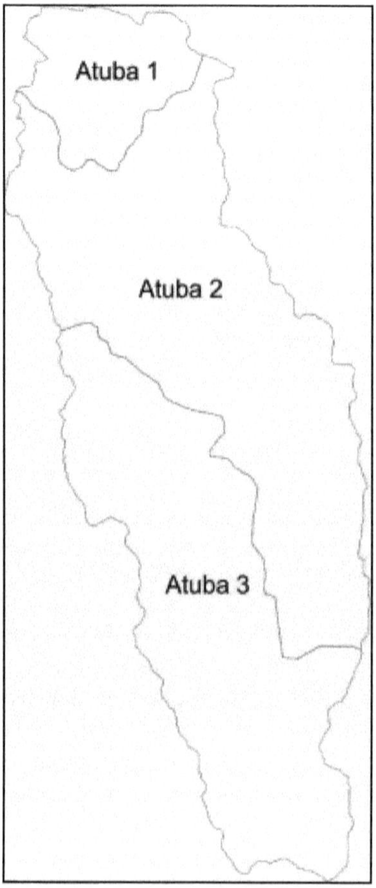

Figure 3: The three catchment areas of River Atuba basin. Source Richa and Schafer, 2014, p. 02.

Despite the values of rivers, in Curitiba, the care for the waterscape was never in the forefront of the urban agenda, as illustrated in Figure 4. Although nowadays the city is known for its ecological projects (Rocha and Smaniotto 2022, Rocha et al. 2023), with the urban growth waterways were used to dump sewage in and, after time, these waterways were straightened, covered and turned into sewers. As canalized sewers, the waterways represent undoubtedly the most typical type of construction of artificial or anthropogenically modified water bodies, and thus the strongest loss of natural richness. Almost all rivers that cross the city, are degraded, and anthropogenically modified. This evidence shows the low environmental quantity and quality of its waters, exposing the urgency of measures to restore and revitalise Curitiba's rivers (Rocha et al. 2023). The situation has improved in the last decade, as significant investments to implement sewage treatment infrastructure serving communities across the city (with the implementation of sewage treatment systems, as of March 2024 almost 98% of the population has access to the sewage collection network, and 100% of the sewage collected is treated (AEN 2024). Such advances, however, do not include the rivers as they still remain the main element

Figure 4: The map of Curitiba of 1894 shows the central area with a street grid running in straight lines. The existing rivers seem to be a hurdle within this street grid arrangement which demonstrates the scant attention paid to topography and hydrologic aspects. Source: Amorin, 2019.

in the sewer network. Rivers still have high rates of water pollution, in particular by organic pollution originating from illegal dumping and undersized treatment plants (IAP 2009). Although the municipality and regional government seek to clean up Curitiba's waterways, as demonstrated by the increase of investment in infrastructure (AEN 2024), the treatment plants do not follow the urban growth. New developments, in particular the newly occupied areas usually do not have a sewer network, and the wastewater is discharged into rivers. As urban areas sprawl into the landscape, the more concrete and asphalt, the higher is the amount of stormwater, surpassing the capacity of sewers, besides increasing the urban heat effect. Additionally, the climate change is bringing heavier rainfall that again overburden sewers.

4. Features of a Linear Park and the Linear Park of Cajuru

A linear park is a particular typology of public open space usually narrow and long, becoming thus a longitudinal shape. Linear parks often result from reusing sites along a linear element of the landscape, such as watercourses, streets and other infrastructure. They form a continuous system of greenspace. Recycling an underused space is generally seen as an opportunity to adapt an old infrastructure to contemporary needs (Faggi and Vidal 2016). In the case of a park along a watercourse, linear parks create new opportunities for nature in cities, to enhance biodiversity and its ecosystem services besides the improvements in water systems. The creation of a linear park goes often together with the restoration of a watercourse, which can include uncovering a lost river, or parts of it, by removing obstructions that are covering it and bringing the rivers back to light, renaturalization of previously canalized water bodies, recovering banks, reestablishing floodplain areas towards buffering the effects of heavy rainfall and in this way protect economic activities and communities further from flooding or heavy rains. Recovery and protecting vegetation on river banks and slopes contribute to soil stability and mitigation of erosion and landslides. The restoration contributes thus to the natural drainage and retention of water, in addition to, of course, increasing the number of green areas in the city. Hence, through an adequate design, nature-based solutions or imitating natural ecosystems, linear parks can help protect, conserve, restore, manage and enhance water courses, reduce flooding in urbanised areas, run-off pollution, and other environmental issues urban areas face (Smaniotto et al. 2023).

The riverain population can enjoy better water and air quality, becoming thus more resilient. With adequate infrastructure, linear parks provide also green access in neighbourhoods (Faggi and Vidal 2016), and make "parks" accessible to more people. Linear parks can thus have a unique and important significance to the urban environment, offering to communities several social, economic, and environmental benefits, as any other greenspace (Smaniotto et al. 2008, EEA 2022). Actually, cities need to take advantage of any space to bring back more nature, in new and innovative ways (Reichwein et al. 2019). Transforming abandoned or underused sites make them useful again and help to protect and restore the environment, while reaping the social, health and economic benefits that they bring. After all, because such an improvement also helps increase nearby land values. Once cities are reinventing their relationship with water courses, a large number of linear parks have been built on riverside, waterfront and riverbanks, creating new connections and a new urban scenery. Citizens can thus enjoy new landscape and recreational functions.

In Curitiba, linear parks as a typology of public open spaces were adopted in 2003, when four areas along rivers were placed under protection (PMC 2024). There are currently five such parks (Table 1). The parks were planned to perform an important drainage function and help prevent floodings within urbanised areas. Another relevant goal is to prevent new land take (called in Brazil *ocupações*, lit. occupations) for informal settlements. The occupation of protected and inappropriate areas for housing in Brazil is a wicked problem, as Rocha and Smaniotto Costa (2022a) explained on the example of gated communities in Curitiba. With these

Table 1: Linear Parks in Curitiba

Name	**Area**	**Region**
Cajuru	104.000m²	South-east
Mairi	43.000m²	South
Mané Garrincha	120.000m²	South-east
Yberê	238.105m²	South
Rio Bonito do Tatuquara	44.400 m2	South-west

Source: PMC 2024

linear parks, the city aims, according to PMC (2024), to solve both social and environmental problems, prevent new occupations, and create for local communities leisure and recreation opportunities while integrating more nature into the city.

For the research on a linear park the Asset-Based Community Development (Haines 2014) offers an interesting research venue as it places emphasis on the wealth of assets present in a community. The PLM Cajuru has the potential to be the asset for unleashing local development both in terms of social and environmental improvements.

4.1 The Cajuru Linear Parque

The Cajuru Linear Parque (PML Cajuru) was created between 2002 and 2003 along a stretch of the River Atuba in the south-eastern region of Curitiba. Located at the border between the municipalities of Curitiba and Pinhais, it thus builds the limits between both municipalities. Protected as a park, however, is only its west bank, on the Curitiba side. The Pinhais riverside is classified by the land-use law as green area (*zona especial verde)* (PMP 2022), which means the area is not suitable for building up and should be managed as a greenspace. While the Curitiba side is predominantly a residential area, the adjacent areas in Pinhais are classified as "Zona Mista Consolidada" (Urbanised Mixed-Use Zone), where retail, service and industrial activities are allowed within a low and medium density residential use (PMP 2022). For the purpose of protecting water resources and following the National Environmental Law, the laws of both municipalities require a buffer strip along the watercourse and lakes. This permanent conservation buffer, whose width could vary between 30 m and 15 m wide, should be created on both sides of the river. In best cases, where space is available the 50m strip should comprise a 30m wide trip for buffering and a 20m wide trip for renaturing. These buffer areas should ensure the drainage of river and rainwater and act as a place of refuge for wildlife and provide connections to other habitats (PMP 2022).

Following the goals of the Curitiba City Council, the PML Cajuru was created aiming to restore the environmental functions of the River Atuba, through restoring the riparian areas and thus prevent riverbank erosion. Because of the main river hazards - flooding and erosion, both sides of the river are considered as depited in Figure 5. Just because of this vulnerability both sides became low-income neighbourhoods, lacking adequate access to basic services, such as drinking water, sewage or electricity. This follows a vicious cycle, as people living in deprived areas

Figure 5: View of the Rive Atuba as a border line between two municipalities, on the left as a linear park, with landscaped greenery, and on the right as an abandoned land. Photo: Authors, 2024.

are more exposed to environmental risks (EEA 2024). Because the area of the park is considered vulnerable, it is thus not appropriate for housing. The municipality does not build full infrastructure in terms of basic sanitation, and the neighbourhood population remains vulnerable. On the flip side, these inappropriate areas are mostly those that the poor population can access, and the cycle resumes.

In order to "gain" space for reestablishing the riparian vegetation in the area of the PML Cajuru, about 400 families have been relocated to other neighbourhoods, and there remains potential for the neighbourhood's vulnerable populations to be displaced in the future (Müller et al. 2012, p. 69). Figures 6 and 7 depict the changes in the area to provide space for re-establishing for riparian vegetation zones. The River Atuba rises in the Serra do Açungui (Açungui hill) in the city of Colombo, and flows south-east crossing the municipalities of Colombo and Pinhais before reaching the Iguaçu River in Curitiba. It flows through gently sloping land which forms an extensive flood plain. Its watershed is undergoing strong urbanization, with new neighbourhoods arising in these three municipalities. The population growth in the river watershed and the resulting rise in soil impermeability, tend to increase flood risk, helping to trigger hazards and increase societal vulnerability.

As abovementioned, the PLM Cajuru was created in 2003, first through a protection act and secondly by implementing equipment commonly used in other public open spaces, such as multi-purpose pitches, outdoor gym, playgrounds, and bike lanes, etc. Due to the limited width these are placed in a row, which in turn, puts these facilities very close to their users. The PLM Cajuru has neither toilets nor parking lots. In June 2022, the park was refurbished by the City Council; the network of bike lanes was increased with 240 m added to the 4,000 m, the multi-purpose pitches were recovered, and new basketball hoops were installed. The skate park was completely renovated and got a new skateboard rail for grinding (PMC 2024).

4.2 Curitiba Drainage Master Plan

The region of Curitiba is blessed with abundant rainfall and is frequently invested by huge flooding due to sever rain fall. The city has a long history of struggling

Figure 6: shows the southern part of the River Atuba, before some houses along the river were removed. Source of Figures 5 to 11: Google Earth, adapted by the authors.

with flooding, and this combined with scant prevention measures plus explosive population growth bring about the change required to ensure urban resilience. In particular, due to the growing land use and the effects of climate change the tendency is to worsen the situation. To answer these requirements, the Metropolitan area of Curitiba has since 2002, a Drainage Master Plan for the River Iguaçu basin, covering an area of 2,500 km² and a population of 3,7 million (AEN 2024). This plan has as a goal to provide technical support for the 14 municipalities on the river, towards mitigating and preventing flooding. The plan consists of a mosaic of different but interconnected actions carried out along the River Iguaçu. Curitiba has significantly in 2024 intensified the prevention programme, with the creation of new parks and investments in the protection and restoration of watersheds, following the concept of the sponge city. Both parks and protected areas are intended to absorb and hold rainwater. The programme also includes the construction of a new water reservoir to also prevent future droughts, building a fresh water reserve for the future.

The plan also foresees the implementation of the Committee of the Bassin of River Iguaçu, encompassing the 14 municipalities. This committee is also a request of the Water Resources Policy as a means to exercise social control on the use of water (AMEP 2025). Even if an overarching and regional framework is paramount, the municipalities, although responsible for local interests and urban planning,

Figure 7: shows the park being created with the implementation of the sport fields and restoration of the riparian vegetation. Also the streets were new designed and paved.

encounter difficulties in implementing local actions, since they do not have political and administrative control over water resources, whose responsibility is shared between the regional and central government. This also makes the call to increase the understanding that environmental issues are not the only ones to impact urban rivers. They are influenced by cultural, political and economic factors, besides the "urban" ones, such as the occupation of the banks and removal of riparian forests. Therefore, an integrated and participative management of a river basin is necessary to guarantee the effectiveness of its multiple dynamics and uses (Reis 2016), as well its ecological performance.

5. Method

Methodologically the research was outlined using different survey instruments. First, a literature review was performed on major scientific databases, such as Web of Science, Google Scholar, SciELO , as well as SIBI UFPR and SIBI USP, the library systems of the Brazilian universities Federal do Paraná and São Paulo. Keywords and keyphrases used were linear parks, greenspaces, sprawl and disordered urbanisation, urbanisation of valley bottoms, green infrastructure, degraded urban areas, socio-environmental vulnerabilities and risks, monitoring indicators, urban equipment, housing policy. Different master plans available on the website of the City of Curitiba were used for further analysis. Based on the findings, it shows a comprehensive and up-to-date analysis of both the landscape and the changes in land use in Curitiba in the last decade, including the positive and negative aspects. Such results were also relevant, when considering indicators for monitoring land cover and land use around the PML Cajuru.

Developed between April and August 2024, this research was constructed using an exploratory method, based on observational field research. This method allows the collection of information that often cannot be provided by the individuals themselves or that they may be reluctant to share. Thus, it represents an effective strategy for identifying behaviours, allowing the collection of responses that would be difficult to access through direct questioning (Gil, 2008). The research uses monitoring indicators to understand the processes that are converting the space into a public park, and to a limited degree the discourses that accompanied these transformations. The aim is to provide an intertwined analysis of the ways in which people and ecosystems interact, thereby creating complex patterns of spatial, social and ecological interdependencies. Fieldwork was the second main survey instrument. Between January and May 2024, using observations, visual techniques and field mapping, the variables concerning the park area and the park features were analysed for both conservation indicators and park use indicators. The field work consisted of 10 visits with different length of stay – between 30 to 90 minutes and in different week days.

5.1 The Rationale for Analysing the Use of the PLM Cajuru

According to Cerqueira (2019), indicators are actionable tools to quantify, simplify and make information visible and thus demonstrate their meaning. Indicators can be used on different scales and fields of knowledge; in urban planning, in particular, they provide background information for public policy and government decision-making. Moreover, the author stresses that indicators have the task of describing the analysed system, be it of environmental, social or economic nature, and the results can show pathways, potential for development or even risks in urban development. Such results are in fact an assessment that enables us to gain insights into the relationships between urbanisation and the natural environment, and shows to decision-makers and society the need for corrective actions and policies. Martins (2014) and Kimic et al. (2023) note that any park and greenspace in an urban environment must contribute to the conservation of natural resources and improve local environmental quality, besides providing the stage for public use, in particular for the leisure and recreation (Smaniotto et al. 2008). Thus, our research question is centred on understanding the role of PLM Cajuru for protecting the local natural environment and providing the structure for recreation and leisure. For these reasons are selected those monitoring indicators that reflect the park use and its function. These indicators, as well as the compatibility among them, are discussed in the forthcoming sections.

5.2 Walking and Image Bank as Visual Analysis Tools

The fieldwork was organised towards using the same variables - the park features and attributes, and this for both nature conservation and park use. This was achieved through surveys and directed field observations. Walking was used as a tool for capturing visual perceptions of the park's use, and its correlations to the local features. According to García-Esparza et al. (2024), walking is a creative tool that related to ethnographic approaches (Menezes and Smaniotto Costa 2024) enables researchers to learn about users' practices, activities and behaviours in the space to be analysed.

Table 2: Selection criteria for the park analysis and their metrics (variables and indicators).

Variable	Short description	
Park features and function	**Nature conservation Indicators**	**Park use Indicators**
River and riverscape	State of the riverbed (natural, canalised or covered)	Users' perception and quality of landscape/riverscape
Ecological structure of the Park	Species diversity and landscape quality, maintenance of riparian habitats	Social profile of residents and land use patterns (residential, retail, industrial, services purposes)
Economic Issues	Investments in water and sanitation infrastructure	Investments in accessibility, mobility, and waste management
Quality of adjacent neighbourhoods	Quality of public open spaces, and built environment (kind and state of repair of buildings, their height and proximity to the riverbank	Use (recreational, cultural, artistic), features and equipment, events (cultural, sporting, religious), visual and physical contact with the river
Emotional and sensory aspects	Environmental Education Projects, risk perception on using or misusing the park	Park ambience (park atmosphere, sounds, visual pollution and smells, aesthetics and identity with the landscape

Menezes (2023) brings into focus the value of walking to transport researchers into a sensory immersion in the territory, providing an opportunity to discover and explore an indigenous perspective. As walking provides both quantitative and qualitative information, it assists us in identifying and visualizing elements that contribute to understanding the usability of the park, and how this hinge on its features.

Along walking, a visual data mining approach was employed to generate an image bank. The images were taken by a third person, who is not familiar with both research and the park. This person was requested to visit it and take scenery pictures of what attracts the attention. The sole requirement is that images of the river should be also captured. This person got only very basic information about the park and its location. The park was visited in December 2024 twice, in a weekday afternoon and on Sunday morning. It resulted in an image bank with over 40 images. The image analysis was intuitively backed by the author's personal knowledge; the images were scanned towards filtering the main components (primary focus) and the background. The analysis enabled a visual representation of data, which is used as highly valuable supplementary information for enriching the understanding of the interplay of the park and its features.

5.3 Building Indicators

As above-mentioned, one of the research goals is to enhance the understanding of the relationship between the quality of the park and how this affects its use. To achieve this, surveys and field observations were addressed using monitoring indicators that consider both nature conservation variables and public use variables. To understand the relationship between both features is fundamental, as urban greenspaces can

Figure 8: The entrance gate to the park, which serves also as a barrier for vehicles. On the left is the football stadium and on the right is an open-air gym. Photo: Authors, 2025.

fulfil both functions when they are adequately designed, implemented and well-maintained. In this context it is recalled that in urban settings the access to nature can be limited, and parks and greenspaces offer for many city dwellers the only opportunity to contact and experience nature (Smaniotto Costa et al. 2008). To verify the performance of the PLM Cajuru in this regard, an index with background variables and indicators are identified - these are listed in Table 2. The framework for the typology of variables and indicators, the methods for data collection and analysis besides addressing several key issues, should also be of easy access and use.

6. Results and Discussion

The literature review demonstrated the interest of the municipal and regional governments concerning the water quality of the rivers that cross the city. In March 2024, coinciding with World Water Day on 22 March 2024, different agencies shine the spotlight on the ecological status of rivers that cross the city. Although this is a positive development, the current situation is far from satisfactory. Rivers covered with scattered rubbish, sewage, and bad smells make up the current scenario in Curitiba. In particular, the unpleasant smell and the pollution prevent people from using riverscapes. Still, in Curitiba there are no signs of removing obstructions (such as concrete or pavement) which cover the rivers across the city and restoring them to a better environmental condition.

6.1 The Spatial Analysis of te PLM Cajuru

Regarding the spatial analysis and indicators, the reslts are as follows. Along the park there is only one shared-use path to accommodate pedestrians and cyclists. This path is also the common element that connects all parts of the park from north to south. During walking tours through the park and the immediate adjacent neighbourhoods, field notes were taken on these relationships. These are analysed and discussed below.

The first trip began at the park's gate (Figures 8 and 9), which it is located near the João Ramos dos Santos Stadium, a soccer field of the "suburban" football club called

Figure 9: The stadium seen from one of the road bridge that connects the municipalities of Curitiba and Pinhais. Photo: Authors, 2025.

Santissima Trindade. Being 22 years old, this club was there before the construction of the park. The stadium is named after the former landowner, who donated the area for its construction. Residents reported that the area around the stadium experiences more floods, for this reason several families were forcibly relocated. It was also reported that some time ago, the residents planted themselves some trees to "make the place more beautiful", since there were no trees there.

The north section has different features, with several sports courts and fields, lawn areas and a small stream that after crossing the neighbourhood flows into the Atuba River. In this section, the shared-use path is very damaged, and the street along the park is not paved. The courts were not a subject of intensive use during the day, only in the evenings there is an intensive use by football players. The playground is very damaged and thus not used during the day. Maybe because of the main entrance, this area is well visited, in particular for dog walking. It is a popular spot on weekends, due to the football matches that take place there. To the north there is one of the few bridges over the River Atuba, this is one of the main connections between the municipalities, just before the bridge there is a playground and two grass courts, which are also widely used during the evenings and on weekends (Figures 10 and 11).

Further two sand courts, a multi-sports court, a skate park and an outdoor gym are facilities that make up the northern end of the park (Figures 12 and 13). This section reaches with 102 m its greatest breadth. Here there are also the most equipment facilities, such as outdoor fitness equipment, skate park, multipurpose and sand courts. This equipment makes this section the most used in the PLM Cajuru, both on weekdays and on weekends and holidays. Along the river, while on the Curitiba side there is little riparian vegetation, on the Pinhais side the wild vegetation features more bushes and trees with shrubby growth habits. On the grassland there are some trees, with new ones recently planted by the city council and local residents. Some are already leafy, providing shade to the sand courts. Unpleasant is an illegal small rubbish dumping close to the bus stop.

Figure 10: A grass sport court. Photo: Martins, 2014.

Figure 11: The northern playground. Martins, 2014.

Figure 12: The refurnished skate park. Photo: Martins, 2014.

Figure 13: The multi-sports court. Photo: Martins, 2014.

During the week, this section is used for sport and physical education classes by a nearby elementary school. This can be considered as a sign that the park is a community resource also used for the benefit of the children. This highlights the central idea of a park in promoting physical activity, social skills, and well-being (Smaniotto et al. 2008).

From the park entrance, heading southwards, the park becomes narrowed again by several houses, making access to it difficult. There are no streets or cycle lanes, only beaten paths. Two small streams reach River Atuba in this section, but there are no bridges to cross them. Maybe due to this inaccessible feature there are not many facilities here, just a bocce ball court and a grass court, both seem to be not in use. The park vegetation along the river is very shrubby, but without trees, except for those recently planted by the residents in the lawn area. Parts of the park are often used as a parking lot, and for dumping or burning rubbish. This may account for the large number of "No Littering" signs in this section.

In the most south part of the park just before the River Atuba flows into the River Iguaçu. The park has not been implemented yet, so there is no equipment or a "sign" of the council intervention. The vegetation is composed mainly of pioneer species. Due to inequities in Brazil, the houses result from a typical practice of land occupation with irregular or illegal nature, reflecting the struggle for land rights.

6.2 Assessing the Quality of the PLM Cajuru

As there is no data about the PLM Cajuru, we decided to investigate the quality of the park through a simple and visual qualitative analysis and focused our attention particularly on visitors' levels of activity supported by the park features and attributes.

Watercourse (analysed is the riverbed channel along the park) - the river runs in an open channel; in areas with sparce vegetation its artificial straightening can be observed. Also noticed is the illegal sewage dumped on both sides of the river, in addition to the irregular garbage disposal, especially on the Pinhais side bank. These often contain construction rubble which can be remains of the demolished houses. In February 2025 in a joint effort of both councils (Curitiba and Pinhais) the riverbed was dredged, sediments and garbage removed towards improving the water flow (PMC 2025).

Vegetation structure - the field analysis indicates the riparian vegetation is severely fragmented, not covering the whole length of both riverbanks. This may lower the benefits a riparian corridor can provide (filtering and absorbing sediments, pollutants, etc.). Both planted, and spontaneously established plants show visible signs of poor maintenance, such as unpruned bushes, poor crown forms, broken limbs, etc. In addition, invasive species, such as Ricinus communis, are becoming dominant. During one of the visits, the lawn was being mowed by the council gardeners, and they also replaced damaged small trees. This is regarded as an indication that the municipality is increasingly caring for the park. The fact that the park has a greater proportion of its surface covered by vegetation (lawn and bushes along the river) is of course a major benefit for the city (in the sponge city context) and residents (less artificial environment).

Neighbourhood qualities - The surrounding buildings are mostly residential, with some local retail. The population belongs to low familiar income and low-wage workers, who build their own homes, mostly on weekends. The businesses are family-run and, in general, the owner lives in the back of the commercial premises. There are few services available, but no industrial use. Besides the park, there is no other public space in the neighbourhood - only an outdoor gym with fitness equipment.

Most of the streets are still unsurfaced, with abundant dust or mud. There are no sidewalks, pedestrians walk along the streets. The only pedestrian paved lane is the shared path in the park. On the streets near the streams, you can see garbage dumped on the banks of these tributaries. There are few collection points for organic and recyclable waste on the streets surrounding the park, even so there is illegal waste dumping in the park.

Emotional and sensorial Aspects - There are no information panels on display in the park. Visitors can not be informed about the park history or about the goals of removing the population. There is also no information about fauna and flora of the park. The commemorative plaque in the park entrance has been removed, probably stolen and has not been replaced yet. Information about park facilities and equipment is also not available, new users do not get an overview of possibilities. The park is equipped with lighting, also for the sports courts and fields so that the participants and visitors can use the park after dawn.

Use of the park - The park is heavily used for a variety of purposes, especially due to the different leisure facilities (picnic tables, benches) and sports equipment distributed along the park; these are connected through the sparce greenery by a shared-use path. Users moving around the park concentrate on this path. The park lacks information boards; rest rooms and litter boxes well distributed throughout it. Furthermore, criticised is the poor maintenance of facilities, in particular many sports and bowling fields became weed-infested and unplayable. Also fencing the areas for vegetation recovery would help a rapid regeneration. Even such problems do not prevent an intensive use of the park, perhaps because of the lack of opportunities to be outdoors in the neighbourhood. The pattern of use shows a well-used park, both on weekdays and weekends, with the number of visitors varying according to the moment of the day. On weekdays, in the morning period, adults and seniors use the park for waking, and school children for school-based physical activities. In the afternoon, children and teenagers use the playgrounds and the sports facilities. In the evening, adults and teenagers use the multi-sports courts and the skate park. Predominantly users live in the neighbourhood, coming usually by foot, bike or skateboard. Often people come to picnic or idly rest in the park, enjoying the silence and contact with nature, others read while sitting on the benches, and still others just walk their pets.

Moreover, there were no cultural, artistic or religious events in the park, although a common practice in Brazil. Sports events, such as championships and tournaments have been organised by the residents, who also funded prizes, medals and trophies. During the field research some residents could be approached to get some views about the park's values. Among them are two ladies who were planting trees in the park. Both did not want to provide any official information because they mean to be acting illegally. However, their efforts in providing trees and thus caring for the park is to be positively understood. The other users who could be directly approached where adolescent skateboarders who explained that they use the park very often, during the day and evenings. Both uses (by the ladies and skateboarders) are considered as a positive action towards the park, as they contribute to improve its natural assets and set an outward sign of the parks' values for the community.

6.3 The Park Management (or the Lack of a Coherent Management)

The situation of PLM Cajuru calls for developing a park management that understands its complexity with intrinsic linkages between the natural, fluvial and man-made systems, the park design, its maintenance, patterns of use, and the surrounding neighbourhood. The concept of 'management' is thus to be drawn broadly to recognise how built, social and natural environments intersect with and influence each other. For the park and neighbourhood resilience the management approach has to bring together these systems. This is not evidenced at the moment. While the Curitiba bank side of the river is developed as a linear park and contains some amenities for the community, the Pinhais side is literally abandoned, showing scant care. In order to reap full benefits for the community and environment and transform the park in a wetland able to act as a natural filter, helping thus both, a concerted and long-term action integrating both councils and communities have to be put in place. The current situation of the Atuba River and the PLM Cajuru are good examples of how the land take in natural floodplains without prevention, the straightening of rivers and suppression of adequate drainage, put citizens at risk besides demonstrating the ineffectiveness of the urbanisation model, which disregards any concern for rivers (Correa, 2018). This model provokes economic and social damage with severe environmental consequences for the whole society.

7. Conclusions

In highly urbanised areas, due to the permanent changes in watershed and catchment areas, a complete recovery of river processes and functions is impossible. This calls for creating opportunities to mitigate the urbanisation impact while creating more nature-like spaces, as well as a service for the residents. Another reason is also due to the fact that changing riverbeds is often expensive, so it is important to prioritise responses that bring the most improvement for everyone - for environmental issues, for social issues and for collaborative governance. Parks, as Duarte (2022) aptly states, are key for a green infrastructure, and the case of the PLM Cajuru shows how a simple solution can bring better benefits. Parks along rivers help mitigate extreme events, such as flooding, while simultaneously linking liveability, restoration potential and biodiversity. By providing a series of ecosystem services, parks integrate spatial, social and natural resources. The research on the PLM Cajuru improved the knowledge base on the impact of river regeneration and, in particular, the role of river management in providing a social space as well as a place for mitigating the impact of urbanisation. Although the regeneration benefits are evident for environmental agencies, for example, they are less widespread among further technical infrastructure agencies and for the society in general. This calls for further research on societal issues within the catchment area, and the engagement of the population to better balance the different needs, as informal recreation, sports and physical activities, water management and biodiversity further limit the land available for river recovery.

The PLM Cajuru is a good example of how public spaces can be a free resource in low-income communities and attract different age and gender groups to socialise, be more active and experience nature. Such benefits, social and environmental, however, do not follow administrative boundaries. Considering the status of the River Atuba banks on the Pinhais' side and completing the idea of a linear park with a river in its core would not only bring nature and residents together but also calls for a collective response from both councils.

This research being anchored in the RUN Project reinforces the experience of this wide network, which shows that all partners in a river basin benefit from healthy aquatic ecosystems, also because they open new opportunities for social engagement and help clean cities up as well as to reduce social and environmental inequities. In conclusion, the research on urban rivers and the proposed pathway to a strengths-based approach for the PLM Cajuru offer a promising framework for fostering inclusive and sustainable community development. To achieve Curitiba's vision of a thriving, sustainable and future-fit city also means that it is imperative to take steps in this regard, and recovering rivers and nature from anthropogenic impact is a substantial first step.

Acknowledgements

This publication is based upon work of the Project RUN | Citizen science and co-creation for regeneration of urban rivers and mitigation of risks, which has received funding from the CYTED Programa Iberoamericano de Ciencia y Tecnología para el Desarrollo, under grant agreement No 420RT0008, https://www.cyted.org/RUN-RIOS-URBANOS-NATURALIZADOS.

References

AEN - Agência Estadual de Notícias. 2024, March 29. Curitiba chega aos 329 anos como a segunda melhor capital do País em saneamento. https://www.aen.pr.gov.br/Noticia/Curitiba-chega-aos-329-anos-como-segunda-melhor-capital-do-Pais-em-saneamento (Accessed on 26 Dec 2024).

AEN. 2024, Aug 28. Com 137,7 mil pessoas a mais, RMC chega a 3,6 milhões de habitantes em 2024. https://www.aen.pr.gov.br/Noticia/Com-1377-mil-pessoas-mais-RMC-chega-36-milhoes-de-habitantes-em-2024 (Aceessed on 26 dec 2024).

AEN. 2022; March 22. Sanepar comemora Dia Mundial da Água com garantia de qualidade para 346 municípios. https://www.aen.pr.gov.br/Noticia/Sanepar-comemora-Dia-Mundial-da-Agua-com-garantia-de-qualidade-para-346-municipios (Accessed on 26 Dec 2024).

AMEP - Agência de Assuntos Metropolitanos do Paraná (n.d.). Municípios da Região Metropolitana de Curitiba. https://www.amep.pr.gov.br/FAQ/Municipios-da-Regiao-Metropolitana-de-Curitiba.

AMEP. Comitê das Bacias do Alto Iguaçu e Afluentes do Ribeira COALIAR. https://www.amep.pr.gov.br/Pagina/Comite-das-Bacias-do-Alto-Iguacu-e-Afluentes-do-Alto-Ribeira-COALIAR (Accessed on 26 Dec 2024).

Amorim, E. 1st August 2019. Curitiba em 1894. Memoria urbana. https://www.memoriaurbana.com.br/curitiba-em-1894 (Accessed on 12 Dec 2024).

Câmara de Curitiba 2024, March 22. Rios de Curitiba: CMC Podcasts debate tema na semana do Dia Mundial da Água. www.curitiba.pr.leg.br/informacao/noticias/rios-de-curitiba-podcast-da-camara-discute-problemas-e-solucoes-na-semana-do-dia-mundial-da-agua (Accessed on 26 Dec 2024).

Corrêa, R. H. A., Vazquez, G. H., and Vanzela, L. S. 2018. Projeto estratégico de ocupação do fundo de vale do córrego da Aldeia no perímetro urbano de Fernandópolis/SP. Urbe – Revista Brasileira de Gestão Urbana, 10 (2). https://doi.org/10.1590/2175-3369.010.002.AO15.

Duarte, L. F. 2022. Parques Lineares implantados em áreas urbanas ambientalmente degradadas. XIII Congresso Brasileiro de Gestão Ambiental. Teresina. http://dx.doi.org/10.55449/congea.13.22.VI-008.

EEA - European Environment Agency. 2024. Environmental inequalities. https://www.eea.europa.eu/en/topics/in-depth/environmental-inequalities (Accessed on 12 Dec 2024).

EEA. 2022. Who benefits from nature in cities? Social inequalities in access to urban green and blue spaces across Europe. https://www.eea.europa.eu/publications/who-benefits-from-nature-in (Accessed on 12 Dec 2024).

Faggi, A., and Vidal, C. Z. 2016. Linear Parks: Meeting People's Everyday Needs for Secure Recreation, Commuting, and Access to Nature. The nature of cities. https://www.thenatureofcities.com/2016/04/14/linear-parks-meeting-peoples-everyday-needs-for-secure-recreation-commuting-and-access-to-nature/ (Accessed on 16 Dec 2024).

García-Esparza, J. A., Smaniotto Costa, C., Hansen, P., Aragão, N., and Nikšič, M. 2024. Walking Methods as a Place-Learning Experience for Participatory Design. In Placemaking in Practice Volume 2. Leiden: Brill, 13–41. https://doi.org/10.1163/9789004691919_004.

Garcias, C. M. 2016. Experiências na revitalização do Rio Belém: Realidades urbanas do Brasil. *Revista Eletrônica de Gestão e Tecnologias Ambientais*, 4(1): 50–68. https://doi.org/10.9771/gesta.v4i1.14910.

Haaland, C., and van den Bosch, C. K. 2015. Challenges and strategies for urban greenspace planning in cities undergoing densification: A review. Urban Forestry & Urban Greening, (14)4: 760–771. https://doi.org/10.1016/j.ufug.2015.07.009.

Haines, A. 2014. Asset-based community development. pp. 67–78. *In:* An introduction to community development. Routledge. https://www.taylorfrancis.com/chapters/edit/10.4324/9780203762639-14/asset-based-community-development-anna-haines.

Hamilton, K., Kaczynski, A.T., Fair, M. L., and Lévesque L. 2017. Examining the Relationship between Park Neighborhoods, Features, Cleanliness, and Condition with Observed Weekday Park Usage and Physical Activity: A Case Study. J Environ Public Health. https://doi.org/10.1155/2017/7582402.

Hermida, A., Cabrera, N., Osorio, P., and Cabrera, S. 2021. Rio urbano: medición, representación espacial y estrategias de diseño para las márgenes de los ríos urbanos. Cuenca, Equador: Universidade de Cuenca.

IAP - Instituto Ambiental Do Paraná. 2009. Monitoramento da qualidade das águas dos rios da Bacia do Alto Iguaçu, na Região Metropolitana de Curitiba, no período de 2005 a 2009. Curitiba: IAP.

IAP and Superintendência de Desenvolvimento de Recursos Hídricos e Saneamento Ambiental. Plano Diretor de Drenagem para a Bacia do Rio Iguaçu na Região Metropolitana de Curitiba, 2002. Retrieved from https://www.iat.pr.gov.br/Pagina/Plano-de-Drenagem-do-Alto-Iguacu. Ascessed on 26 dez 2024.

IBGE - Instituto Brasileiro de Geografia e Estatística. Cidades. Pinhais. https://cidades.ibge.gov.br/brasil/pr/pinhais/panorama (Accessed on 26 Dec 2024).

Kimic, K., Smaniotto Costa, C., Bocci, M., and Aragão, N. T. 2023. How People Change Public Parks by Using: Notes on Before and After the Covid-19. *In*: Smaniotto et all. (Eds.). Dynamics of Placemaking Volume 1, 272-289. Leiden: Brill. https://doi.org/10.1163/9789004542389_016.

Liziero, L. B. da S., and Carvalho, F. 2018. Federalismo e centralização no Brasil: contrastes na construção da Federação Brasileira / Federalism and centralization in Brazil: contrasts in the construction of the Brazilian Federation. *Revista de Direito da Cidade*, 10(3): 1483–1503. https://doi.org/10.12957/rdc.2018.32661.

Martins, L. F. V. 2014. Monitoramento de parques urbanos em fundos de vale: análise das funções de conservação e uso público – estudos de casos múltiplos em Curitiba, Paraná. PhD thesis in Geography. São Paulo: Universidade de São Paulo. https://teses.usp.br/teses/disponiveis/8/8135/tde-07052015-164053/pt-br.php (Accessed on 26 Dec 2024).

Menezes, M. 2023. A framework to grasp socio-territorial relationships: a conceptual path. *In:* Smaniotto Costa & Aragão (Eds.). Understanding and transforming the territory. New Approaches and Perspectives. Culture & Territory, Vol. 5. Lisbon: Lusófona University Press. https://doi.org/10.60543/ecati/2qb0-kc79.

Menezes, M., and Smaniotto Costa, C. 2024. The Role of Collaborative Ethnography in Placemaking. Humans, 4: 284–297. https://doi.org/10.3390/humans4030018.

Müller, C. R. C., Oliveira, K. L. de., and Motta, J. T. W. (Orgs.) 2012. Plano Municipal de Recuperação e Conservação da Mata Atlântica. Prefeitura Municipal de Curitiba.

PMC - Prefeitura Municipal de Curitiba 2025, Febr 20. Prefeituras de Curitiba e Pinhais intensificam desassoreamento do Rio Atuba. https://www.curitiba.pr.gov.br/noticias/prefeituras-de-curitiba-e-pinhais-intensificam-desassoreamento-do-rio-atuba/76144 (Accessed on 26 Dec 2024).

PMC - Secretaria Municipal de Meio Ambiente. 2024. Parques Lineares Municipais de Curitiba. https://www.curitiba.pr.gov.br/conteudo/parques-lineares-municipais-de-curitiba/3327. (Accessed on 26 Dec 2024).

PMP - Prefeitura Municipal de Pinhais. Mapa de zoneamento. n.d. https://geo.pinhais.pr.gov.br/geo/mapa/pinhais/.

PMP - Dispõe sobre o Zoneamento, o Uso e a Ocupação do Solo no Município de Pinhais. 2022. https://leismunicipais.com.br/plano-de-zoneamento-uso-e-ocupacao-do-solo-pinhais-pr (Accessed on 26 Dec 2024).

Popelka, S. J., and Smith, L. C. 2020. Rivers as political borders: a new subnational geospatial dataset. Water Policy, 22(3): 293–312. https://doi.org/10.2166/wp.2020.041.

Querol, M. 2019. Revisiting the Concept of Water Tenure: Filling the gap between water rights and water governance. FAO. https://www.waterlex.org/wp-content/uploads/2021/05/191130-Water-Tenure-Paper-Final-w-logo.pdf.

Reichwein, S., Smaniotto Costa, C., and Westerholt, D. 2019. Fostering biodiversity in densely built-up areas. A contribution to the debate on research needs. *In:* Menezes, M. and Smaniotto Costa, C.(Eds.) Neighbourhood & City - Between digital and analogue perspectives. Lisbon: Lusófona University Press: 167–184.

Richa, C. H., and Schafer, A. G. 2014. Correção de Inconsistências nas Geométricas de uma Bacia Hidrográfica Utilizando Funcionalidades da Extensão Espacial de um Sistema Gerenciador de Bancos de Dados. https://www.researchgate.net/publication/334222919_correcao_de_inconsistencias_nas_geometrias_de_uma_bacia_hidrografica_utilizando_funcionalidades_da_extensao_espacial_de_um_sistema_gerenciador_de_bancos_de_dados (Accessed On 03 Nov 2024).

Rocha, E. 2019. Intervenções, reinvenções e subjetividades sociais em Curitiba: análise da criação da Praça de Bolso do Ciclista sob a ótica da Antropologia do Espaço. *In:* Menezes, M., Smaniotto Costa, C. (2019). Neighbourhood & City – Between digital and analogue perspectives. Lusófona University Press: 199–208.

Rocha, E., and Smaniotto Costa, C. 2022a. The gated communities and their socio-spatial configurations in the Metropolitan Region of Curitiba, Brazil. *In:* Yapicioglu B. and Lalenis, K. (Eds.). Boundaries and Restricted Places. The Immured Space. Elgar Studies in Planning Theory, Policy and Practice: 155–168. https://doi.org/10.4337/9781800884083.00020.

Rocha, E., and Smaniotto Costa, C. 2022b. O efeito metrópole de Curitiba: as configurações espaciais e os reflexos socioambientais da periferização. *e-cadernos CES*, 36: 131–148. https://doi.org/10.4000/eces.6793. http://journals.openedition.org/eces/6793.

Rocha, E. P., Demarch, M. C. R., and Piola, T. S. 2023. Análise preliminar da vulnerabilidade ambiental e sustentabilidade social na região da sub-bacia do Rio Bacacheri em Curitiba (Brasil). Press: 293–307 *In:* Smaniotto Costa, C., Menezes, M., Pallares-Barbera, M., Pastor, G., Rocha, E. do P. and Villalba

Condori, K. O. (Eds.). Rios Urbanos na Ibero-América: Casos, Contextos e Experiências / Ríos Urbanos en Iberoamérica: Casos, Contextos y Experiencias. Cultura & Território, Vol. 6. Lusófona University. https://doi.org/10.60543/ecati/zngj-t014.

Smaniotto Costa, C. 2016. A Framework for Guiding the Management of Low-impact Mobility towards Making Room for Sustainable Urban Green Infrastructure. Journal of Traffic and Logistics Engineering, 1(4): 74–82. https://doi.org/10.18178/jtle.4.1.74-82.

Smaniotto Costa, C., and Hoyer, J. 2014. Why Invest in Urban Landscapes? Impacts of European Research on Paradigm Shift in Urban Planning, Focus, Journal of the City and Regional Planning Department (11): 60–68. https://doi.org/10.15368/focus.2014v11n1.7.

Smaniotto Costa, C., Menezes, M., Pallares-Barbera, M., Pastor, G., Rocha, E. do P., and Villalba Condori, K. O. (Eds.). 2023. Rios Urbanos na Ibero-América: Casos, Contextos e Experiências / Ríos Urbanos en Iberoamérica: Casos, Contextos y Experiencias. Cultura & Território, Vol. 6. Lisboa: Lusófona University Press. https://doi.org/10.60543/ecati/4tk9-1a66.

Smaniotto Costa, C., Šuklje-Erjavec, I., and Mathey, J. (Ed.) 2008. GreenKeys @ your city - A Guide for Urban Green Quality, IOER, Dresden.

Soares, M. C. 2024. Parques lineares em São Paulo: uma rede de rios e áreas verdes que conecta lugares e pessoas. Master's thesis. Universidade de São Paulo. São Paulo. Retrieved from https://www.teses.usp.br/teses/disponiveis/16/16135/tde-23062015-162603/pt-br.php (Accessed on 26 Dec 2024).

2

Development of Riverfronts in Belgrade

From Recreation Area to Megaprojects

Aleksandra Djukić, Jelena Marić, Nikola Mitrović* and *Emilija Jović*

1. Introduction

In contemporary urbanism, riverfronts are increasingly recognized as prominent spatial segments. Although the functional significance and perception of riverfronts have been transformed over time, both the professional community and citizens now recognize their potential for ecological and social engagement, as well as overall public benefit. This perception aligns with the principles of sustainable development as we know them today. While initially considered to hold equal importance, the three aspects - environmental, social, and economic - have since been reevaluated to prioritize the environmental aspect above the social and economic aspects. (United Nations 2015, Santos et al. 2023). However, considering that the trend of riverfront development began when financial factors were still of great importance, it is clear that in certain megaprojects, profit continues to emerge as a central purpose. This approach frequently provokes public reactions, and this chapter explores how and through which media the public has responded to a specific megaproject involving the Belgrade riverfront, to reveal the complex layers of public participation and professional engagement.

Faculty of Architecture, University of Belgrade, 11000 Belgrade, Serbia.
Emails: jelena.maric1989@yahoo.com, 0000-0002-3906-850X;
nikola.mitrovic@arh.bg.ac.rs, 0000-0002-3721-8842;
emilija.jovic@arh.bg.ac.rs, 0009-0002-1451-1285
* Corresponding author: adjukic@afrodita.rcub.bg.ac.rs, 0000-0002-7815-6588

DOI: 10.1201/9781032711737-2

In the first half of the 20th century, riverfronts served primarily as ports—central hubs for transportation and commerce (Petrović-Balubdžić 2017). Their importance was crucial for the urban economy, and these areas held utilitarian value throughout technological development. Following the economic crisis caused by the global wartime context in the mid-20th century, port development stagnated (Stupar 2009). In the modern context, there is a renewed need to restore the value of these urban sites, manifested through the opening of urban fronts towards the water. Urban renewal and regeneration often involved transformations that altered the physical appearance and contributed to the development of new visual identities, as well as changed the functional perception of space (Petrović-Balubdžić 2017).

The relationship with waterfront areas in Europe has evolved significantly throughout history. Initially, rivers were used for transportation, energy production, and waste disposal during the Industrial Revolution and subsequent periods of stagnation. This resulted in a disconnect between rivers and the citizens, with these areas functioning more as separate entities rather than as public spaces. However, in recent years, there has been a shift in perspective, viewing riverfronts as valuable urban assets with significant potential for development. Scholars such as Pinto and Kondolf (2020) highlight the importance of riverfronts as central elements in the urban landscape, positioning them as key features of city skylines. During the latter part of the 20th century, there was a noticeable trend toward revitalizing riverfront areas in response to the decline of industrial activities. This shift led to the implementation of major redevelopment projects along the waterfronts of different cities, including Oslo, Hamburg, Liverpool (Desfor and Jørgensen 2004), and Tampere (Rek-Woźniak 2023). While riverfront development is complex, successful examples often involve a diverse range of activities and active engagement from various social and age groups. This approach is based on both social and economic aspects. Pinto and Kondolf (2020) discuss this theory in their work, aligning it with the contemporary principles of sustainable development, previously mentioned at the beginning of the chapter. In considering a megaproject, it is essential to rely on a wide range of functions, thus creating content for various user groups. When a complex does not depend on a single source of funding, such as the sale of residential or private spaces, but instead of a variety of revenue streams, the economic sustainability of the project is ensured. In other words, when multiple investors are involved in the megaproject, including the participation of the city itself, the complex gains value on multiple levels.

HafenCity in Hamburg represents this idea. The initial plan, which involved the development of ten distinct districts interconnected by the riverfront, ensured functional diversity (Hanakata and Gasco 2018). Another key aspect of a successful waterfront megaproject is citizen participation. In planning *the Five-Star City Centre* in Tampere, Finland, civil organizations were involved, and the plans were executed transparently and systematically (Rek-Woźniak 2023). However, in certain contexts, reality can differ. While certain cityscapes may eventually become ingrained in a city's identity, it does not imply that the professional community or citizens endorsed the initial transformations. In their project on the Copenhagen waterfront, Desfor and Jørgensen (2004) argue that the euphemism *quality construction* was used as a political tool to justify the construction of inaccessible housing, rather than creating a project designed for public benefit. In this way, the authors suggested the potential issue of social segregation in building complexes that primarily rely on fulfilling economic aspects.

Reflecting on such examples also raises the question of whether time catalyzes a somewhat romantic view of certain cases. The extensive return of rivers to urban cultural life seems to be a byproduct of other initiatives. Given that awareness of the importance of riverfronts arose during an era when sustainable development emphasized the financial aspect as a primary benefit, it is clear that the riverfront, even after industrial expansion, continued to serve as a *financial resource.* In light of this, Goss's (1996: 237) claim that "*some cities have attempted to create a waterfront where none had previously existed*" becomes fully comprehensible and aptly conveys the trend where urban development leaned on riverfront development and branding. Nevertheless, prototypes are designed to analyze and extract essential lessons that require attention and action. In other words, it should be acknowledged that certain cities truly treated the waterfront as a zone connecting water and land for the first time, rather than as a zone of separation between them. Accordingly, pioneering projects have their flaws, and we cannot view these mistakes in the same way as we would the mistakes of later examples that failed to implement previously learned lessons.

In a context somewhat closer to the subject of this chapter, the case of Zagreb in Croatia opens a new discussion. For a long time, the city maintained an unclear relationship with its river regarding the urbanization of the waterfront. Although initial steps toward a megaproject were taken as early as 2014, this momentum became evident from 2017 to 2019. The authority's proposal to transform the riverfront into Zagreb Manhattan faced significant backlash from citizens and the professional community (Čamprag 2024). While the project proposal included recreational spaces, cultural facilities, and open public areas, the main argument against it was the lack of continuity with the existing urban fabric and design. During the project presentation, the process was inadequately explained and lacked transparency, reminding citizens of the very example that will be the focus of this text: the Belgrade Waterfront megaproject (Čamprag 2024). In their research, Bulatovic and Djukic (2022) demonstrate that planning documentation plays a crucial role in developing a project from concept to realization, as it enables the assessment of the needs for executing a particular idea. It is precisely due to the neglect of this principle that citizen reactions emerge. Public protests and debates led to the suspension of the Zagreb project but also illustrated a strong interest in the participatory process of urban planning. This example underscores the importance of lessons learned from history and highlights the active involvement of citizens in urban planning decisions.

All these examples highlight how complex historical, economic, and social layers shape today's riverfronts. In doing so, they provide an overview of the various aspects involved in the development of megaprojects and set the foundation for a more detailed analysis of Belgrade's unique waterfront development.

2. Belgrade—From Plain Shore to the Urban Riverfront

2.1 Dynamics of Waterfront Development from the Origins of the City to the Modern Era

Throughout history, rivers have often posed significant threats, as shown in the timeline. For centuries before the modern era, Belgrade was built nearby on higher ground but kept a safe distance from the water. With the start of the industrial era in the 19th and early 20th centuries, an industrial zone, a cargo port, and railway lines were developed along the Danube River (Đukić et al. 2014). This development integrated the river belt but also served as a buffer zone for residents (Djukic et al. 2018). Rivers were seen as resources but also carried stigmas. However, between the two World Wars, when global economic growth slowed, a strong river culture appeared in the area known as Bara Venecija (Mitrović, 2023), which will be explored further later in this chapter (Janic Ljubisavljevic et al. 2024). This part of the city was first included in the General Plan of Belgrade in 1923 (Figure 1). In the 1930s, Bara Venecija, an undeveloped area along the Sava River, became an unofficial beach (Figure 2). Reflecting this trend, the public complex Belgrade Fair was established in 1937 to serve this purpose (Petrovic-Balubdzic 2017). This project marked the start of construction and the development of an urban identity based on it.

After World War II, the city embarked on a planning and reconstruction process. Informal urban forms like beaches no longer had a place in the city. Along the river, a

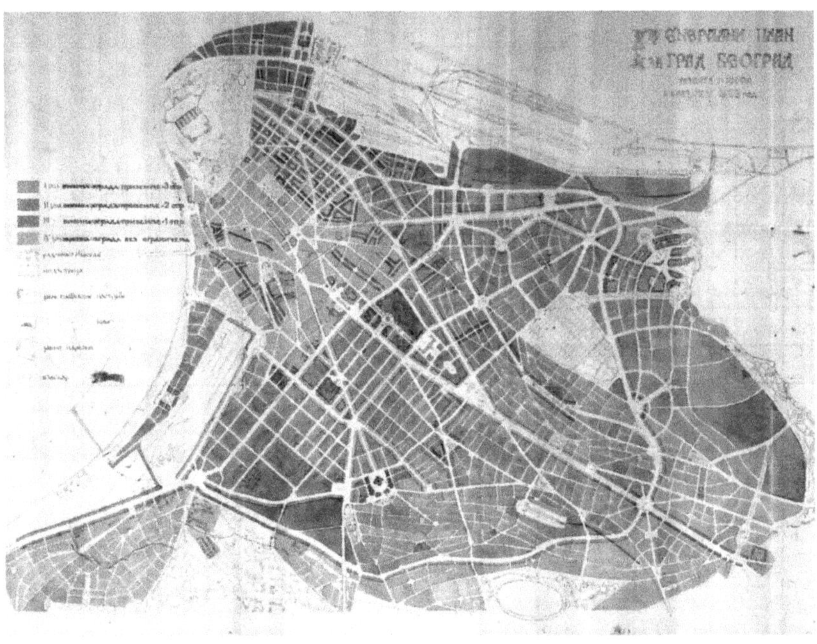

Figure 1: Đorđe Kovaljevski, General Plan of Belgrade, 1:10.000, 1923. Source: Djukic et al. 2014.

Figure 2: Informal urban life on the Belgrade riverfront. Source: Janic Ljubisavljevic et al. 2024.

promenade was laid out, extending a significant urban project - the development of a new settlement on the opposite bank. The size and importance of this settlement were such that it felt like a parallel Belgrade was emerging on the other bank, aptly named New Belgrade. The settlement was built according to C. I. A. M. principles[1] with key features including an orthogonal grid and the layout of monumental structures, which established a pattern that influenced the city as a whole (Bogdanovic 1986, Djukic 2015). This new context, with Belgrade now extending on both banks of the two rivers, established a completely new position for the riverfront, not only spatially but also functionally and culturally.

Through the centuries, the role of the river for Belgrade evolved from a threat and boundary into an important space for economic and cultural activity. The construction of New Belgrade ultimately established the river as a key element of the city's urban identity, laying the foundation for its further development as an integrated part of the city's structure.

2.2 Dynamics of Waterfront Development in the Modern Era

The dawn of the last decade of the 20th century brought an economic crisis due to internal conflicts. Planned urban development of the waterfront stagnated until the beginning of the 21st century. Conversely, unregulated construction emerged as

[1] C.I.A.M. (Congrès Internationaux d'Architecture Moderne) was founded in 1928 in La Sarraz (Switzerland) by 25 leading European architects of the time. The principles applied in the construction of New Belgrade (Ristić Trajković, Stojiljković, Ćirić 2013) are based on the Athens Charter, which is believed to have been developed from the C.I.A.M. IV Congress held in 1933, focused on the analysis of the Functional City (Gold 2019) .

a phenomenon across the city, with the riverfront also being informally inhabited (Petrovic-Balubdzic 2017).

The location previously known as Bara Venecija is now called *The Sava Amphitheater*. The fact that this area, with the construction of the *Belgrade Fair*, was the birthplace of the riverfront that would later become the city's most recognizable vista places, places its historic significance in context. In the second decade of this century, competitions were announced for urban planning solutions directly connected to this area. However, the processes related to competitions, plans, and strategies lack sufficient transparency when it comes to the Sava Amphitheater site.

This serves as an introduction to the megaproject that soon began developing at this location. At the time it was presented, the area to be planned varied 177.27 ha. The project envisioned 2,000,000 m^2 of construction, with exactly half for residential buildings. The second most common use would be commercial, and the main architectural landmark, an office building, was planned to reach a height of 210 m. The megaproject would be implemented in three phases: the first to be completed in eight years, and the final phase in 30 years (Zeković et al. 2018).

3. Riverfront as Belgrade Waterfront

One of the urban projects whose formal planning process has been linked to numerous controversies is the megaproject that currently defines the city's identity – Belgrade Waterfront (Čukić et al. 2022). Although urban dynamics at the end of the 20th century were unsatisfactory, efforts were made to start a design competition for the Sava Amphitheater area. However, unresolved issues such as land ownership, among other challenges, further delayed progress (Djukić et al. 2020). By 2012, public competitions were held for nearby areas around the main focus of this chapter, attracting notable figures like Daniel Libeskind and Zaha Hadid. The year 2012 marks the official start of the Belgrade Waterfront project (Djukić et al. 2020).

In 2014, the megaproject was presented to the citizens for the first time, when the then Prime Minister of Serbia, Aleksandar Vučić, together with the investor from the United Arab Emirates, Mohamed Alabbar, organized a project presentation and press conference. In the following days, the project was showcased to the broader public within the sales center of the company, Belgrade Waterfront. With no plan defining its purposes, capacities, or essential urban parameters, the only medium used was a visual representation (rendering) of the final stage (Figure 3) (Petrovic-Balubdzic 2017).

The Belgrade Waterfront project is a joint venture between the Government of Serbia and Eagle Hills, a private development company based in Abu Dhabi. It is valued at €3 billion and was officially intended to transform a neglected urban area into a luxury district (Bartlett et al., 2015). As a country dependent on oil revenues, the UAE established a strategy of investing its income abroad in preparation for a post-oil future (Bartlett et al. 2015). This project fit perfectly into that strategy, as Serbia, being a developing market, offered relatively low land prices, while the potential for value growth was tied to urban transformation (Bartlett et al., 2017). In this way, revenue is generated, and capital is increased. Eagle Hills is known for this type of business model. Moreover, there are claims that such projects are a covert means of expanding the UAE's global influence through the construction of its now recognizable architectural style, thereby strengthening its international position through a soft power approach (Koelemaij and Janković, 2020).

The collaboration between the UAE government and Serbia was legislatively approved as a basis for developing the BW project (OG RS 3/2013); however, *through Lex Speciales*, Serbia was required to adapt all legal amendments to suit the foreign investor's interests, creating an imbalance of power (Zeković et al. 2018). This context created an environment of near-complete exclusion of citizens as well as domestic professionals, which later proved accurate. Only two local institutions participated in the planning process. The first one is the Belgrade Urban Planning Institute, which created the BW Master Plan Concept in July 2014 based on graphic documentation prepared by SOM (Skidmore, Owings & Merrill), an international architectural firm, deviating from standard planning procedures. The second is the Agency for Spatial and Urban Planning of the Republic of Serbia, which prepared the plan fully following the master plan (Perić and D'hondt 2020). In September 2014, despite numerous complaints, amendments to Belgrade's General Urban Plan were officially adopted. Furthermore, in June of the same year, the Serbian government passed a formal decision to develop a Spatial Plan for a Special Purpose Area, a planning instrument of higher legal authority intended to bypass local coordination mechanisms and accelerate planning procedures (Perić and D'hondt 2020).

The civil sector's response included the activities of certain non-governmental organizations that monitored government decisions and translated all irregularities into colloquial language for the public. Over 3,000 complaints were filed against the proposal to amend the general plan, but neither this nor the hundreds of citizens actively participating in public discussions could halt the development process. The plan was officially adopted in 2014 (Perić and D'hondt 2020). In 2015, construction works commenced, and as Djukić et al. (2020) concluded, significant logistic facilities and nodes, such as the railway and main bus stations, were relocated to accommodate the project, challenging the city's overall development and establishing new power dynamics regarding central area development in Belgrade. The final resistance by citizens before execution manifested in protests during 2016 and 2017, but these, like previous attempts, were unsuccessful. This is a prelude to Belgrade's Waterfront tale: the Sava River, with the megaproject rising above, and defining the entire riverfront skyline (Figure 3).

3.1 Public and Professional Engagement

Public and professional engagement in urban development is essential for fostering inclusive, sustainable, and well-integrated urban spaces. Effective public engagement ensures that community voices shape development projects, aligning them with local needs and fostering a sense of ownership and trust in the process. Legitimacy in such processes depends on transparency, meaningful public consultation, adherence to legal and regulatory frameworks, and the equitable distribution of impacts throughout the community. Meanwhile, professional engagement introduces technical expertise and strategic planning, guiding projects toward viable and impactful outcomes.

Figure 3: Presentation render of the megaproject Belgrade Waterfront from 2014 (source: https://www.danubeogradu.rs/2014/06/projekat-beograd-na-vodi-otvoren-za-posetioce/).

Insights into public and professional engagement within the BW megaproject are based on the following sources:

(1) Official Governmental and Planning Documents,

(2) Civil Society Reports and Activist Publications,

(3) Media Coverage, such as newspapers and internet blogs,

(4) Scientific Journal Articles and Conference Papers,

(5) Public Insights and Debates (mainly simulated or unrealized),

(6) Court Records and Legal Proceedings,

These sources together illuminate the diverse and intricate layers of public and professional engagement, presenting multiple perspectives on how the BW project has been perceived, contested, and implemented. Rather than reflecting a balanced or participatory process, the BW megaproject reveals a consistent marginalization of public engagement, professional autonomy, and citizen input, underscoring significant procedural deficiencies and governance failures in the development of urban riverfront spaces.

Positioned on Belgrade's Sava River, the BW project promised substantial redevelopment but instead showcased a top-down approach that largely disregarded public and professional input. This disconnection reflects broader systemic issues, where planners, citizens, and even local government often find themselves sidelined by powerful investor-state alliances. The BW project is a paradigmatic example of trends in urban megaprojects across transitional societies, where private interests often override public participation and democratic processes. The BW project is frequently cited as a case where public engagement was simulated rather than authentically facilitated, demonstrating the broader systemic issues within urban planning in Serbia (Perić 2020).

From a professional standpoint, urban planners and architects in Serbia have faced significant limitations when voicing concerns about the BW project. Unlike in some European contexts where urban planning incorporates close participation of citizens, Serbia's post-socialist setting remains marked by strong centralized control and limited participatory practices. Planners, traditionally responsible for balancing public and private interests, became marginalized in favor of a top-down decision-making process led by national-level politicians and investors. This approach involved minimal public participation and limited institutional oversight. Notably, the BW plan was adapted by SOM Architects before consulting with local experts or residents, and later amended into Belgrade's master plan, bypassing typical planning processes and public input. This lack of meaningful professional engagement fueled criticism from the local architects' community, which questioned both the project's ethical and spatial impacts (Perić 2020, Perić and Maruna 2022). The authoritarian approach driving the BW project has also been marked by the misuse of legal instruments. In order to expedite the project, the Serbian government adopted a Special Spatial Plan for the area, bypassing the usual regulatory processes that involve rounds of public debate and public input. This legal maneuvering not only accelerated the project's timeline but also significantly limited opportunities for public intervention. Additionally, the introduction of the *Lex Speciales* in 2015, which allowed for the fast expropriation of land for the BW project, further sidelined the public interest in favor of the private investor, Eagle Hills (Transparency Serbia 2015). This law, despite public protests and legal challenges, was passed, showcasing the imbalance of power between the government, private developers, and the public (Perić 2020).

In July 2014, the International Network for Urban Research and Action penned an open letter addressing concerns about the BW project. This group of 50 urban development experts, visiting Belgrade for a conference, emphasized the city's unique urban qualities and potential for a vibrant waterfront (Pescanik 2014). However, they caution against the economic risks of entrusting the entire project to a single investor, which could leave Belgrade with an incomplete and underutilized space. They argue that the current plans follow a standard corporate model that does not reflect the city's cultural or social fabric and could exclude the local community from both the design and possible benefits. Also, the experts highlighted the project's potential to harm Belgrade's ecology, warning against building on flood-prone riverbanks without measures for climate resilience. They advocate for more inclusive urban planning, involving the local community to create a development path that meets local needs and honors the city's distinctive character, identity, and provides continuity with the existing urban matrix. The letter closes with a call for a phased, community-driven approach that prioritizes environmental sustainability and aligns with Belgrade's unique identity rather than imposing a generic, profit-driven design.

Citizens' responses have been notably shaped by their limited measures for influencing the BW project's direction. Public opposition included initiatives where activists highlighted flaws in the project's planning and execution processes. These groups voiced concerns over inadequate public transparency and the perceived neglect of community needs in favor of developer interests. Despite some efforts at public involvement, such as a public insight session with over 200 participants in November 2014 (Čukić, Sekulić, Slavković, Vilenica 2015), residents felt that their input was ultimately disregarded. Over 3,000 citizen objections were either dismissed or superficially considered, reinforcing public skepticism about the authenticity of participatory opportunities in Serbia's urban development context (Perić 2020). Reports and statements from activist groups such as *"Ne davimo Beograd"* [Don't Let Belgrade D(r)own][2] and the *"Ministarstvo prostora"* [Ministry of Space][3] offer a detailed perspective on citizen-led engagement. These organizations have documented their objections, organized protests, and facilitated public discussions, helping to shape the discourse around insufficient transparency and accountability in the BW project, emphasizing the phenomenon of simulation of public debate.

The blog, *"Serbian Protesters Say Authorities Were Behind Demolitions,"* (2016) details a controversial incident in Belgrade that ignited significant public outcry. In April 2016, around 30 masked men armed with sticks and equipped with construction machinery demolished buildings in Belgrade's Savamala district - the site of the BW project. The demolition, which allegedly included the assault and unlawful detention of bystanders, triggered allegations that the Serbian authorities were involved or complicit. The above-mentioned activist group *"Ne davimo Beograd"* [Don't Let Belgrade D(r)own] organized a protest with around 4,000 people, calling for the resignation of several city officials, including the Police Minister and the Mayor of Belgrade. According to an *Annual Report of the Protector of Citizens (Ombudsman) of Serbia*, police records and conversations suggest that authorities deliberately ignored calls for help and likely coordinated with those responsible for the demolitions (Annual Report, 2016). Protesters argue that the authorities' silence and denial of involvement raise deeper concerns about transparency and accountability.

Each of the aspects presented reflects underlying concerns about legitimacy and collectively highlights the failure to fulfill public expectations concerning fairness, justification, and procedural integrity in the planning and execution of the BW project. Legitimacy in urban development rests on principles of transparency, authentic public consultation, adherence to legal frameworks, and equitable outcomes for all stakeholders. However, the BW project has been characterized by a lack of transparency, with key agreements and project details kept from public scrutiny. There are notable discrepancies between the investment figures promoted publicly and the actual contractual commitments (Grubbauer and Čamprag 2018). Public insights and debates were often bypassed or conducted superficially, serving more as formalities than genuine participatory exercises, thereby undermining the credibility of public involvement (Cvetinović et al. 2016). Furthermore, legal instruments were strategically manipulated—such as the introduction of a *Lex Speciales* to expedite expropriation and construction permits—which circumvented standard procedures and limited opportunities for public objection (Grubbauer and Čamprag 2018). This approach not only marginalized professional and civic voices but also resulted in

[2] The official website of this organization: https://nedavimobeograd.rs
[3] The official website of this organization: https://ministarstvoprostora.org

an uneven distribution of impacts, favoring the investor while neglecting broader social and environmental concerns. Civil society groups and urban movements have consistently critiqued this imbalance, framing it as a violation of the "right to the city" and a striking example of state-corporate collusion that excludes citizens from shaping the future of their urban environments (Božilović 2018). Consequently, the BW project serves as a clear example of a breakdown in the normative principles that should underpin legitimate urban governance.

This discussion highlights that urban riverfronts have increasingly become battlegrounds between public interests and private development agendas, particularly in transitional urban contexts where governance frameworks are either fragile or evolving. In Belgrade, local communities have advocated for a riverfront that functions as an inclusive public space rather than a commercial development intended for elite use, especially in a landscape where global megaprojects typically prioritize private interests. This tension between private development and public accessibility in riverfront initiatives underscores the necessity for systemic reforms in Serbia's planning and regulatory practices. By prioritizing open public participation in future riverfront projects, Serbia could cultivate more inclusive and resilient urban environments that respect both the ecological significance of river systems and the social needs of riverine communities (Perić and Maruna 2022). For authentic public engagement to occur, a shift towards a more collaborative and transparent decision-making process is essential, one that involves a broader spectrum of stakeholders, including civil society and local communities.

4. User Perception Based on Surveys

To expand the information on public and professional engagement with the riverfront, especially the BW megaproject, offering insights into how people experience and value urban spaces, how projects impact everyday life (bridging the gap between design intentions and actual use), surveys from two previous research studies will be discussed.

The aim of the two surveys (conducted in 2018 and 2024) was to gather user perspectives and provide an overview of local opinions regarding the BW project and the ongoing changes to the Sava and Danube riverfronts. The traditional method for obtaining qualitative data was employed, conducted through an online questionnaire. Eleven thematic semi-open questions about the local population's perceptions of the continuing changes to the Sava and Danube riverfronts were part of the survey's structure. Before these, three introductory questions were asked to determine the respondents' age, gender, and living area in Belgrade.

4.1 The Structure of the Surveys

The survey was based on Kevin Lynch's theory (Lynch 1960), alongside user activities, thermal comfort, perception, and satisfaction analysis. The structure of the survey was designed to capture a broad understanding of the public's engagement and attitudes toward Belgrade's riverfronts, covering demographic information, general use patterns, specific preferences, and perspectives on the BW project. The survey comprised eleven questions, organized into three sections (Table 1 provides a comparison between 2018 and 2024 studies):

Demographic Profile (Question 0): gathering basic demographic information about respondents (gender, age, and living area of each participant),

General Riverfront Use and Preferences (Questions 1–8): aimed to understand the general usage patterns and preferences for the Sava and Danube riverfronts (how often and with whom respondents visited these spaces, their main reasons for visiting, and which specific parts of the riverfronts were most frequented, as well as evaluation of spatial qualities),

Landmarks and the BW Project (Questions 9–11): investigated respondents' views on prominent riverfront landmarks and the BW project.

This structured approach allowed for a comprehensive understanding of user demographics, general riverfront usage, specific area preferences, and attitudes toward recent developments, with a focus on the role of the BW project in shaping public engagement with Belgrade's riverfronts. Furthermore, the results from the 2024 survey are compared with those from the survey conducted in 2018 to examine how user perception has changed from the beginning of the BW construction until today. The aim is to understand shifts in public perception of the project from its early construction phase in 2018 to the near completion of the riverfront in 2024. Understanding user perception is essential for public engagement, as it empowers citizens and ensures that their needs and preferences shape urban development projects. When people see their feedback integrated into planning, they are more likely to trust and support the outcomes, leading to more vibrant and accessible public spaces (Marić and Djukić 2018). These two studies, conducted six years apart, include responses from 281 participants in 2018 and 120 participants in 2024.

4.1.1 Public Perception of Belgrade's Riverfronts in 2018 (survey with 281 respondents)

The survey with an e-questionnaire was conducted among the Belgrade population in July and August of 2018. In total, 281 respondents participated (Djukić et al. 2020). The survey aimed to gather insights into residents' use and perception of Belgrade's riverfront areas. Demographic data showed a slight majority of female respondents (56%), with most participants living in Old Belgrade (60%), followed by New Belgrade (27%). The largest age group was 25–34 (32%), with other age groups fairly evenly represented.

The survey revealed high engagement with the riverfront, as 95% of respondents actively used these spaces, primarily for socializing, with most visits occurring several times per month. People often visited the waterfront with friends or partners, underscoring its role as a social space. Entertainment (46%) and recreation (37%) were the most popular reasons for visiting.

Usage patterns varied between the Sava and Danube riverfronts. The Sava Riverfront, including Ada Ciganlija and New Belgrade Quay, was favored by 50% of respondents, while one-quarter mainly used the Danube riverfront, especially around

Zemun. The survey also indicated that Old Belgrade's waterfront areas were less frequented.

Accessibility was the top priority for respondents when choosing riverfront areas, followed by path quality and greenery. Important landmarks included the "Jugoslavija" Hotel in Zemun (47%) and Kalemegdan Fortress (41%). When asked about the new BW promenade, opinions were mixed, with 42% expressing a positive view, though only 23% said it encouraged them to visit the Sava riverfront more often.

4.1.2 Public Perception of Belgrade's Riverfronts in 2024 (survey with 120 respondents)

The survey was conducted among the Belgrade population in Jun and July 2024. The survey results included a total of 120 respondents (Djukic et al. 2024), where female respondents were slightly more prevalent than males (58% female). New Belgrade and Old Belgrade were the most common residential areas, with 34% and 43% of respondents respectively. The age group 25–34 was most prevalent (41%), followed by 35–49 (29%).

An overwhelming majority (98%) actively used Belgrade's riverfronts, with 39% visiting almost daily. Social interactions remained a primary motivation, with respondents visiting mainly with friends (39%) or partners (28%). Leisure (48%) was the top reason for visits, with family activities, like walking with children or pets, also notable.

The Danube riverfront was most popular, favored by 42% of respondents, with Zemun and New Belgrade as frequently visited areas. On the Sava, participants evenly distributed their use among New Belgrade, Old Belgrade, the BW area, and Beton Hala.

Accessibility and greenery were the most valued qualities of riverfront spaces, with urban furniture, lighting, and pavement quality also contributing to area selection. Kalemegdan Fortress remained a significant landmark (41%), while the BW Tower and "Jugoslavija" Hotel also held importance, reflecting the growing recognition of newer developments.

Attitudes toward the BW project showed an increase in approval. A majority (53%) liked the new BW promenade, and 81% indicated they visited the Sava riverfront more often due to the BW project, illustrating its role in increasing engagement with the waterfront.

4.1.3 Comparative Considerations from the Surveys

The 2018 and 2024 surveys on public engagement with the Belgrade riverfronts reveal important shifts in user demographics, usage patterns, and attitudes toward the Belgrade Waterfront (BW) project, offering a comparative insight into the evolving relationship between citizens and the city's urban spaces (Table 1).

In terms of demographic characteristics, both surveys showed a slight prevalence of female respondents. In 2018, 56% of respondents were female, while in 2024, this figure slightly increased to 58%. A notable shift occurred in the residential distribution of respondents: in 2018, 60% of respondents lived in Old Belgrade,

Table 1: Comparison between the 2018 and 2024 studies.

Q No	Questions	2018 responses (281 participant)	2024 responses (120 participants)
0	Demographic Profile: Age Gender Location	25 to 34y – 32% Female – 56% Old Belgrade – 60%	25 to 34y – 41% Female – 58% Old Belgrade – 43%
1	Do you use the Belgrade riverfront areas?	95% YES	98% YES
2	How often do you walk along the riverfronts?	almost every day 16% once or twice per week 17% several times per month 43% several times per year 24%	almost every day 39% once or twice per week 18% several times per month 23% several times per year 20%
3	When you walk along the riverfront, you are with?	alone 12% with partner 36% with family 15% with friends 35% with pet 2%	alone 2% with partner 28% with family 23% with friends 39% with pet 8%
4	State the main reason for using the riverfronts.	recreation 37% entertainment 46% walking with children 8% walking with a pet 2% other reasons 7%	recreation 48% entertainment 21% walking with children 12% walking with a pet 15% other reasons 4%
5	Which area of Belgrade riverfront do you use mostly?	The Sava riverfront 37% The Danube riverfront 29% Ada Ciganlija 13% Both are equally 21%	The Sava riverfront 26% The Danube riverfront 42% Ada Ciganlija 12% Both are equally 20%
6	Which paths along the Sava riverfront do you usually use?	New Belgrade quay 46% Old Belgrade Beton Hala 19% Old Belgrade Indoor Arena 19% Ada Ciganlija 16%	New Belgrade quay 42% Old Belgrade Beton Hala 43% Old Belgrade Indoor Arena 8% Ada Ciganlija 7%
7	Which paths along the Danube riverfront do you usually use?	Old – Zemun 37% New Belgrade, Ušće district 30% Old Belgrade, Marina Dorćol 25% Ada Huja 9%	Old – Zemun 38% New Belgrade, Ušće district 30% Old Belgrade, Marina Dorćol 26% Ada Huja 6%
8	Which are the spatially related reasons you chose a certain area of the Belgrade riverfront? Choose the level - % of influence	Quality of pavement 31% Quality of urban furniture and lighting 18% Greenery 28% Accessibility 46% Nearby recreational activities 10% Nearby leisure activities 27%	Quality of pavement 23% Quality of urban furniture and lighting 22% Greenery 29% Accessibility 31% Nearby recreational activities 18% Nearby leisure activities 33%

Table 1 contd. ...

...*Table 1 contd.*

Q No	Questions	2018 responses (281 participant)	2024 responses (120 participants)
9	Which of the following are the most important landmarks along the Belgrade riverfront areas? Choose the level - % of influence	Hotel "Jugoslavija" 47% Kalemegdan park and fortress 42% Museum of contemporary arts 22% "Ušće" business district 15% BW tower 6% Ada bridge 21%	Hotel "Jugoslavija" 23% Kalemegdan park and fortress 41% Museum of contemporary arts 12% "Ušće" business district 11% BW tower 25% Ada bridge 12%
10	Do you like a new promenade along the Sava River (BW promenade)?	Yes 42% No 22% I don't know 36%	Yes 53% No 21% I don't know 26%
11	Do you use the area of the Sava riverfront along the BW project more often after the beginning of the implementation of the BW project?	Yes 78% No 22%	Yes 81% No 19%

while New Belgrade followed at 27%. By 2024, the proportion of respondents living in Old Belgrade decreased to 43%, and New Belgrade rose to 34%. Zemun and the Third Belgrade maintained their lower respondent rates in both surveys. Age demographics also shifted between the two surveys. In 2018, the largest group of respondents was aged 25–34 (32%), followed closely by younger respondents aged 16–24 (24%). By 2024, the dominance of the 25–34 age group increased to 41%, with fewer participants in the younger 16–24 category (unreported for 2024 but presumably lower). The 35–49 age group also grew in importance, from 22% in 2018 to 29% in 2024, reflecting a slight ageing of the population engaging with the waterfronts.

The usage of the riverfronts also evolved. In 2018, 95% of respondents actively used the waterfront, with 44% visiting several times per month, and only 5% of respondents indicating they did not use the riverfront. By 2018, the active use of riverfronts increased to 98%, with 39% of respondents visiting almost daily, signaling a rise in the regularity of riverfront usage. Social activities remained a key reason for visiting, with participants in both surveys primarily using waterfronts with partners or friends. However, the 2024 survey saw a shift in company types, with 39% of respondents visiting with friends (up from 35% in 2018), while those visiting with a partner declined to 28% from 36%. The growing popularity of walking with children or pets was also evident in 2024, accounting for 12% and 15% of responses, respectively, suggesting an expanded use of the riverfronts for family-oriented activities.

Patterns of use between the Sava and Danube riverfronts showed considerable changes between the two surveys. In 2018, the Sava riverfront, particularly New Belgrade Quay (Marić 2021), was the most popular, with 50% of respondents favoring it, while the Danube riverfront was used by about one-quarter of respondents. In contrast, by 2024, preferences had shifted in favor of the Danube riverfront, with 42% of respondents favoring it, while only 26% preferred the Sava riverfront, and 12% favored Ada Ciganlija. The data suggests that the Danube waterfront's appeal grew over time, possibly due to increased development or improvements in accessibility and amenities.

The perception of urban landmarks and spatial elements also evolved (Figure 4). In 2018, respondents highlighted accessibility (46%) and the quality of pathways and greenery as the key reasons for choosing specific waterfront areas. By 2024, accessibility remained a top priority (31%), but the importance of greenery rose to 29%, indicating a growing appreciation for natural elements in urban spaces. Landmark preferences also shifted slightly. In both surveys, the "Jugoslavija" Hotel in Zemun and Kalemegdan Fortress in central Belgrade were prominent, with 47% and 41% respectively in 2018. However, by 2024, Kalemegdan Fortress remained significant (41%), while the importance of the "Jugoslavija" Hotel declined to 23%, and newer landmarks such as the BW Tower (25%) and the Ušće business center (11%) gained recognition, reflecting the impact of new developments on the urban landscape.

The public's attitude toward the BW project experienced notable changes between the two surveys. In 2018, the public was divided regarding the new BW promenade along the Sava riverfront, with 42% expressing appreciation for it. By 2024, support had increased, with 53% sharing a positive view of the promenade. Additionally, while in 2018 only 23% of respondents reported using the Sava riverfront more frequently due to the BW project, by 2024, this figure had risen sharply to 81%, indicating that the project's completion had significantly influenced the public's use of the Sava riverfront.

In summary, the comparative analysis of the 2018 and 2024 surveys reveals a clear increase in public engagement with the Belgrade waterfronts, particularly in the use of the Danube riverfront, and a growing acceptance of the BW project. This evolution highlights the dynamic relationship between urban development and public space use in Belgrade, as well as the importance of accessible, green, and landmark-rich spaces for fostering citizen engagement with the city's waterfronts.

5. Lessons Learned and Conclusion

The complex question regarding the urban riverfront area has been tackled in this chapter, offering several important conclusions that could be summarized in the form of lessons learned. The development of Belgrade's riverfronts, particularly the whole process of BW megaproject, from the planning documents to the construction sites offers significant insights into these challenges and focus on the key messages: (01) The importance of public participation and cooperation; (02) Balancing private and public interests; (03) Focusing on environmental sustainability and multifunctionality; (04) The importance of historical urban context; (05) Transparency in the urban development process and (06) Importance of media, activism and public perception. In addition to these lessons, it is crucial to highlight the significance of research based on public perception to ensure citizen satisfaction and a deeper understanding of their wishes and needs.

Figure 4: BW megaproject 2025. Source: A. Djukić, 2025.

The most important lesson from the BW project is the necessity of robust public engagement throughout the planning and implementation processes. Initial resistance from citizens and grassroots activism highlighted a widespread necessity for inclusion in decision-making. Engaging communities not only helps to identify local needs and aspirations but also fosters a sense of ownership over the public realm. This is the case with the majority of top-down projects that are usually related to the riverfront areas in the European context. Future developments must prioritize inclusive strategies that empower residents to voice their concerns and contribute to the vision for their urban environment.

The BW development could be considered an example of prioritizing private over public interests. Today, the government additionally insists on expanding the territory of the megaproject with the explanation that it is a project of special interest for the Republic of Serbia (Petrušić 2024) (Figure 5). While private investments are vital for urban renewal, striking a delicate balance between financial considerations and community needs should be the top priority in urban riverfront developments.

While private investments can drive urban revitalization, they should not come at the expense of community welfare or accessibility. Therefore, the profit should not come before the social sustainability approaches. Also, the environmental implications of urban riverfront developments cannot be overstated. The BW project, while ambitious, encountered many issues regarding its environmental impact and failure to adequately address issues of sustainability, especially regarding the heat island effect emphasized in many studies done for this area (Simić 2023) Megaprojects developed in the area with greenery and natural benefits must consider sustainable design principles that harmonize with the natural landscape, enhancing both ecological health, physical comfort, outdoor thermal comfort and aesthetic value.

Additionally, in order to ensure adequate usage of the riverfront areas, the diversity of activities and features along the whole area is important. There is a need for riverfronts to be multi-functional and accessible to a wide range of users, creating a vibrant and livable public open space that has the potential to become a playground for most people in urban areas. The BW project initially faced criticism for its lack of inclusive public spaces that cater to varied demographics. Also, one of the major issues regarding the BW project is the neglect of historical values, historical context, and spatial and architectural heritage, thus neglecting Belgrade's complex identity. The biggest mistake of the project, without a doubt, is the lack of transparency of the planning process and the lack of adequate public participation. Establishing trust between authorities, politicians, investors, and citizens is essential for fostering cooperative and healthy relationships that support sustainable urban riverfront development. Finally, the powerful role of media and activism should be addressed. The BW project showcased the importance of NGOs and citizen groups' activism in shaping the overall public opinion regarding an urban development project. This opinion later shapes the characteristics of emotional connection to the place and *genius loci* of the area, which can lead to misuse of the space. However, the survey done among the BW users showed that significantly more users are in the BW area in 2024 compared to 2018, when the activism and criticisms over BW were stronger.

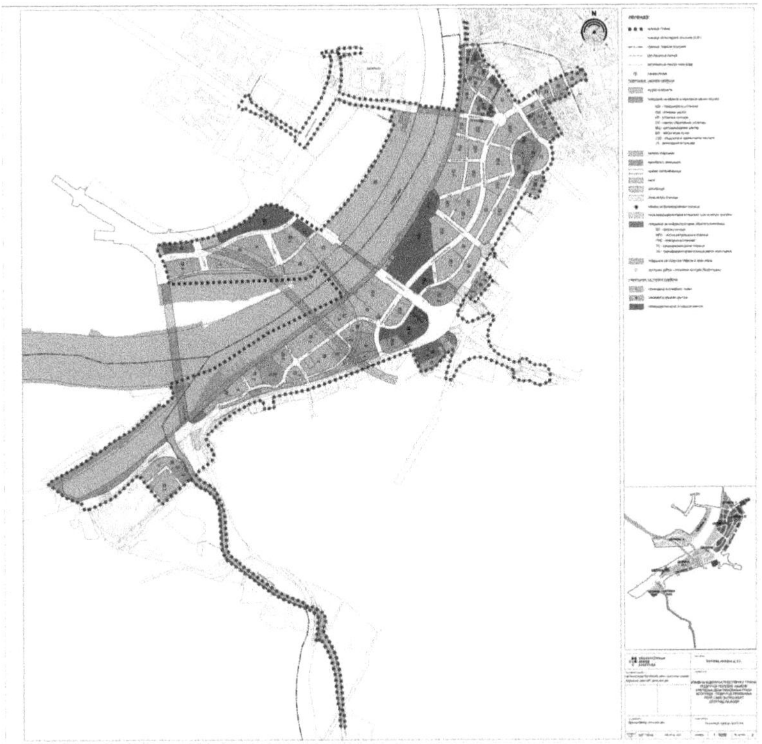

Figure 5: Further expansion of the BW megaproject. Source: https://www.mgsi.gov.rs/lat/dokumenti/rani-javni-uvid-povodom-izrade-izmena-i-dopuna-prostornog-plana-podrucja-posebne-namene-8.

In summary, the lessons learned from BW development underscore the need for a holistic, community-oriented approach to riverfront regeneration by prioritizing public engagement, balancing private and public interests, emphasizing sustainability, and fostering diverse uses.

References

Bartlett, W., Ker-Lindsay, J., Alexander, K., and Prelec, T. 2017. The UAE as an emerging actor in the Western Balkans: the case of strategic investment in Serbia. Journal of Arabian Studies. 7(1): 94–112.

Bogdanović, B. 1986. Budućnost Novog Beograda, Međunarodni konkurs za unapređenje urbane strukture Novog Beograda [The Future of New Belgrade, International Competition for the New Belgrade Urban Structure Improvement]. Arhitektura Urbanizam. Special issue: 3–13.

Božilović, J. 2019. Right to the city: Urban movements and initiatives as the pulse of civil society in Serbia. pp. 151–170. *In:* Petrović, J., and Backović, V. [eds.]. Experiencing Postsocialist Capitalism: Urban Changes and Challenges in Serbia: University of Belgrade – Faculty of Philosophy, Institute for Sociological Research, Serbia.

Bulatović, S., Djukic, A. 2022. Modern Transformations of City Riverfront - Comparison of the Riverfronts of Belgrade and Zagreb. ICUP 2022 International Conference on Urban Planning, Urban Planning Cluster, Niš, Serbia: 1–8.

Cvetinović, M., Maričić, T., and Bolay, J.-C. 2016. Participatory urban transformations in Savamala. Belgrade – Capacities and Limitations. Spatium, 36(1): 15–23.

Čamprag, N. 2024. Effects and Consequences of Authoritarian Urbanism: Large-Scale waterfront redevelopments in Belgrade, Zagreb, and Novi Sad. Urban Planning. 9. Article Number: 7589.

Čukić, I., Stojić, B., Ćatić, A., and Mićanović, M. 2022. Izrada planova iza pritvorenih vrata: Izveštaj o stepenu učešća građana u procesima urbanističkog planiranja u Beogradu za 2021. Godinu [Planning Behind Closed Doors: A Report on the Level of Citizen Participation in Urban Planning Processes in Belgrade in 2021]. Institut za urbane politike / Ministarstvo prostora.

Desfor, G., and Jørgensen, J. 2004. Flexible urban governance. The case of Copenhagen's recent waterfront development. European Planning Studies. 12(4): 479–496.

Djukic A. 2015. New Belgrade: visions, plans and realizations pp. 1947–2014. *In:* G. Doytchino, A. Djukic, C. Ionita [eds.]. Planning Capital Cities: Belgrade, Bucharest, Sofia: 160–173.Verlag der Technischen Universität. Graz, Austria. http://dx.medra.org/10.3217/978-3-85125-398-6.

Djukic A., Vujičić, T., Špirić, A. 2018. Urban design competition and megaprojects in a context of identity of cultural heritage: case study Belgrade`s riverfronts. pp. 59–73. *In:* Ifko, S., and Stokin, M. [eds.]. Protection and reuse of industrial heritage: Dilemmas, problems, examples. Institute for the Protection of Cultural Heritage of Slovenia and ICOMOS Slovenia, Ljubljana, Slovenia.

Djukic, A., Maric, J., Joković, J., and Antonic, B. 2024. Re-Evaluation of Urban Renewal Waterfront Development: The Comparison of user perception regarding the Riverfront in Belgrade, Serbia. XVI. Conference of European Association of Urban History EAUH 2024. Cities at the Boundaries. Ostrava, Czech Republic. (in preparation).

Djukić, A., Marić, J., Antonić, B., Kovač, V., Joković, J., and Dinkić, N. 2020. The evaluation of urban renewal waterfront development: the case of the Sava Riverfront in Belgrade, Serbia. Sustainability. 12(16): 6620.

Đukić, A., Vukmirović, M., and Vaništa-Lazarević, E. 2014. Planski okvir, konkursi, projektna rešenja i vizije za uređenje Savskog amfiteatra [The planning framework, projects, competition designs and visions for planning the Sava Amphitheater]. Izgradnja. 68(3–4): 103–120.

Gold, J. R. 2019. Athens Charter (CIAM), 1933. The Wiley Blackwell Encyclopedia of Urban and Regional Studies. 1–3.

Goss, J. 1996. Disquiet on the waterfront: Reflections on nostalgia and utopia in the urban archetypes of festival marketplaces. Urban Geography. 17(3): 221–247.

Grubbauer, M., and Čamprag, N. 2019. Urban megaprojects, nation-state politics and regulatory capitalism in central and Eastern Europe: The Belgrade Waterfront project. Urban Studies, 56(4): 649–671.

Janić Ljubisavljević, J., Mrišuk, S., Nakaniši, B. Đ., Đivković, Z., Vuk, Ž., and Gajić, D. 2024. Plažama predaka [Along Ancestral Beaches]. Likos Auto. Belgrade.

Koelemaij, J., and Janković, S. (2020). Behind the frontline of the Belgrade Waterfront: A reconstruction of the early implementation phase of a transnational real estate development project. Experiencing postsocialist capitalism: Urban changes and challenges in Serbia: 45–65. University of Belgrade – Faculty of Philosophy.

Lynch, K. 1960. The image of the city. M.I.T. Press, Cambridge.

Marić J., 2021. Towards More Resilient City: Improving Public Health by Increasing the Usage of Urban Green Open Space - a Case Study of New Belgrade. DOCONF/2021 - Facing Post-Socialist Urban Heritage: proceedings, 8-9th October 2021: 274–283. Budapest, Hungary.

Marić J., and Djukić, A. 2018. Connecting the Centre of Belgrade with the Danube and Sava Riverfronts to Increase Attractiveness. Periodica Polytechnica Architecture. 49(1): 23–28.

Mitrović, N. 2023. Urban Highway Areas: Reconsidering Morphological Elements of the Approach to Urban Studies. In: XXIX International Seminar on Urban Form. ISUF 2022 Urban Redevelopment and Revitalisation. A Multidisciplinary Perspective. 6th June – 11th September 2022, Łódź–Kraków, edited by A.A. Kantarek et.al.: 1125-1131. Lodz University of Technology Press. https://doi.org/10.34658/9788367934039.90.

Peric, A., and hondt, F. D. 2020. Squandering the territorial capital in the Balkans? Urban megaprojects between global trends and local incentives. Urban Design International. 27(4): 288–306.

Perić, A. 2020. Public engagement under authoritarian entrepreneurialism: the Belgrade Waterfront project. Urban Research & Practice. 13(2): 213–227.

Perić, A., and Maruna, M. 2022. Post-socialist discourse of urban megaproject development: From City on the Water to Belgrade Waterfront. Cities. 130. Article Number: 103876.

Petrovic-Balubdzic, M. 2017. Creating the Belgrade waterfront identity through a prism of architecture and urban planning competitions. Spatium. 37: 74–81.

Pinto, P. J., and Kondolf, G. M. 2020. The fit of urban waterfront interventions: matters of size, money and function. Sustainability. 12(10): 4079.

Rek-Woźniak, M. 2023. Discourses of growth in megaproject-based urban development: a comparative study of Poland and Finland. Policy and Society, 42(2): 245–258.

Ristić Trajković, J., Stojiljković, D., and Ćirić, D. 2013. Transformation of Modern Urban Space of New Belgrade in Conditions od the Post-Socialist Translation and Sustainable Development. Docomomo/2013 – The Survival of Modern Form Coffee Cup of Plan: proceedings, 7-10 August 2013: 381-387. Espoo, Finland.

Santos, L. C. T. D., Frimaio, A., Giannetti, B. F., Agostinho, F., Liu, G., and Almeida, C. M. V. B. 2023. Integrating environmental, social, and economic dimensions to monitor sustainability in the G20 countries. Sustainability, 15(8): 6502.

Stupar, A. 2009. Grad globalizacije – izazovi, transformacije, simboli. [The City of Globalization – Challenges, Transformations, Symbols]. Orion art, Belgrade.

United Nations. 2015. Transforming our World: The 2030 Agenda for Sustainable Development. United Nations. https://sustainabledevelopment.un.org/post2015/transformingourworld/publication.

Zeković, S., Maričić, T., and Vujošević, M. 2018. Megaprojects as an Instrument of Urban Planning and Development: Example of Belgrade Waterfront. In S. Hostetter, S. Najih Besson, and J.-C. Bolay [eds.]. 153–164. Technologies for Development: From Innovation to Social Impact, London: Springer. https://link.springer.com/chapter/10.1007/978-3-319-91068-0_13.

Additional sources - internet blogs

Bartlett, W., Ker-Lindsay, J., Alexander, K., and Prelec, T. 27 November 2015. UAE policies towards the Western Balkans: Investment motives and impacts. LSE Middle East Centre Blog. Retrieved 22 June 2025. https://blogs.lse.ac.uk/mec/2015/11/27/uae-policies-towards-the-western-balkans-investment-motives-and-impacts/.

Beograd na vodi: Predizborni trik ili realnost [Belgrade Waterfront: Pre-election trick or reality]. 17 February 2014. Radio Slobodna Evropa. Retrieved 25 October 2024. https://www.slobodnaevropa.org/a/beograd-na-vodi-predizborni-trik-ili-realnost/25266956.html

Čukić, I., Slavković, L. j., Vilenica, A., and Sekulić, D. 6 May 2015. Report from Belgrade Waterfront. Eurozine. Retrieved 22 June 2025. https://www.eurozine.com/report-from-belgrade-waterfront/ .

Dragojlo, S. 12 May 2016. Serbian Protesters Say Authorities Were Behind Demolitions. Balkan Insight. Retrieved 27 October 2024. https://balkaninsight.com/2016/05/12/thousands-serbians-protest-accusing-authorities-for-masked-demolition-05-12-2016/.

'Ne davimo Beograd' uz Toplodolce – pozivaju na odbranu reka ['Don't Let Belgrade D(r)own' with the citizens of Topli Dol – they call for the defence of the rivers]. September 9 2019. Južne Vesti. Retrieved 25 June 2025. https://www.juznevesti.com/drustvo/ne-davimo-beograd-uz-toplodolce-pozivaju-na-odbranu-reka/.

Open letter to the people of Belgrade. 01 August 2014. Pescanik. Retrieved 27 October 2024. https://pescanik.net/open-letter-to-the-people-of-belgrade/.

Petrušić, L. 17 August 2024. Dalje širenje Beograda na vodi - Državni projekat u privatnom interesu. ["Further expansion of Belgrade on the water - State project in private interest"]. Radar. Retrieved 29 October 2024. https://radar.nova.rs/ekonomija/beograd-na-vodi-sirenje/.

Simić, I. 11 July 2023. Prva mapa urbanih toplotnih ostrva Beograda: gde su najčešće tropske noći? ["The first map of Belgrade's urban heat islands: where are the most tropical nights?"]. Klima101. Retrieved 29 October 2024. https://klima101.rs/toplotna-ostrva-u-beogradu-tropske-noci/.

Additional sources - legislation

Annual Report 2016. Ombudsman of Serbia. Protector of Citizens, 2016. 13 June 2017, Belgrade, Serbia. Retrieved 25 June 2025. https://www.ombudsman.org.rs/attachments/article/134/Introduction_2016%20Annual%20Report.pdf.

Documentation for Early Public Insight regarding the preparation of Amendments on the Spatial Plan for the Area of Specific Use for the Development of the Coastal Area of the City of Belgrade - The Coastal Area of the Sava River for the "Belgrade Waterfront" Project (Dokumentacija za Rani javni uvid povodom izrade Izmena i dopuna Prostornog plana područja posebne namene uređenja dela priobalja grada Beograda – Područje priobalja reke Save za Projekat „Beograd na vodi"). 07 August 2024. Ministry of Construction, Transport and Infrastructure of Republic of Serbia. Retrieved 25 June 2025. https://www.mgsi.gov.rs/lat/dokumenti/rani-javni-uvid-povodom-izrade-izmena-i-dopuna-prostornog-plana-podrucja-posebne-namene-8.

'Lex specialis' – for exceptional cases only." 16 March 2015. Transparency Serbia. Retrieved 25 June 2025. https://www.transparentnost.org.rs/index.php/en/110-english/naslovna/7473-lex-specialis-for-exceptional-cases-only.

OG RS 3/2013 (Official Gazette of the Republic of Serbia 3/2013). The Act confirming the Agreement on Cooperation between the Government of the Republic of Serbia and the Government of the United Arab Emirates (Zakon o potvrđivanju Sporazuma o saradnji između Vlade Republike Srbije i Vlade Ujedinjenih Arapskih Emirata). Retrieved 25 June 2025.

3

Exploring Stakeholder Engagement in Urban River Recovery

An Ibero-American Perspective

Ana Vera,[1,*] *Antoni Mas-Ponce,*[1] *Montserrat Pallares-Barbera,*[1] *Nagayamma Aragão,*[2] *Sandrine Simon,*[3] *Sònia Sànchez-Mateo*[4] and *Carlos Smaniotto Costa*[5]

1. Introduction

Throughout history, rivers have been of fundamental importance in the development and success of human settlements. Despite their historical importance, rivers have frequently been severely altered, often straightened, channelised, and covered, with riparian vegetation replaced by concrete and roads, leading to their diminished visibility and disappearance from urban landscapes. Land use changes and urbanisation have demonstrably led to severe urban problems, including increased

[1] Universitat Autònoma de Barcelona, Geography Department, Barcelona, Spain.
Emails: tonimasponce@gmail.com, 0000-0003-3116-8423;
Montserrat.pallares@uab.cat, 0000-0002-1595-3248
[2] Universidade Lusófona, Department of Architecture and Urban Planning and Interdisciplinary Research Centre for Architecture, Urbanism and Design (University of Lisbon), Lisbon, Portugal.
Email: nagayamma.aragao@ulusofona.pt, 0000-0002-2207-9587
[3] University of the West of England, School of Architecture and Environment, Bristol, UK.
Email: sandrine.simon@uwe.ac.uk, 0000-0002-3769-9834
[4] Fundació RIVUS, Granollers, Spain. Sonia.Sanchez.
Email: Mateo@uab.cat, 0000-0001-6544-2967
[5] Universidade Lusófona, Department of Architecture and Urban Planning and Interdisciplinary Research Centre for Architecture, Urbanism and Design (University of Lisbon).
Email: smaniotto.costa@ulusofona.pt, 0000-0002-1896-4663
* Corresponding author: Ana.vera@uab.cat, 0000-0001-7865-1610

DOI: 10.1201/9781032711737-3

flooding, erosion, landslides, the proliferation of unhealthy settlements, extensive waste pollution, and the spread of disease. Thus, rivers are frequently, and unjustly, reduced to their negative consequences (Smaniotto et al. 2023). To respond to climate change, cities need to formulate strategies capable of facilitating paradigmatic shifts and generating solutions to the environmental and social deterioration of urban environments. One such shift resides in giving more importance to stakeholder participation and, in particular, to examining its role in river recovery under a climate change scenario (Khatibi et al. 2021).

River recovery is a complex challenge and requires the involvement of numerous stakeholders because certain stakeholders have the potential to influence urban decision-making (Yang 2014). Thus, an open, transparent, and participatory process is essential to ensure that river recovery benefits nature, drainage systems, urban morphology, and local communities. Consequently, a river recovery initiative must encompass and incorporate ecological, social, economic, and urban dimensions. Only an integrated programme can help achieve sustainable, resilient, and inclusive outcomes.

Anthropogenic interventions in rivers, affecting both their bed and water composition, are a primary cause of reduced ecological balance (Cao et al. 2020). This degradation extends to other dimensions, such as the decline of biodiversity, the artificial transformation of urban environments, and the impoverishment of the urban fabric. In the European context, the European Water Framework Directive (EWFD) furnishes political support for a paradigmatic transformation (EC 2000). It establishes a regulatory framework for water body protection and restoration, aiming to achieve 'good status' in all inland waters, both chemically and ecologically. In addition to relying on such hard scientific data, a change in perspective is also required, encompassing both those involved in ecosystem-based river management and the broader group of concerned stakeholders. This change in mindset is vital, as historical river management prioritised controlling nature through top-down strategies (Zingraff-Hamed et al. 2022). The EWFD mandates a comprehensive shift in environmental governance and management, moving from a top-down to a collaborative approach. This approach aims to enhance community-government interactions, exemplified by community-led and government-fed initiatives in planning practice. Therefore, the RUN Project, focusing on urban rivers, requires a dynamic approach to systematically identify, explore, and integrate stakeholder perspectives in a human and ecosystem-centred river restoration. Consequently, a methodology for mapping stakeholder perspectives, needs, and interests is required to ensure their incorporation into subsequent planning phases and to organise and support co-creation processes. This study employs comparative stakeholder mapping to analyse stakeholder engagement in regeneration activities. Instead of providing generic stakeholder recommendations, this study offers strategic insights for project leaders to determine appropriate stakeholder involvement at each regeneration stage.

This study analyses, in an exploratory way, the stakeholder participation in river recovery, giving particular importance to how knowledge, as well as the relationship between urban rivers and society, are being enhanced. The analysis encompasses the RUN Project's network, which involves fifteen partners from eight countries, of which six are in Latin America, alongside partners in Spain and Portugal. This analysis examines the eleven cases brought to the network by these partners (listed in Table 1). The wide range of cases allows us to illustrate the diverse stakeholder processes employed in collaborative solution-finding and their variation across local contexts. Despite significant variations in geographic, institutional, and demographic contexts across the case study regions, the analysis reveals substantial similarities across all eight countries, particularly in the nature of river management problems.

2. Integrating Participatory Processes into Urban River Restoration and Governance

Given that nature is a public good, ecosystems like rivers require support beyond single groups, such as governments or private companies. As Zingraff-Hamed et al. (2022) state, resilient and sustainable environmental governance requires collaborative planning that includes both institutions and concerned citizens. Within such governance process, citizens are no longer considered as passive recipients of governmental planning but, instead, as active contributors (Meijer 2019). Doing so will be particularly useful in addressing one major shortcoming in the government's approaches to prioritising legibility in their projects – that is, the fact that they often overlook valuable local knowledge and social structures (Scott 1998). Citizens and their local knowledge provide, in fact, a vital resource for comprehending the urban environment as a dynamic space that resists complete control and management. As a consequence, using local knowledge requires first identifying and then involving stakeholders in public goods management. The effectiveness of planning and implementation indeed depends on collaborative strategies, co-creation procedures, and self-organisation.

The necessary political and strategic backing for collaborative strategies and community engagement is provided by the UN's 2015 Sustainable Development Goals, in particular SDG 11 (which focuses on inclusive, safe, resilient, and sustainable cities), the New Urban Agenda (UN 2016, 2022), the Paris Agreement and the European Commission's Green Deal (EC 2021). All focus on and highlight the particularly topical dimension of urban issues. In addition, the New Leipzig Charter (EC 2020) places great trust in what it calls the "transformative power of cities for the common good". All together, these key documents provide a policy framework for implementing these agreements by defining crucial principles for urban governance, which include integration, participation, co-creation, multi-level collaboration, and a place-based approach (EC 2020, p. 6-8).

The cumulative effects of decades of economic development and rapid urbanisation have resulted in the progressive degradation of urban rivers, characterised by increased impervious surfaces (reducing groundwater recharge), diversion (channel modification), disconnection of rivers and floodplains, elevated water demand, and increased contaminant levels. Everard and Moggridge (2012) highlight the considerable deterioration of urban rivers, resulting in the loss or severe limitation of their beneficial ecosystem services for human communities. Legal frameworks like the EWFD, the Habitats Directive (EC 1992), and the Millennium Ecosystem Assessment (EC 2005), which detailed river ecosystem services, significantly increased awareness of issues concerning rivers in human settlements, both within and beyond Europe.

The imperative for river restoration, particularly pronounced in developing countries, is exemplified by Latin America, where, as elucidated by Walteros and Ramirez (2020), the majority of urban watercourses are impacted or degraded, with contaminant inputs and untreated wastewater discharges identified as the primary agents of river degradation. Moreover, Mondragon-Monroy and Honey-Roses (2016), in their review of successful Latin American urban river restoration, found that available documents were sparse and dispersed, and lacked comprehensive planning process descriptions. According to the authors, their study revealed far fewer restoration projects than expected. However, a range of motivating factors for the restoration projects was identified by the authors, such as the enhancement of fluvial transportation, commerce, and trade, exemplified by the River Magdalena (Colombia). Other drivers included restoring ecological and landscape quality (Bogota Project, 2012, due to a public legal claim) and creating public spaces to address segregation (Medellín, Colombia). D'Orey et al. (2022) found that large restoration projects were scarce in Latin America, with most analysed cases being community-led initiatives aimed at improving local environments.

From this, some prime objectives for the restoration projects being explored by the RUN Project emerged and were outlined by Smaniotto et al. (2023). A key premise, which is evident in the analysed cases, is the need for the engagement of riverine communities and broader society in river restoration. In addition, three initial lessons that could help to understand how and why urban rivers transit from 'problem spaces' to 'opportunity spaces':

1. Polycentric governance and co-creation, involving diverse stakeholders, are considered more effective for managing public goods than traditional top-down approaches (Zingraff-Hamed et al. 2020, D'Orey et al. 2022).
2. Community-based participation, defined as "the engagement of residents in decision-making processes concerning issues that impact their lives" (Hoffman 2022, p.1), as corroborated by Stewart and Lithgow (2015), remains in need of refinement to ensure inclusivity and the provision of meaningful citizen input into decision-making processes.
3. Co-production and co-creation, as social processes, are growing as a means of unlocking innovation and justice towards more sustainable and resilient development. Sideris (2021) and Hoffman (2020) agree that clarifying participation goals and identifying structural inequalities in planning are essential for challenging power imbalances and creating more effective cooperation.

3. The Rationale of Stakeholder Mapping

Stakeholder involvement in problem-solving and decision-making is increasing and becoming a permanent feature (Scholes and Clutterbuck 1998). In river restoration, this calls for exploring the collaborative potential of concerned communities and institutional stakeholders. In the realms of policymaking, urban governance and urban river restoration, a discernible trend towards stakeholder engagement and collaborative participation has emerged (Koontz 2005, Smaniotto et al. 2023). Meaningful stakeholder contributions require a clear understanding of who the relevant stakeholders are. In the literature, as Reed and Curzon (2015) point out, stakeholder definitions are broad and varied, encompassing employees, customers, competitors, communities, special interest groups, and society. Stakeholders can also include the general public (Carroll 1993), local, national, or international pressure groups (Scholes and Clutterbuck 1998), as well as managers, suppliers, and creditors (Hill and Jones 1992). In view of categorising stakeholders, certain scholars, such as Zingraff-Hamed et al. (2020), have adopted an approach based on roles. According to this approach, 'stakeholders' can then be described as decision-makers, implementers, coordinators, providers of expert knowledge, funders, lobbyists, or mediators. Another way of categorising stakeholders relates to the specific project stage at which they engage. For instance, they might be involved in the assessment and planning phase, or else in the design, implementation, operation and maintenance, monitoring or evaluation stages. Skarlatidou et al. (2019), on the other hand, prioritise stakeholders' interest and influence levels throughout the process, whilst Lopes and Videira (2015) and Sedlacko et al. (2014) utilise participatory systems mapping to engage stakeholders in creating causal loops, thereby providing insights and fostering knowledge exchange.

In view of such diversity of approaches aimed at defining stakeholders, Zingraff-Hamed et al. (2020, p.3) advised that using a "systematic method to identify relevant stakeholders seems to be essential to a) enable higher planning efficiency and b) reduce bottlenecks and time needed for planning, deciding, and implementing (restoration) projects". Stakeholders' Mapping and analysis can provide a framework for structuring these efforts. It consists of three key steps: identifying affected stakeholders (including ecological entities and future generations), defining relevant aspects of the phenomenon, and prioritising these stakeholders for decision-making involvement (Carayannis and Campbell 2009, Reed et al. 2009). While traditionally used in management, education, and corporate social responsibility, stakeholder mapping has been expanded to natural resource management (Skarlatidou et al. 2019). According to Reed and Curzon (2015, p.21), stakeholder mapping can be categorised into three distinct approaches:

- normative, which prioritises stakeholder legitimacy and empowerment,
- instrumental, which aims to manage stakeholder behaviour for specific outcomes, and
- descriptive, which focuses on analysing stakeholder relationships with a given phenomenon.

Stakeholders' mapping would allow a project to lead to a better understanding of people's and institutions' roles, engagement levels and benefits of being part of the process. Thus, ultimately, the aim of using such analysis has been to optimise long-term participatory governance in urban river restoration projects by identifying the influence and authority distributed across multiple layers, and this, in the specific cases of the project, as well as in river restoration projects in general. Understanding these dynamics is crucial for the success of a restoration project, such as the

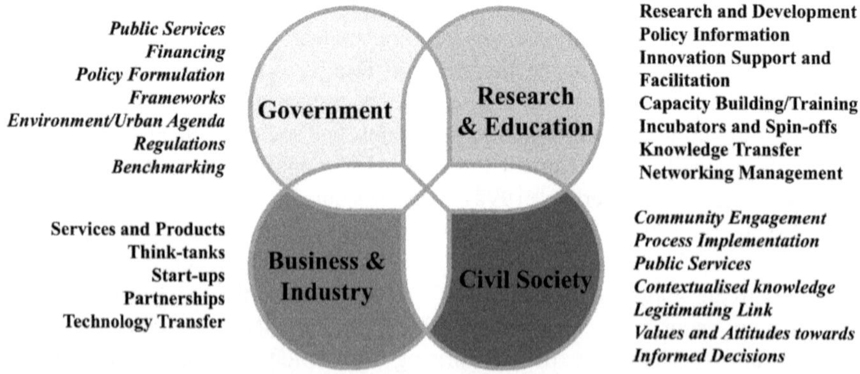

Figure 1: Quadruple Helix Stakeholders: Categories and Potential Input for River Restoration. Source: Adapted from Carayannis et al. (2012), EC (2015) and Carayannis et al. (2018).

RUN Project, since partners rank stakeholder management as one of their biggest challenges (Smaniotto 2025).

To understand the role of stakeholders in river restoration, this study uses the Quadruple Helix Model of Innovation (Figure 1) (Carayannis and Campbell 2009), which identifies four key stakeholder groups - Business & Industry; Research & Education; Government; and Civil Society. Grouping diverse stakeholders helped us to better understand their interests and how they can influence an initiative. The Quadruple Helix Model stresses that diverse knowledge and innovation potentials are of equal importance and must be addressed collaboratively. The togetherness fostered by this approach facilitates the changes needed to achieve environmentally friendly, socially equitable, and economically efficient outcomes (Schütz et al. 2019). The model emphasises the significance of societal participation in research, innovation, and decision-making.

The European Union adopted the Quadruple Helix Model in its 2015 Open Innovation 2.0 policy for a digital single market. Consistent with the Quadruple Helix Model, EC-sponsored projects and policies, alongside a growing number of governments, are placing a high priority on increased public engagement in innovation processes (Schütz et al. 2019). Because greater public involvement entails greater demand for inter- and trans-disciplinary processes, the scientific community is facing increasing pressure to elaborate and refine existing understandings of knowledge and the methods of its production and transfer.

4. Methodology

4.1 Study Design

To analyse and understand stakeholder involvement in the case studies of the RUN Project, the Quadruple Helix Model is used as a way to approach stakeholder mapping (based on the rationale explained earlier). This analysis aims to: (i) map stakeholders

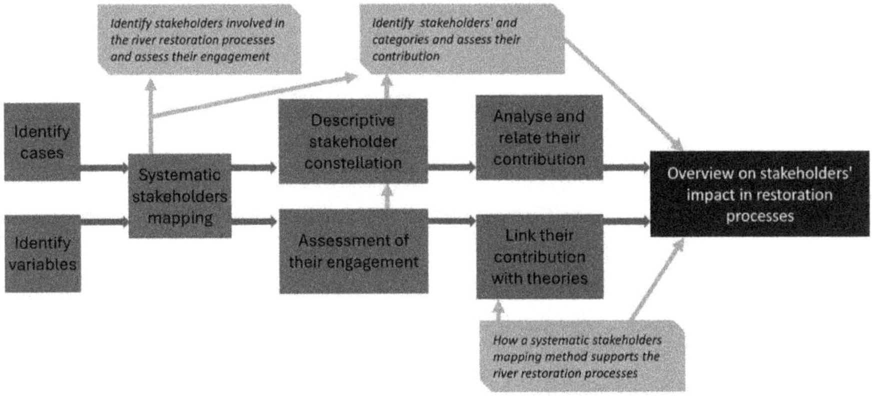

Figure 2: The methodological approach. Source: Adapted from Zingraff-Hamed et al. (2022).

involved in recovery projects provided by the RUN network, (ii) understand their contributions to restoration development, (iii) extract lessons from their contributions in achieving restoration goals, and (iv) identify potential stakeholders for future involvement (Figure 2).

The study was designed collaboratively by the RUN partners UAB and ULUSO. It was intensively developed during a collaborative workshop in Lisbon (January 2023), where the questionnaire content, the organisation of the questions, and the analysis methodology were determined. A questionnaire was developed in the project's three languages (Portuguese, Spanish, and English) and sent to the head of each of the RUN restoration project.

4.2 Data Collection and Analysis

Partners were invited via email to complete an online survey. The data collection process consisted of a questionnaire with two parts. In Part A, the respondents identified their current stakeholders and selected a corresponding category from the Quadruple Helix Model groups. For each stakeholder listed in Part A, respondents were asked to then complete Part B, which included the following nine questions:

1. **Name** of the stakeholder,
2. **Category group** – to be selected from the four Quadruple Helix groups
3. **Since when has this stakeholder been involved** – open question to be stated in month/year.
4. **Evaluation of the degree of involvement** – to be answered on a five-point Likert rating scale, ranging from 1 (low grade of involvement) to 5 (high degree of involvement).

5. **The activities performed by the stakeholder** – a multiple-choice question with 13 possible answers provided. The answers exemplify typical, common activities assumed by the stakeholders.
6. **Are there any issues or problems with this stakeholder?** In this open question, the partners were asked to very briefly describe the difficulties encountered in interacting with stakeholders.
7. **Degree of influence based on attitude, interest and power**; This question had three sub-questions:

   7a) ***What is this stakeholder's attitude about the restoration process?*** Partners were asked to assess this on a five-point Likert rating scale, ranging from (very negative) to 5 (very positive).

   7b) ***Does this stakeholder have a positive or negative interest in the regeneration process?*** – Multiple-choice, to select between "Negative interest" or "Positive interest".

   7c) **Does this stakeholder have power in the regeneration process**? A five-point Likert rating scale, ranging from 1 (has little power) to 5 (Has a lot of power).
8. **What is the impact and potential of the collaboration with this stakeholder?** This was an open question.
9. **Are there active interactions between this stakeholder and the others? If yes, with which stakeholder?** This was an open question.

The questionnaire gathered both numerical data (via Likert scales and multiple-choice questions) and descriptive insights (through open responses). The analysis used a multi-method approach, triangulating data to gain a nuanced understanding of stakeholders and their roles in river restoration. The survey questions were answered by the lead researcher for the local case study, given that the responses were based on subjective assessment.

Since this study relies on the researchers' subjective evaluations, the stakeholders' perspectives become a uniquely important quality criterion. As emphasised by the EC (2015), the Quadruple Helix Model is useful for analysing the role of communities in the innovation process: it provides a solid basis to analyse how stakeholders engage, particularly within innovation ecosystems and in tackling complex societal issues like urban river recovery. Partners identified and classified the local stakeholders according to their primary helix category. They found out that the stakeholders' core roles and interests in river recovery are connected to their underlying motivations. The analysis also assessed how their potential contributions, in terms of resources, knowledge, skills, and influence, can be translated into both tangible contributions (e.g., funding, technology) and intangible contributions (e.g., local knowledge, public support). Furthermore, it analysed the stakeholders' potential impact, the extent of their motivation, influence, and power, and whether their involvement ultimately affected the project's outcome positively or negatively.

It is worth stressing that the use of this approach went beyond simply identifying stakeholders. It also highlighted the dynamic interactions and relationships among the four helix groups, and helped us in understanding patterns of active collaboration, cooperation, conflicts, and dependencies within the helices and with other stakeholders.

5. Results and Interpretation

By analysing stakeholder participation across the international RUN Project network (in 11 urban river recovery initiatives), this study offers locally relevant insights into enhancing the relationship between urban rivers and society. Quadruple Helix Model enabled the categorisation of the identified stakeholders in each case, thereby establishing a framework for understanding their involvement, influence, power, and perceived benefits.

Table 1: Nodes, case studies and number of stakeholders. Source: the authors.

Node	**Case Study**	**City (Country)**	**Number of Stakeholders**
Watercourse			
Universidad de Cauca	Molino river	Popayán (CO)	6
Universitat Autònoma de Barcelona	Besòs and Tordera basins	Barcelona (SP)	5
Universidade Federal de Minas Gerais (UFMG)	Ribeirão Arrudas	Belo Horizonte (BR)	5
Instituto Argentino de Investigaciones de las zonas áridas	Desaguadero river	Mendoza (AR)	5
Universidad de Cuenca	Tomebamba river	Cuenca (EC)	5
Universidade Lusófona (ULU)	Ribeira das Jarbas	Lisbon (PT)	4
Claretiano Centro Universitario (CEUCLAR)	Bacacheri and Atuba rivers	Curitiba (BR)	2
Universidad Católica de Santa María	Chili river	Arequipa (PE)	-
Universidad Nacional de San Martín	Reconquista river	San Martín (AR)	-
Wetlands			
Universidad de Buenos Aires (PIRNA-IG, FFYL, UBA)	Wetlands in the Metropolitan Region of Buenos Aires	Buenos Aires (AR)	10
Universidad Nacional de Asunción	Vertedero Cateura	Asunción (PY)	-
Memories of vanished landscapes			
Laboratório Nacional de Engenharia civil (LNEC)	Memories of vanished landscapes	Lisbon (PT)	2
Maps and stools to study the landscape			
Universidad Nacional de la Plata (LIFIA)	Citizen sicence	La Plata (AR)	13
Universidade de São Paulo (USP)	urbVerde	São Carlos (BR)	4
Universidad de Valladolid	Valladolid (GPS)	Valladolid (SP)	-

5.1 Overview of Participating Nodes and Case Studies

Data for this analysis were provided by 11 out of 15 of the total RUN Project's participating nodes (73%), as four nodes did not submit responses. The partners' case studies and the corresponding stakeholder numbers (Table 1) demonstrate the significant breadth and diversity of urban river recovery initiatives, along with the varying scales of stakeholder involvement in different local settings.

5.2 Categorisation of Stakeholder Engagement Patterns

Based on the distribution of identified stakeholders across the four categories of the Quadruple Helix Model (Business & Industry, Research & Education, Government, and Civil Society/Users), Figure 3 offers a general overview of the stakeholder groups involved in urban river recovery initiatives. The distribution of stakeholders shows that urban rivers, as a common good, concern primarily governments (41%) and

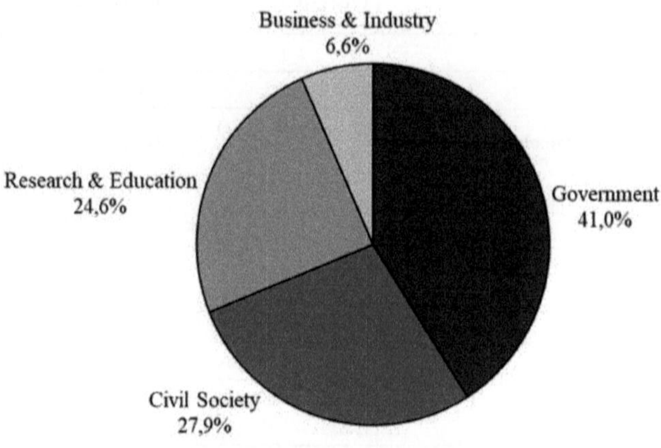

Figure 3: The percentage distribution of stakeholders according to the Quadruple Helix Model.

civil society/community groups (24.6%). The significant involvement of Research & Education stakeholders (27.9%) reflects the fact that academic partners coordinate these cases. In contrast, the proportion of stakeholders from the category Business & Industry remains minimal (6.6%), and is not evident in every case.

5.3 Assessing Stakeholder Involvement: Attitude, Influence, and Power

Table 2 quantitatively summarises the perceived engagement and impact of different stakeholder groups in river recovery efforts. It presents the average reported involvement, derived from their attitude, influence, and power, as reported by partners (nodes) for each Quadruple Helix category, across all case studies. A Likert scale, ranging from 1 (low) to 5 (high), was used for these assessments (Likert 1932, Bryman 2016). Our influence assessment reveals that all stakeholders demonstrate a very positive interest in the river restoration projects. Across all case studies, every

Table 2: Stakeholder Analysis: Involvement, Attitude, and Power.

Quadruple Helix	Involvement	Attitude	Power
Business & Industry	33% – 36%	33% – 56%	11% – 34%
Research & Education	20%– 50%	33% – 63%	10% – 33%
Government	18% – 46%	25% – 50%	17% – 45%
Civil Society	25% – 45%	30% – 59%	11% – 30%

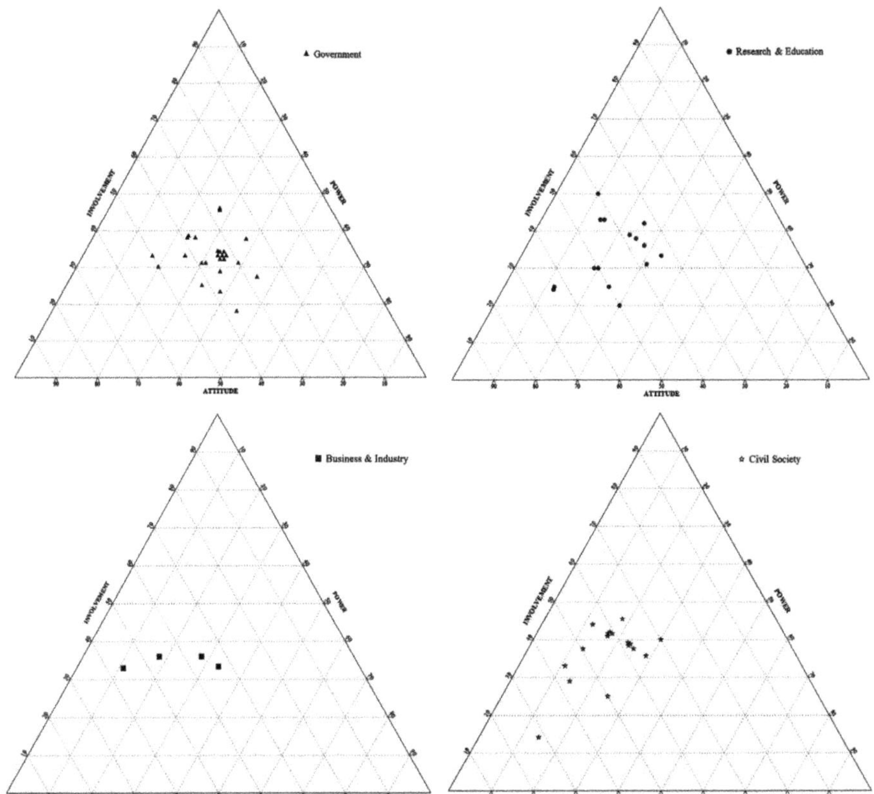

Figure 4: Involvement, influence and power of the stakeholders, displayed as Likert Scale Triangles.

stakeholder category expressed a keen interest in the regeneration process. This insight into their willingness to participate is vital for achieving river recovery goals. Indeed, the unanimous positive attitude and interest of stakeholders toward these recovery projects are considered crucial for making significant advancements.

As depicted in Figure 4, the perceived engagement and impact across stakeholder groups present a diversified picture. Comparing the results for four categories, the Business & Industry category occupies a relatively central position within the Likert scale triangle, showing a tendency towards the 'Attitude' axis, ranging from 40%

to 60%. This outcome signifies a homogeneous performance with low internal variability among the analysed stakeholders in this category.

These results indicate that Business & Industry partners significantly influence decision-making processes and wield considerable power, due to their economic or strategic importance. This enables them to actively intervene in recovery initiatives. However, their involvement and commitment to participatory actions, such as co-creation or community management, are relatively low. This pattern suggests that their participation is primarily instrumental or conditional, driven by economic interests or a desire for increased institutional visibility, rather than a genuine, long-term commitment to environmental or social goals.

The Research & Development group provides a more heterogeneous behaviour. Although being a highly participatory and collaborative type of stakeholder, this group faces limitations in terms of authority or formal power in the decision-making process. Its main contribution is related to knowledge production, awareness raising and providing support for innovation.

Stakeholders from the Government category show a relatively concentrated distribution, with a slight predominance in the influence and power dimensions, maintaining intermediate levels of involvement. This behaviour reveals that institutional stakeholders can significantly shape decision-making and have the organisational or normative power to implement measures. On the flip side, their direct involvement in collaborative, community-based or co-creation actions is moderate, suggesting that a more institutionalised and formal engagement will not be committed to more horizontal participatory practices.

Finally, the Civil Society category shows a more diffuse distribution, with a strong concentration towards the involvement dimension, with little influence and low power. This pattern highlights a high commitment to local and participatory actions, namely in promoting collaborative processes, community events and grassroots activities. However, its capacity to influence strategic decisions or mobilise structural resources is limited, reflecting an asymmetry in power sharing and formal decision-making. Such heterogeneity also suggests that the mobilisation and intervention levels vary significantly between different civil society entities, and this may be justified by factors such as available resources, organisational experience or specific territorial contexts.

5.4 Analysing Stakeholder Activities in River Recovery

The performed activities were analysed and displayed by radial diagrams (Few 2012), as depicted by Figure 5. The diagrams display the frequency of reported activities performed by identified stakeholders across all case studies, revealing their practical contributions to the urban river recovery initiatives. The 13 possible answers exemplify typical, common activities assumed by the stakeholders and are the same for the four categories. This analysis allows a critical reflection on the specific roles of each category, their limitations and potential in the collaborative co-creation processes.

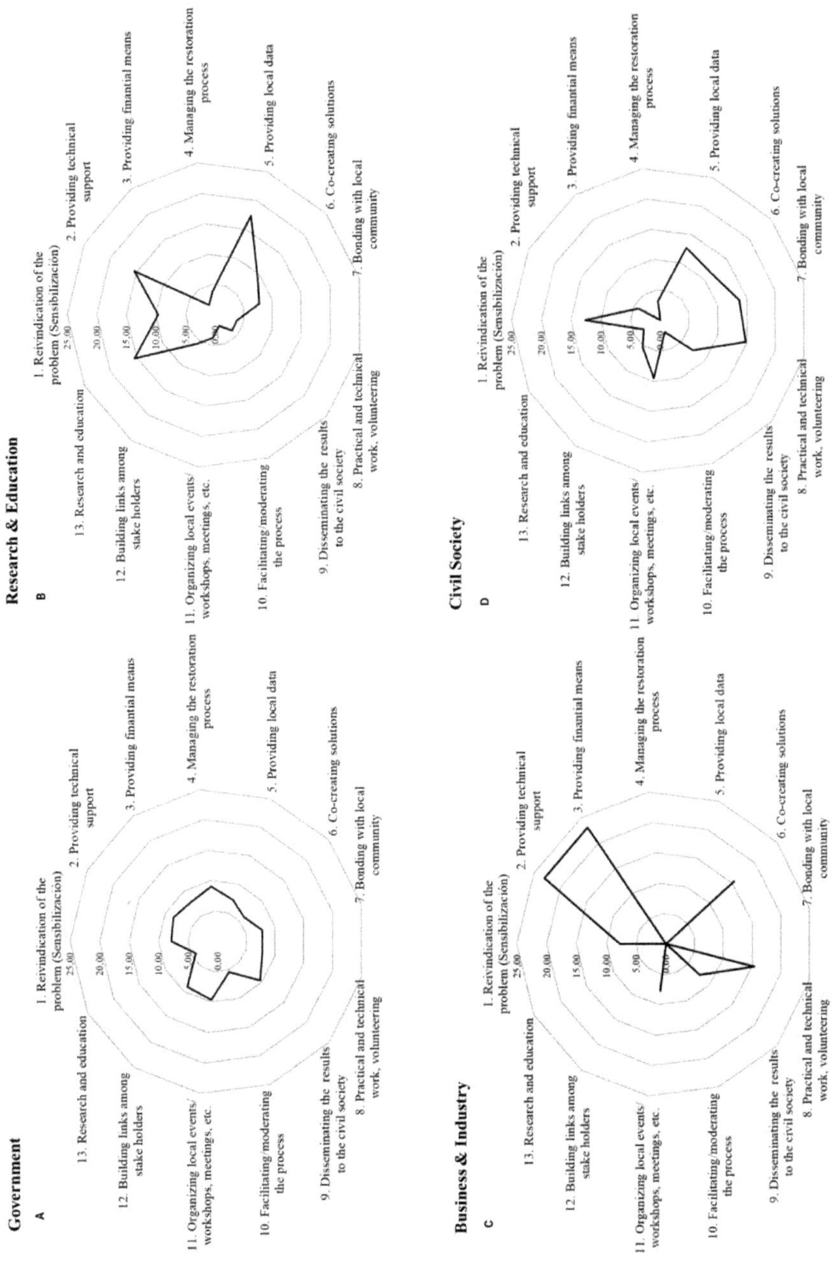

Figure 5: Stakeholders' involvement in local activities.

Regarding the category Government, graph A shows a relatively balanced action profile, with moderate participation in almost all activities, as well as contribution to the organisation of local events, building stakeholder networks and managing processes. Such activity richness is consistent with governments' key institutional role in intersectoral coordination. The category Research & Education (B), on the other hand, shows a strong commitment to providing local data, technical support and knowledge exchange. Among the main performed activities are systematising information and producing scientifically based evidence that helps in strategic decision-making, risk reduction, and supporting the design of interventions. The category Business & Industry (C) shows a greater involvement in activities related to the provision of technical support and financial resources for the recovery processes. This trend is consistent with the category's economic vocation and investment capacity. However, there is a weak participation in community-related activities, such as enhancing awareness, moderating the participatory processes and engaging in educational or scientific activities. Finally, the category Civil Society (D) shows a significant involvement in fostering relationships and participatory processes, such as awareness-raising, volunteering and co-creation of solutions. On the flip side, its participation in more technical or financial activities is only marginal.

5.5 Interactions and Perceived Issues in Interactions among the Stakeholders

The analysis of stakeholders' active interactions among themselves revealed a wide cooperation network. Observed patterns of interaction include, for instance, the creation of strong networks within certain categories, in particular Civil Society and Research & Education, between the key bridging organisations, but also limited interaction between specific groups. Understanding these interaction patterns is crucial for fostering more integrated and effective collaboration within the categories as well as with the other categories.

The qualitative data from the open-response question regarding 'issues encountered in the interaction among the different stakeholders' revealed several recurring patterns. The frequency of these issues varied across stakeholder categories and case studies, suggesting context-specific challenges in stakeholder engagement. The problems related to interacting with stakeholders were shared by the four categories, but with a divergent extent between the groups (as shown in Figure 6). The most frequently reported issues came from the Civil Society category, which highlighted challenges such as significant dependence on government and academic partners, and a lack of dedicated financial resources for local recovery projects. Government stakeholders frequently reported issues such as conflicting priorities, poor communication, bureaucratic obstacles, and delays in decision-making due to political interests. The Research & Education stakeholders often cited a lack of consistent interaction and insufficient funding, noting that funds often had to be secured through competitive calls, with technology development not consistently receiving financial support. The Business & Industry stakeholders primarily

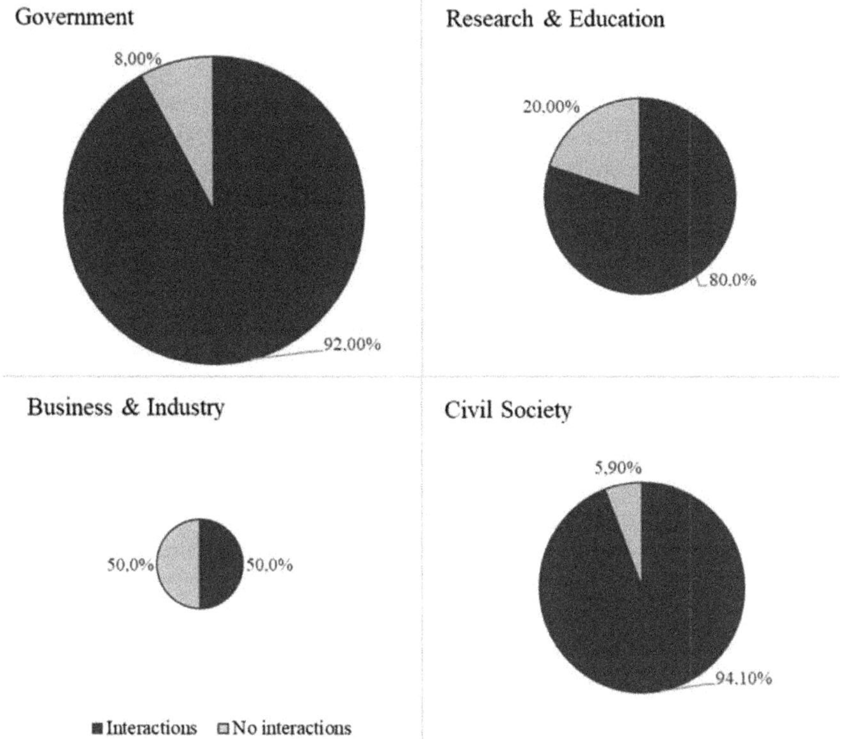

Figure 6: Number of issues identified among the stakeholders – the size of the circles reflects the total number of stakeholders in the study.

interacted with local authorities, risk management and disaster response institutions, and emergency relief organisations, including local grassroots movements.

5.6 Impact and Potential of Collaboration

The open-response data underscored the critical role of partnerships and effective collaboration in achieving the multifaceted goals of river recovery. Key reported benefits of collaboration include knowledge sharing, resource mobilisation, enhanced legitimacy of processes and projects, and improved outcomes. Additionally, the responses highlight the significant potential of the category Research & Education to boost citizen participation in public policy design.

Government stakeholders are crucial for river recovery initiatives, as they are usually the owners of the natural resources and thus provide invaluable experience and knowledge of the area. They are significantly involved in actions to reduce wastewater impact on rivers, engage in research and restoration efforts, and foster connections with other stakeholders. Their participation extends to awareness campaigns and citizen science initiatives. Their authority over the river environment means that their authorisation is required for all related activities. It also facilitates new projects by offering technical and financial aid. This presents a vital opportunity for wetland conservation and the restoration of all watercourses and bodies within the basin. These stakeholders are essential to the overall response plan, delivering critical training and community preparation for specific programs. Their wide-ranging contributions include generating and sharing information, launching new social, technical, and educational awareness campaigns, and actively participating in urban river regeneration projects. Furthermore, they help to formulate policies with specific criteria for waterway revitalisation and champion community intervention projects.

Key stakeholders directly influence the urban water network. Their collaboration involves coordinating efforts across institutions and actors, planning and implementing pilot projects, executing interventions, and providing continuous monitoring. They also work to integrate research into public policy and disseminate project information in crucial decision-making forums. As vital local actors, initially linked with the municipal project, they facilitate its implementation; their involvement is indispensable. They can also replicate environmental care projects across the entire province and build networks with actors often inaccessible to civil society. Ultimately, the main goals are to implement comprehensive public policy and enhance areas adjacent to urban rivers.

Research & Development stakeholders are actively engaged in research and development of innovative techniques to improve water quality (specifically nutrient reduction) and enhance river ecosystems. They are also crucial for generating foundational information about the basin's wetlands, which is vital for informed decision-making by other actors. The knowledge generated by academic groups would be even more impactful with continuous support. Significant efforts are dedicated to transferring knowledge about the importance of these topics to future educators in basic education and to partners who champion environmental issues. Close collaboration with the local community is essential, especially as direct contact and participation in group meetings are being established. A key challenge identified is the deficiency in health research data, which is crucial for the project's success. Despite this, the public views these efforts positively. While some actors primarily provide valuable information, others could potentially channel funds (albeit limited) to support community work. There is strong potential for contributions in interdisciplinary research and knowledge production, particularly in the social, environmental, and landscape aspects of the work. This promotes a valuable connection between scientific and technical organisations and government entities. Key outcomes include generating scientific information for publication, expanding the project, and developing technology and research for citizen participation. This includes the execution and expansion of the project, further contributing to scientific knowledge.

The Business & Industry's strategy is dependent on the current local administration. There is a strong possibility of implementing re-naturalisation actions in the developing restoration initiatives. Key concerns include ensuring the continuity of, and securing funding for these initiatives, as well as financing for the project's ongoing needs.

The Civil Society stakeholders play a multifaceted and crucial role in river recovery efforts, offering connections with various other stakeholders and the potential to attract the business and private sectors to support river conservation. Their involvement extends to environmental education and outreach, fostering citizen engagement through volunteering, and driving both research and river restoration projects. A key strength of Civil Society lies in its local knowledge and broad community support. This sector has the potential to advance citizen participation in public policy management, particularly in integrating community involvement into environmental policy for wetlands. It is instrumental in articulating measures with local residents that balance wetland restoration in places like Laguna de Rocha, characterised by local housing needs.

Civil society is vital in recognising that residents around a basin are vulnerable to environmental threats affecting their social, labour, cultural, environmental, and material well-being. Its involvement ensures that communities take ownership of technically suggested measures. This includes acknowledging the river's current and future state and actively participating in the design, execution, care, and monitoring of restoration processes. Civil Society groups provide access to local communities residing along riverbanks and possess strong community organising capabilities for proposed initiatives. Their involvement ensures sustained community engagement and long-term mobilisation. Furthermore, civil society members collaborate in data collection, co-designing tools, and in establishing restoration priorities. They contribute to data generation, co-constructing tools, and keeping residents actively involved when scientific teams are not on site. Educators, in particular, participate in the co-creation of specific classroom tools (e.g., manuals, games, activities for children) and help to transmit the project's issues within families. They also manage water resources and organise outreach events. Finally, and most importantly, civil society excels at transforming technical data into urban quality indicators and proposals for public authorities and demonstrates a strong capacity for social mobilisation.

The next section of this chapter, focused on the results of our research, provides a structured overview of the stakeholder engagement landscape within the RUN Project. The subsequent Discussion section will delve deeper into these findings, relating them back to the literature review and addressing the specific analytical points outlined earlier.

6. Discussion of Results

By examining the 11 cases and stakeholders mapping results, this study aims to compare and understand the general benefits of engaging stakeholders and whether such engagement provides a greater variety of quality decisions, as concluded by Reed and Curzon (2015), thereby increasing the sustainability of the recovery cases. The analysis reveals that the objectives of river recovery initiatives vary according to the interests and approaches of different stakeholder groups. While the category Government tends to focus on institutional and technical coordination goals, the Civil Society group demonstrates a strong commitment to ecological regeneration and the increase of urban qualities. The Business & Industry group, in turn, shows a more instrumental participatory approach. Its strategic objectives exercise is often carried out aiming at economic return, which limits its contribution to the long-term environmental or social goals. This sets river recovery as an issue at the core of Government and Research & Education.

Such heterogeneity of motivations has a direct influence on the degree of involvement and on the type of contribution of each stakeholder in the different phases of the processes. These patterns are consistent with recent studies that highlight the variability of priorities and barriers faced by different stakeholders in urban river recovery projects (Saikia et al. 2022, Yan and Ibrahim 2023). In all analysed cases, an unsustainable type of development is caused by human interventions in the riverbed or water composition, by the effects of unsustainable urbanisation processes, as well as by changes along the riverbeds and banks, leading to the fragmentation of river ecosystems. The perceived problems range from biodiversity loss to degradation of the urban fabric and missing links between the community and the riverscapes. This perception pattern highlights a widespread recognition of the need for interventions that promote ecological requalification and the restoration of the socio-environmental functions of rivers. Recent literature highlights the importance of integrating place-based approaches and multi-level governance to address the complex challenges of river recovery in urban contexts (Novalia et al. 2024).

The overall number of stakeholders and their share in the four categories of the Quadruple Helix demonstrates the fact that urban rivers, as common goods, concern mainly stakeholders from government (41% of total stakeholders) and Civil Society (24,6%). The group of Government plays a crucial role in environmental protection, acting through regulatory legislation, policies and programmes, while civil society is one of the main beneficiaries, with improved urban qualities (Zingraff-Hamed et al. 2020) and social ties (Meijer 2019, Itsumi et al. 2025). The cooperation between both categories, as well as encouraging local communities to take part in decision-making and in managing environmental resources, has proven to be effective and long-lasting (D'Orey et al. 2022). The main contribution of Research & Education lies in knowledge production and exchange, awareness raising and bringing research evidence into policies. Thus, it is crucial to strengthen the relationships of the Research & Education group with other stakeholders, towards translating its capabilities into concrete impact.

When scrutinising the levels of influence, power, and involvement of the stakeholder groups, the comparative analysis reveals a structural asymmetry. While the Business & Industry group has high levels of influence, it shows low engagement in collaborative or co-creation processes. Its activities, focused on financial and operational management, demonstrate little integration with the educational, social, or community aspects of recovery. Research & Education, on the other hand, is highly participatory and motivated, in particular in the production and exchange of technological and scientific knowledge. However, its *formal* influence on decision-making processes is limited, which undermines translating technical knowledge into direct impact. These findings are in line with research that highlights the importance of understanding the factors that influence stakeholder participation in river restoration initiatives (Yan and Ibrahim 2023). The Government group, as expected, holds a relatively centralised place, with institutional power to actively intervene in the territory. However, it shows moderate engagement in more horizontal or community-based actions. Its engagement seems to be limited by institutional and bureaucratic factors, which hinder more agile and collaborative management. The Civil Society group shows a strong involvement in relational and participatory dimensions, such as awareness-raising and volunteering. However, it lacks formal power and access to strategic decision-making mechanisms; this could be understood as a failure in the planning and governance framework (Menezes and Smaniotto 2024). These patterns reflect common challenges identified in studies on sustainable management of urban environments, where effective stakeholder participation is crucial for the success of local interventions (Rao et al. 2024).

There is also a strong relation between stakeholders' levels of influence, power, and involvement and the activities performed in recovery initiatives. The results suggest that the Business & Industry group, by adopting the abovementioned goals, adds little value to the social and educational aspects of the recovery. This behaviour could compromise the sustainability of the initiatives in the long term, although this group can be crucial for efficient technical-operational and financial support. Research & Education, by fostering innovation, creativity and knowledge sharing, improves the chances of knowledge influencing a direct application in local contexts. Another key issue of Research & Education is its close ties to communities, which can result in more robust citizen science and participatory models. This issue is also related to the group of Governance, although its involvement does not reach high levels, presumably due to institutional, bureaucratic or budgetary constraints. These calls for improving its performance range and capacities through an approach of flexibility and administrative agility, both crucial parameters of agile cities and sustainable (natural) resources management (Anwar et al. 2025). Moreover, combining these approaches (robust citizen science, participatory models and agile management) with a strong connection to places and high social capital of the Civil Society group, can result in a better-tailored context and more legitimate recovery efforts.

The interactions between knowledge production, resource efficiency and social and environmental improvements point to the importance of designing policies across sectors and stakeholders, besides demonstrating the importance of strategic partnerships. As suggested by Lopes and Videira (2015) and Sedlacko et al. (2014), gaining a deep understanding of stakeholders' roles, interests, and contributions is crucial for effectively constructing causal loops, as it illuminates the interdependencies and feedback mechanisms within a system. Likewise, understanding better the motives for engagement and the perceived problems may improve prevention strategies, reinforcing the call for collaboration and co-creation to address challenges that riverine communities face (Smaniotto et al. 2023). Human and environmental-centred governance must engage multiple stakeholders and combine multidisciplinary knowledge, participatory approaches, co-creation of solutions and multi-institutional collaboration.

The expected benefits vary among the different category groups, and so does their view on, and impact of, specific influential factors. Civil Society expects fewer risks and more improvements in the environment and the quality of life, and, when well informed, expresses a stronger symbolic tie with the river. Research & Education seeks to validate scientific knowledge and strengthen its practical applications and impacts. While the Government group looks to ensure the institutional viability of interventions, Business & Industry mainly aims, as mentioned above, at increasing financial returns, but also at enhancing its brand image as a responsible entity that appeals to social and environmental values. Such diversified visions show, on the one hand, how variegated inputs a river and the riverine ecosystems can deliver, and how many 'interests' can be involved in its improvements and recovery. On the other hand, these diversified visions also have an influence on ecological, social, economic or urban goals, conditioning the way in which these stakeholders engage in recovery processes. As Enu et al. (2025) rightly state, to overcome barriers and facilitate effective implementation of recovery projects, there is a need to align priorities. Co-creation and collaboration, e.g., with the engagement of stakeholders from the four categories, open new avenues for development and operational efficiency of recovery projects (Fenten and Dieperink 2024, Aragão et al. 2025).

Yet, integrating local knowledge held by Civil Society into formal planning processes remains a challenge. Better knowing the needs of the community affected and tapping local knowledge and resources for project planning and implementation can improve not only the local planning framework but also deliver more contextualised results. Such results can be relevant for those communities facing environmental and health risks (Corburn 2003), as is the case with rivers under urbanisation pressures. Mobilising local knowledge is one of the main activities performed by the Research & Education category. Making use of it helps in diminishing the frustration of community stakeholders, as they witness their own contributions being reflected in public policies or implemented actions (Itsumi et al. 2025). In addition, studies of the RUN Project demonstrate that the (re)naturalisation of urban rivers can foster a sense of ownership and community engagement, essential for adaptation to climate change and for achieving long-term sustainability (Aragão and Smaniotto 2023, Aragão et al. 2025). The literature also highlights that the success of river restoration depends on a process-based approach, which considers sustainable objectives, stakeholder involvement and learning over a long period of time. These challenges corroborate with other studies on (multi-level) governance for the successful recovery of river ecosystems (Apilánez et al. 2022, Fenten and Dieperink 2024), agile cities (Anwar et al. 2025), and to address complex challenges of river restoration in urban contexts (Novalia et al. 2024).

7. Conclusions and Future Research

To face climate change, biodiversity loss and environmental degradation, cities must adapt and evolve to safeguard and improve their environmental qualities, and thus become more safe, sustainable, resilient and inclusive. The success of change initiatives in complex urban landscapes relies on the implementation of effective co-creation and participatory processes. This, in turn, calls for a multi-level governance approach. Tackling rivers restoration in urban landscapes, as an integral part of a transition plan, allows cities to improve agility and resilience, crucial for success in the climate action age. This means, focussing on cultivating a mindset that embraces continuous learning, strong partnerships and civic empowerment towards building resilience.

The analysis of the restoration cases and the engaged stakeholders' evidence of competing claims on natural resources (such as rivers and their landscapes) and this in turn calls for establishing collaborative processes. The different stakeholders in each case show diverse roles and the power they share and exercise, which, when well-coordinated, can reinforce the effectiveness, complementarity and legitimacy of interventions in recovering rivers. Effectively integrating stakeholders in all phases of the process – from problem identification to implementation and monitoring – is essential to ensure not only technological-environmental success, but also social ownership and sustainability. The engagement of diverse stakeholders is based on the belief that all stakeholders hold different types of knowledge and skills, which are complementary to science and public management. As stated by Reed and Curzon (2015), such complementarity is critically important for context-specific solutions towards sustainability. Such transformative knowledge must be prioritised as, besides providing legitimacy to collaborative processes, it is an instrument for empowerment. Stakeholders' contributions can enhance the qualities of environmental policymaking, as well as guide more effective and durable decisions. Effective and nature-based river recovery hinges on good governance and the engagement of institutions and stakeholders involved in managing territorial and social development and environmental risks. Understanding the dynamics between the river, the riverine community and the urban space, as well as the key stakeholders, is crucial for a restoration success, and a fragile and threatened ecosystem, as rivers call for sharing responsibilities and knowledge.

This chapter provides a perspective on stakeholder engagement from different real-world cases. By examining their engagement in diverse contexts, it showcases the achievements and difficulties faced by different entities. It needs to be noted that this chapter reports the findings of an exploratory study. Transforming these insights into action is the consequent step, which now has to be set by the municipalities with local stakeholders.

Acknowledgements

This publication is based upon work of the Project RUN | Citizen science and co-creation for regeneration of urban rivers and mitigation of risks, and was funded by CYTED Programme grant agreement No. 420RT0008 (https://www.cyted.org/run-rios-urbanos-naturalizados), and is also partially funded by the projects PID2020/112734RB-C32 (MCIN/AEI/10.13039/501100011033) and by 2021 SGR 00522 (Generalitat de Catalunya).

References

Apilánez, I., Shouler, M., Ward, R., and Jiménez, A. 2022. City Water Resilience Framework: A governance-based planning tool to enhance urban water resilience. Sustainable Cities and Society, 77: 103497. https://doi.org/10.1016/j.scs.2021.103497.

Anwar, S., Elshater, A., Afifi, S., Ismail, A., and Ghalib, H. 2025. Retrieving parameters of agile cities: A systematic literature review. Ain Shams Engineering Journal, (16): 1. https://doi.org/10.1016/j.asej.2024.103230.

Aragão, N., and Smaniotto Costa, C. 2023. A regeneração de rios urbanos. Paralelos entre Portugal e África – Desafios e práticas. In Smaniotto Costa, C. et al. (Eds.), Rios Urbanos na Ibero-América: Casos, Contextos e Experiências / Rios Urbanos na Iberoamérica: Casos, Contextos y Experiências. Series Culture & Territory, Vol. 6: 27-45. Lisbon: Lusófona University Press. https://doi.org/10.60543/ecati/8tgb-ym88.

Aragão, N., Bragança, L. S., and Reis, G. 2025. Racismo Ambiental, Participação Pública, Capacitação e Sistemas de Accountability: Experiências do Recanto das Nascentes na Negociação da Água para o Território. pp. 115-144. *In:* M. R. Alves, C. et al. (Eds.), Rios urbanos na Ibero-América: Propostas teórico-conceituais e metodológicas / Ríos urbanos em Iberoamérica: Propuestas teórico-conceptuales y metodológicas. Portal de Livros Abertos da USP. https://doi.org/10.11606/9788566624410

Bryman, A. 2016. Social Research Methods (5th ed.). London: Oxford University Press.

Carayannis, E. G., Barth, T. D., and Campbell, D. F. 2012. The Quintuple Helix innovation model: global warming as a challenge and driver for innovation. J. Innov. Entrep. 1, 2. https://doi.org/10.1186/2192-5372-1-2.

Carayannis, E., and Campbell, D. 2009. Mode 3 and Quadruple Helix: toward a 21st century fractal innovation ecosystem. International journal of technology management, 46 (3/4): 201-234. https://doi.org/10.1504/IJTM.2009.023374.

Cho, S.-w., Shorna B. A., and Stedman, R. 2024. The Social Fabric of Watershed Management: Comparison of Citizen-Based and Agency-Based Organisations. Water 16(1): 111. https://doi.org/10.3390/w16010111.

Corburn, J. 2003. Bringing Local Knowledge into Environmental Decision Making: Improving Urban Planning for Communities at Risk. Journal of Planning Education and Research, 22(4): 420–433. https://doi.org/10.1177/0739456X03022004008.

Davids, R. 2012. Restoring Bogotá's Waterscapes. Center for Latin American Studies, UC Berkeley. https://clacs.berkeley.edu/sites/default/files/publications/brlas-fall2012-davids-small.pdf.

D'Orey, A., Aragão, N., and Smaniotto Costa, C. 2022. A review of urban river restoration projects. Practices and challenges. *In:* Smaniotto Costa, C., and Aragão, N. (Eds.). Understanding and transforming the territory. New Approaches and Perspectives. Series Culture & Territory, Vol. 5: 113–153. Lisbon: Lusófona University Press. https://doi.org/10.60543/ecati/zc65-5z76.

EC - European Commission. 1992. The Habitats Directive. EU measures to conserve Europe's wild flora and fauna. https://environment.ec.europa.eu/topics/nature-and-biodiversity/habitats-directive_en.

EC. 2000. Water Framework Directive. https://environment.ec.europa.eu/topics/water/water-framework-directive_en.

EC. 2015. Digital Single Market Strategy. https://ec.europa.eu/digital-single-market/en/shaping-digital-single-market.

EC. 2020. New Leipzig Charter- The transformative power of cities for the common good. https://ec.europa.eu/regional_policy/whats-new/newsroom/12-08-2020-new-leipzig-charter-the-transformative-power-of-cities-for-the-common-good_en.

EC. 2021. The European Green Deal. A growth strategy that protects the climate. https://ec.europa.eu/stories/european-green-deal/.

Enu, K. B., Zingraff-Hamed, A., Lupp, G., Raum, S., Moors, E., and Pauleit, S. 2025. Stakeholder priorities and navigating barriers in urban river restoration: Comparative insights from Germany and Ghana. Environmental and Sustainability Indicators, 26: 100683. https://doi.org/10.1016/j.indic.2025.100683.

Everard, M., and Moggridge, H. L. 2012. Rediscovering the value of urban rivers. Urban Ecosyst. 15: 293–314. https://doi.org/10.1007/s11252-011-0174-7.

Fenten, T., and Dieperink, C. (2024). Governance Conditions for a Successful Restoration of Riverine Ecosystems, Lessons from the Rhine River Basin. Water, 16(20). https://doi.org/10.3390/w16202983.

Few, S. 2012. Show Me the Numbers: Designing Tables and Graphs to Enlighten. Analytics Press, Montgomery.

Guzelj, M., Hauer, C., and Egger, G. 2020. The third dimension in river restoration: how anthropogenic disturbance changes boundary conditions for ecological mitigation. Sci. Rep. 10: 13106. https://doi.org/10.1038/s41598-020-69796-0.

Hoffman, J. 2022. Toward Civic Co-production: Using Worldbuilding to Go Beyond Participation in Urban Planning and Enact More Equitable Cities. Front. Sustain. Cities, 4. https://doi.org/10.3389/frsc.2022.907541.

Itsumi, Y., Chibana, T., and Watanabe, S. (2025). Cultivating a sense of ownership over climate change adaptation through the expansion process of local community activities stimulated by urban river restoration in Okazaki City, Japan. Frontiers in Water, 7. https://doi.org/10.3389/frwa.2025.1537235.

Khatibi, F. S., Aysin Dedekorkut, H., Howes, M., and Torabi. E. 2021. Can Public Awareness, Knowledge and Engagement Improve Climate Change Adaptation Policies? Discover Sustainability 2(1). https://doi.org/10.1007/s43621-021-00024-z.

Koontz, T. M. 2005. We Finished the Plan, So Now What? Impacts of Collaborative Stakeholder Participation on Land Use Policy. Policy Studies Journal, (33) 3: 459-481. https://doi.org/10.1111/j.1541-0072.2005.00125.x.

Likert, R. 1932. A Technique for the Measurement of Attitudes". Archives of Psychology. 140: 1–55. https://psycnet.apa.org/record/1933-01885-001.

Meijer, M. 2019.Community-led and government-fed: Comparing informal planning practices in depopulating regions across Europe. The Journal of Rural and Community Development, 14(4): 1–26.

Menezes, M., and Smaniotto Costa, C. 2024. The Role of Collaborative Ethnography in Placemaking. Humans, 4: 284–297. https://doi.org/10.3390/humans4030018.

Millennium Ecosystem Assessment. 2005. https://www.millenniumassessment.org/en/index.html.

Mondragon-Monroy, R., and Honey-Roses, J. 2016. Urban River Restoration and Planning in Latin America: A systematic review. Water Planning Lab. University of British Columbia.

Novalia, W., Suwarso, R., and Nurdin, I. (2024). Connecting place and multilevel governance for urban river restoration. Territory, Politics, Governance, 0(0): 1–26. https://doi.org/10.1080/21622671.2024.2336608.

Nico Polman, P., de Blaeij, P., and Slingerland, M. 2014. Knowledge of Competing Claims on Natural Resources: Toward Institutional Design and Integrative Negotiations. In Padt, F., Opdam, P., Polman, N., and Termeer, C. (Eds.) Scale-sensitive Governance of the Environment. Willey. https://doi.org/10.1002/9781118567135.ch4.

Rao, R., Kansal, A., and Tarannum, F. (2024). Role of Stakeholders in Sustainable Management of an Urban Waterbody and Wetland. Environment and Urbanization ASIA, 15(1): 175–191. https://doi.org/10.1177/09754253241236850.

Reed, M. S., and Curzon, R. 2015. Stakeholder mapping for the governance of biosecurity: a literature review. Journal of Integrative Environmental Sciences, 12(1): 15-38. https://doi.org/10.1080/1943815X.2014.975723.

Scholes, E., and Clutterbuck, D. 1998. Communication with stakeholders: An integrated approach. Long Range Planning, (31)2: 227–238. https://doi.org/10.1016/S0024-6301(98)00007-7.

Schütz, F., Heidingsfelder, M. L., and Schraudner, M. 2019. Co-shaping the Future in Quadruple Helix Innovation Systems: Uncovering Public Preferences toward Participatory Research and Innovation. The Journal of Design, Economics, and Innovation, 5(2): 128–146. https://doi.org/10.1016/j.sheji.2019.04.002.

Scott, J. C. 1998. Seeing Like a State. Yale University Press.

Shunglu, R., Köpke, S., Kanoi, L., Nissanka, T. S., Withanachchi, C. R., Gamage, D. U. et al. 2022. Barriers in Participative Water Governance: A Critical Analysis of Community Development Approaches. Water 14(5). https://doi.org/10.3390/w14050762.

Smaniotto Costa, C. 2025. Final report – RUN Network. Madrid: Cyted.

Smaniotto Costa, C., Menezes, M., Pallares-Barbera, M., Pastor, G., Rocha, E. do P., and Villalba Condori, K. O. (Eds.). 2023. Rios Urbanos na Ibero-América: Casos, Contextos e Experiências / Ríos Urbanos en Iberoamérica: Casos, Contextos y Experiencias, Cultura & Território, Vol. 6. Lisbon: Lusófona University Press. https://doi.org/10.60543/ecati/4tk9-1a66.

Stewart, J., and Lithgow, S. 2015. Problems and Prospects in Community Engagement in Urban Planning and Decision-Making: Three Case Studies from the Australian Capital Territory. Policy Studies 36 (1): 18–34. https://doi.org/10.1080/01442872.2014.981061.

UN. 2015. The Paris Agreement. https://unfccc.int/process-and-meetings/the-paris-agreement.

UN-HABITAT. 2016. The New Urban Agenda. https://habitat3.org/the-new-urban-agenda/.

UN-HABITAT. 2022. The New Urban Agenda. https://unhabitat.org/about-us/new-urban-agenda.

Walteros, J. M., and Ramírez, A. 2020. Urban streams in Latin America: Current conditions and research needs. Revista de Biología Tropical, 68: 13–28. https://dx.doi.org/10.15517/rbt.v68is2.44330.

Yan, S., and Ibrahim, R. B. (2023). Exploring the Factors That Influence Stakeholder Participation in Decision-Making for the Moat System Restoration Project in Tianchang City, China. The Scientific World Journal, 2023, 1–15. https://doi.org/10.1155/2023/9969589.

Yang, R. J. 2014. An investigation of stakeholder analysis in urban development projects: Empirical or rationalistic perspectives. International Journal of Project Management (32)5. https://doi.org/10.1016/j.ijproman.2013.10.011.

Zingraff-Hamed, A., Hüesker, F., Lupp, G., Begg, C., Huang, J., Oen, A. et al. 2020. Stakeholder Mapping to Co-Create Nature-Based Solutions: Who Is on Board? Sustainability, 12: 8625. https://doi.org/10.3390/su12208625.

4

Extraction vs. Heritage

Promoting Riverine Fishing Communities in Arctic Villages

Stefano Tornieri[1,*] and *Agatino Rizzo*[2]

1. Introduction

Fishing along rivers and water bodies has shaped not only the livelihoods but also the identities and heritage of riverine communities across generations. This close interaction with water ecosystems has nurtured a profound sense of place attachment, evident in every aspect of their lives—from the architecture of their homes to the tools and traditions that define their daily practices (Senanayake 2006). This connection transcends mere subsistence; it reflects a deeply-rooted cultural bond that manifests in culinary practices, seasonal gatherings, and the spiritual value attributed to water sources.

From a historical viewpoint, the interwoven relationship between water ecosystems, fishing technologies, food preservation techniques, and local customs has been documented across diverse regions such as the Atlantic Coast (Tavares 2024), Australia (Shamsi et al. 2020), and Italy (Bullo 1940, Ravagnan 1980, Tornieri 2022). Long before the advent of industrialized fishing practices, traditional methods were both sustainable and emblematic of a strong, lasting connection between communities and their environment. This relationship was so fundamental that ecologist Berkes (2012) describes it as "sacred," emphasizing the value of indigenous knowledge, particularly in fishing, as a complement to scientific understanding. Berkes provides

[1] Jade University of Applied Sciences, Architecture Department, Oldenburg, Germany.
[2] Luleå University of Technology, Architecture Research Group, Luleå, Sweden.
Email: agatino.rizzo@ltu.se, 0000-0001-6831-8857
* Corresponding author: stefano.tornieri@jade-hs.de, 0000-0003-2887-0414

DOI: 10.1201/9781032711737-4

an insightful exploration of indigenous fishing knowledge in his chapter on Cree fishing practices in James Bay, Canada, which serves as a key inspiration for this study. Under similar climatic conditions, Berkes describes how Cree fishers have long demonstrated an acute understanding of fish behavior and seasonal patterns: "*fishers knew where the pre-spawning aggregations of fish were in August, and they knew that in September whitefish was best harvested over a sand-gravel bottom at a specific depth*" (Berkes 2012, p. 148).

In many traditional fishing communities, the proximity to water was not just practical but existential, with the community's entire way of life centered around the resources provided by rivers, lakes, and coastlines. Within this framework, it is essential to highlight salmon fishing along the Teno River by the Sami community, as documented by N. Kent (2018). Kent's study details the significance of fishing to the Sami people and traces the evolution of their techniques in response to fish stocks and regulatory changes across Lapland. Key fishing areas such as the Teno River, Inari Lake, and locations near Sjokksjokk in Sweden's Lule Sami region were central to Sami livelihoods from roughly 1550 to 1750. Kent's research illustrates how fishing served as a critical factor in transitioning the Sami from semi-nomadic lifestyles to more permanent settlements, noting that "reindeer herding was only a complement to fishing and hunting" (Kent 2018, p. 240). This historical context underscores the primary role that fishing played, not only as a sustenance activity but as a foundation for social and economic stability in Sami communities.

Today, this wealth of knowledge, community practices, and traditions is undergoing major changes due to new industrial developments driven by the energy transition in the resource industry, mass tourism, and climate change, leading to a significant loss of indigenous knowledge (Hedlund 2017, Rizzo and Sordi 2020, Morata et al. 2020, Tornieri et al. 2024). Modern pressures like environmental changes, new regulations, urbanization, and economic shifts are disrupting long-standing practices. As traditional ways evolve or fade, valuable ecological insights and cultural heritage risk being lost, weakening the cultural identities tied to these fishing traditions. The pressure to find sources of revenues to maintain the local welfare system (schools, healthcare, etc.) has resulted in looking for quick ways to make money and in this context, it has meant to welcome mining industries to prospect for valuable minerals. This over time has sidelined the vernacular, heritage values of village communities in the Torne River. As Ransley (2011) notes, these places serve as rich repositories of cultural heritage, encapsulating the skills, tools, and techniques that sustain daily life and define community identity. These values could spark a more diffuse and sustainable form of local development. However, the region's shrinking population means that reaching a critical mass will take long-term commitment and greater effort from local authorities.

In previous work, we have demonstrated how green and blue infrastructure (e.g., natural water and vegetation systems) positively influences health and well-being (Tornieri et al. 2024). Building on that foundation, this study expands to assess the architectural, urban, and social frameworks within these villages, analyzing how these elements contribute to the cultural heritage value of the area. This chapter aims to argue for the potential of an alternative development model to short-term mining, one that is long-term, sustainable, and locally driven. To do so, we will develop an inventory of local landscape elements such as fishing tools, community infrastructure, architectural forms, and community protocols, to illustrate how these villages' heritage and resilience are interwoven with their environment and history (Figures 1,2,3,4,5). To understand the intricate web of social, urban, and architectural values within Kukkola and Korpikylä, we conducted a series of detailed

Name	Category	Characteristic	Description	Picture
Lippo Dipnet	Fishing Tool	Wood, Plastic Net, Natural Fiber. Lenght: 6-7 m	Lippo, called also dipnet, is a fishing tool consisting of a long pole with a net attached to the end. It dates back to the Middle Ages.	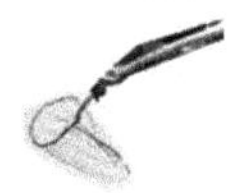
Lamprey Basket	Fishing Tool	Wood, Natural Fiber Height: 1,5 m Length: 1 m	The trap is constructed from wood and resembles an elongated basket. It is used fixed on the wooden pier.	
Burbot basket	Fishing Tool	Wood, Metal Net Height: 1,5 m Length: 1 m	Similar in shape to a lamprey basket, this trap is made of metal netting and is placed underwater beneath the ice. Water currents can be diverted to guide the fish into the trap, enhancing its effectiveness.	
Gaff	Fishing Tool	Wood, Metal Length: 0,6 m	A small and simple tool equipped with several iron nails is used to catch large, heavy fish and lift them out of the water.	
Leister	Fishing Tool	Metal Length: 0,6 – 1 m	Similar to a harpoon, this tool functions like a gaff, designed to extract large, heavy fish from strong currents.	

Name	Category	Characteristic	Description	Picture
Jakopuoti Fish storage	Building	Wood Length: 6,50 m Footage: 8 m Height: 2,5 m	The shed called Jakopuoti was built in 1889. It replaced a similar building which was destroyed by fire in 1886. In this shed the collective fishing tackle of the village was kept.	
Clocktower Sharing Shed	Building	Wood Length: 5,50 m Footage: 8 m Height: 6 m	This shed serves as a space for fish sharing. It features a clock tower that signals the shift schedule for the various fishermen.	
Fishing Shed	Building	Wood Length: 3,50 m Footage: 5 m Height: 2,5 m	The shed is used to store fishing equipment, and these sheds are constructed in a sequence, likely forming a collective space for the fishermen to manage and organize their tools efficiently.	
Grilling shed	Building	Wood Length: 5,50 m Footage: 6 m Height: 3,5 m	The shed consists of a single room with a fireplace located in the center, and benches arranged along the walls. A chimney is visible on the roof.	
Lower mill	Building	Wood Length: 12 m Footage15 m Height: 8 m	The shed is a multifunctional space used by the community for various purposes, including woodworking, ironworks, and food storage.	

Name	Category	Characteristic	Description	Picture
Upper mill Powerstation	Building	Wood Length: 13 m Footage: 16 m Height: 8 m	The building is situated directly on the riverbank, designed to generate electricity through a water mill. Its façade is marked by a distinct, rapid-covered ladder. The roof, featuring a mansard shape, is also unique within the village.	
Aitta	Building	Wood Length: 2 m Footage: 3 m Height: 3 m	An "aitta" is a raised larder, a storehouse used for keeping foodstuffs such as grain, bread, and meat.	
Cattle shed	Building	Wood Length: 5,5 m Footage: 12 m Height: 3 m	The building was constructed at the end of the 19th century as a cattle shed for some cows. It had a section for sheep and pigs. The cattle shed was converted to a garage during the 1930s,	
Fisk Museum	Building	Wood Length: 20 m Footage: 28 m Height:7 m	The old two-story storage building has been refurbished and now serves as a fish museum, preserving and showcasing the region's fishing heritage and practices.	

Name	Category	Characteristic	Description	Picture
Krenkku	Urban facility	Timber logs Length: 2 – 50 m Footage: 2-5 m Height: 0,5 - 2 m Spring to autumn	The structure is a wooden pier, constructed in the spring and dismantled in the autumn, only to be rebuilt the following spring. It is used by fishermen to fish and access different areas along the river, facilitating their seasonal fishing activities.	
Table and bench	Urban facility	Wood Length: 2 m Footage: 3 m Height: 0,8 m	The table and benches serve as a gathering spot for the community, where people meet to chat, observe the river and the fishing activities, and wait for their turn to fish.	
Grilling fish	Community protocol	When organized	Grilling and eating together is a long-standing tradition, typically done outdoors near the grilling hut or around a communal table.	
Sharing Fish	Community protocol	Every day 6 p.m	The sharing takes place on the rapids field in front of the old fish shed when the dipnetting shifts change at 6 p.m. Each farm then divides its catch into smaller portions, distributing them among the member households based on the historical tax value.	

Name	Category	Characteristic	Description	Picture
Pier construction	Community protocol	Twice a year	Building the piers is physically demanding work that requires a great deal of skill. Effective cooperation between the fishermen is crucial, both when constructing and dismantling the piers, due to the strong currents of the river.	
Sheds maintenance	Community protocol	When required	Shed maintenance involves inspecting the overall structure, but the primary focus is on rebuilding and repairing the roof.	
Dipnetting	Fishing Technique	Every day	Dipnetting is done either by standing on the riverbank or by accessing the water from specially built piers. The choice of method depends on the water level of the river	
Lamprey fishing	Fishing Technique	Spring to autumn	Using the krenkku as a framework, the basket is positioned between the main structure and the water. The current then drives the fish into the basket.	
Burbot ice fishing	Fishing Technique	Winter season	the trap is positioned in the right spot underwater. Water currents can be subtly redirected by constructing wooden fences or similar structures beneath the surface.	

Figure 1,2,3,4,5: Inventory of tangible and intangible heritage elements. Source: S. Tornieri.

field surveys in 2022—including photography, filming, and mapping of fishing-specific infrastructure—we gained insights into the profound connection between local knowledge and the environment, reflected in every detail of the villages' spatial and social organization. We were drawn to these villages due to the fascinating contrast they present. On one hand, there is a movement toward developing large, industrialized cities in the Arctic , while on the other, the small and delicate ecosystems of rural villages like Kukkola and Korpikylä, which have endured for centuries, now face the threat of depopulation and aging demographics. Tourism, it seems, may be one of the strongest avenues to preserve these unique communities, helping to sustain their cultural practices and lifestyle in the face of change. By documenting the physical environment and everyday practices of these communities, we aim to trace the historical and seasonal connections between the river and the people. This analysis also includes archival research to contextualize the lived experiences and historical evolution of these villages. Through this approach, we hope to uncover the vital role of the river as an ecological infrastructure that sustains both the tangible and intangible heritage of these unique communities.

2. The Arctic Villages of the Torne River: A Living Heritage at Risk

The Torne River, Sweden's northernmost river, originates near the Norwegian border, just west of Torne Lake, the largest lake in Swedish Lapland is an interesting case study. Stretching 354 miles (570 km), it winds its way to the Gulf of Bothnia, where it meets the sea between the towns of Haparanda in Sweden and Tornio in Finland. Along its lower reaches, the Torne River serves as a natural boundary between Sweden and Finland, with a vast drainage basin covering 15,537 square miles (40,240 square km)—about one-third of which lies within Finland and Norway. In its upper reaches, the river features scenic waterfalls, while the lower course is scattered with rapids, rendering it largely non-navigable. (Figure 6)

Despite this, the river has historically played a vital role in regional industries, once serving as a major route for timber transport and now renowned for its thriving salmon populations. Over the centuries, Arctic communities have adapted ingeniously to the challenges of their harsh surroundings. Along the Torne River, specific fishing villages have thrived within this extreme environment, forming unique architectural styles, landscapes, and social practices intimately tied to their ecological surroundings (AA.VV 1965) (Figure 7). These villages, embedded within

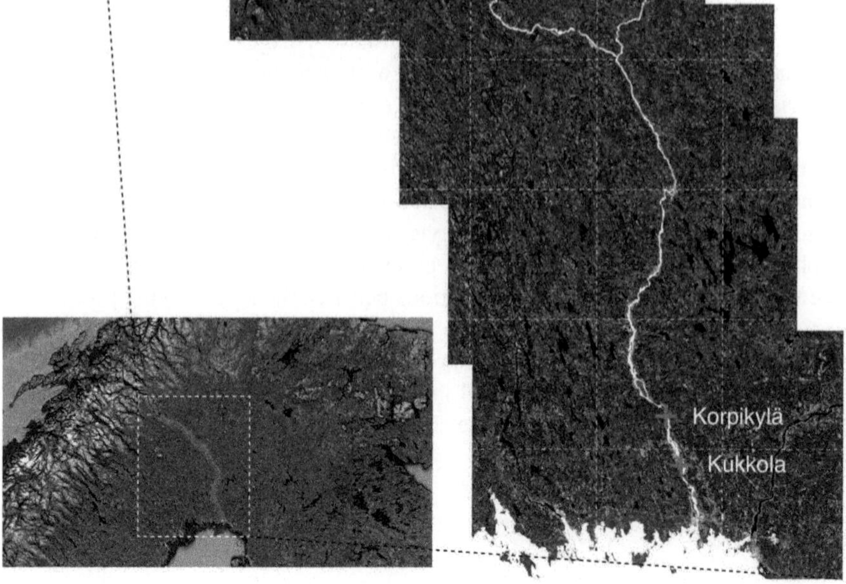

Figure 6: Geographical maps and localization of the Villages. Source: S. Tornieri.

a distinctive Arctic ecosystem, provide an invaluable case study on the resilience and adaptability of human communities in extreme climates. Today, the cultural heritage embodied by these fishing practices, structures, and social traditions is under consideration for inclusion in UNESCO's World Heritage List of Living Cultural Heritage, a testament to the profound and enduring connection these communities hold with their environment.

In the current discourse on heritage and place-based identities, aspects like rural landscapes and residents' sense of belonging are increasingly recognized as key components of maintaining vibrant communities (Soini et al. 2012, Allen et al. 2021, Belanche et al. 2021). The heritage significance of a site or community lies not only in its historical and aesthetic values but also in its symbolic, spiritual, and

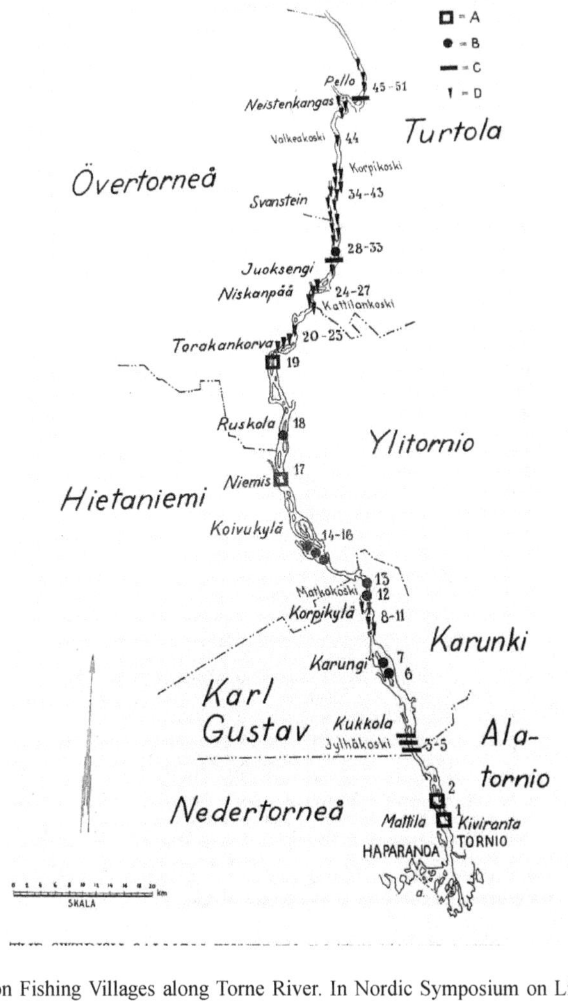

Figure 7: Salmon Fishing Villages along Torne River. In Nordic Symposium on Life in a Traditional Hunting and Fishing Milieu in Prehistoric Times and Up to Present Day. Edited by H. Hvarfner. Luleå: Norrbotten Museum.

social meanings (Azzopardi et al. 2023). UNESCO's definition of intangible cultural heritage acknowledges the complexity of recognizing non-material heritage, such as traditions and practices, which often lack a physical form but are deeply embedded in the community's identity (UNESCO 2003, Walter 2023). In the Arctic regions of Scandinavia, the skills and traditional practices of fishing communities contain invaluable insights that could address contemporary challenges, such as climate change and rural depopulation (Kaslegard 2011, Tornieri et al. 2024).

The emphasis on identifying and supporting local communities stems from their opposition to the mining industry, which remains the dominant extractive activity in northern Sweden. While mining brings significant investment into the region, it simultaneously reveals a profound disregard for the long-term territorial and social consequences during and after extraction activities. As Suopajärvi et al. (2017) highlight, mining operations threaten traditional livelihoods such as reindeer herding, fishing, and small-scale farming, while simultaneously creating a heavy economic dependence on mining, rendering communities highly vulnerable to fluctuations in global market conditions. Given that the siting of mines is determined by the location of underground resources, mining towns frequently emerge in remote rural areas rather than being shaped by environmental considerations such as access to water or favorable microclimates. With the renewed focus on extracting metals and rare minerals, mining activities are increasingly disrupting rural development trajectories across Europe (Mármol and Vaccaro, 2020). The demolition of entire old city centers, such as the well-known case of Kiruna, represents an extreme manifestation of this process, where extractive economic interests override local social, cultural, and territorial values.

The process of the de-localization of people as a consequence of mining development highlights the absence of genuine concern for the local populations, reinforcing the broader detachment from the sense of place. This detachment is further accelerated by technological advancements, as new digital apparatuses increasingly disconnect workers from their workplaces, ultimately leading to the gradual disappearance of human presence within these spaces. Such dynamics are evident in the automation processes across various industries, including data centers (Donoghue, 2017), fish farming (Shingal et al. 2023, Gayday, 2024), and mining, where automation has been shown to produce both income growth and significant inequalities (Paredes and Fleming 2021). Other consequences of the heavy extractivism refer to the post-mining activity. Closing a mine can bring serious socio-economic consequences for nearby communities. These impacts can persist long after the mine has ceased operations, highlighting the need for a more comprehensive understanding of the social ramifications of mine closure (Demajorovic et.al. 2022). Moreover, following the cessation of mining activities, a comprehensive post-closure environmental management plan is essential, particularly in areas at risk of ongoing contamination or ecological degradation. Such a plan should aim to mitigate environmental harm and enhance the potential for alternative land uses, especially for agriculture. In addition, closure planning should consider the repurposing of existing infrastructure—such as housing, facilities, and equipment—and implement policies that support the preservation of social networks and community cohesion in the post-mining phase (Warhurst et al. 2024).

3. The Case Study: Fishing Tools, Techniques, and Rights

This study delves into the fishing traditions of two preserved fishing villages along the Torne River—Kukkola and Korpikylä. Our case study area was considered for the completeness of historical documentation completeness, architectural heritage value, and recent touristic interest. Ingrid Vallin, in collaboration with Ingalill Granlund (AA.VV 1965), created a detailed map of the Torne River fisheries, drawing on extensive documentation by Ingemar Sorensen, a fisheries inspector from the Norrbotten Museum in Luleå's upper northern district. This map provides a rich illustration of village locations along the river and categorizes traditional salmon fishing techniques into four types, as outlined by Grandlund: "Karsina-weirs", the seine method, "Strandpata", and "Mockpata". By overlaying village positions with the unique characteristics of the river's flow, the map reveals a clear connection between fishing techniques and the river's hydrodynamics. For instance, methods like dipnetting have long thrived near rapids, where fish naturally pause in pockets of slower water behind stones as currents intensify. In this study, we highlight Kukkola and Korpikylä—vibrant communities where dipnetting, along with a diverse array of fishing-related architectural structures, has been practiced continuously for centuries. These villages serve as living examples of how traditional fishing techniques and associated architectural features remain an integral part of the cultural landscape today.

Fishing in arctic river valleys, particularly along the Torne River, is more than just a livelihood, it is a legacy steeped in skill, tradition, and cultural significance. The traditional fishing tools, techniques, and rights shape this living heritage, blending practical skill with the unique natural and cultural landscape of northern Scandinavia. From the use of specialized tools crafted from local materials to the seasonal allocation of fishing rights through community auctions, the fishing practices of these villages offer a window into the enduring connection between people and river.

Central to fishing along the Torne River is the lippo, a traditional fishing tool with deep roots in Finnish and Swedish communities. Historical records suggest its use dates back to the 1200s, and the tool has likely existed in various forms even earlier. The lippo is a long wooden pole with a net, or "liina," forming a pouch designed to catch whitefish (siika) and salmon as they pause in the river currents (Figure 8). Its design is both practical and sustainable, with a rounded wooden frame

Figure 8: Fishermen hanging a Traditional Lippo. Source: Norrbotten Museum.

and soft netting traditionally crafted from natural materials - typically birch for the pole and frame and organic fibers for the netting. This traditional structure allows the lippo to dip into the riverbed gently, preventing noise and disturbance that modern metal frames might cause upon contact with rocks.

The art of using the lippo, known locally as lippous, is passed down through generations. Each fisher develops a deep familiarity with the river's contours, the behavior of fish, and the ideal spots for catching them. Fishers often make slight modifications to their fishing sites, adjusting for seasonal shifts in water levels, and some communities even build small wooden platforms and stone anchors along the riverbank. The lippous method requires patience, precision, and knowledge of the river ecosystem, as it involves locating where currents slow and fish rest. This skillful technique is more than a method of catching fish; it is a cultural rite and communal tradition, celebrated as a link between contemporary fishers and their ancestors through shared practices, stories, and seasonal gatherings at the riverbank. Dipnetting, is the fishing method which uses lippo, is a tactile fishing method, requires the fisher to trace the river floor with the net to feel for whitefish resting behind rocks or in pits. Unlike other techniques, dipnetting allows fishers to stay attuned to the subtle rhythms of the river. Whitefish, which return to the river to spawn, follow specific migratory paths influenced by water levels, currents, and the natural formations along the riverbed. Fishers adjust their positions, accordingly, sometimes accessing spots by boat or pier, to maximize their catch. Many of these fishing spots have been used for centuries, with names passed down through generations that signify their historical and cultural importance to the community.

As autumn settles in and the waters cool, fishers along the river shift their focus to catching lamprey. The lamprey is a unique species, caught in September and October

Figure 9: Burbot basket. Source: Norbotten Museum.

when it migrates before the spawning season in spring. For this catch, fishers use wooden traps resembling elongated baskets (Figure 9). Each fisher has a designated location for setting up these traps, an arrangement governed by personal fishing rights. These rights, an integral aspect of the local fishing community's structure, are allocated through an auction each August. Only those with an authorized fishing right may set up lamprey traps in specified spots. This auction system not only allocates fishing locations but also reinforces community ties and respect for traditional fishing rights. Those same rights extend to burbot fishing in the winter, ensuring that the natural resources are sustainably managed and community bonds are reinforced.

Burbot, an elusive fish active during the coldest months, becomes the focus of mid-winter fishing. Along the Kukkolankoski rapids, fishers employ traditional wooden traps specifically designed for burbot, setting them in precise spots beneath the icy water. To guide the fish towards the traps, fishers sometimes use underwater wooden fences or stone structures, carefully placed to subtly alter the current. The traps are checked daily, and burbot can be kept in submerged containers for short periods before use. Similar to the lamprey traps, the burbot fishing spots are allocated through the August auction, with each fisher able to secure several designated spots for the season. This highly regulated, skill-intensive practice demonstrates the deep knowledge and endurance of the fishermen, who work through freezing conditions to uphold a tradition passed down for centuries.

As winter recedes and spring arrives, fishing rights holders engage in lippous without the constraints of a strict timetable, taking advantage of this period to inspect and repair their fishing equipment and prepare the riverbed's fish pits for the busy season ahead. Spring dipnetting, when only a few whitefish have arrived, offers a valuable time for younger, less experienced fishers to practice their skills under the guidance of seasoned community members. This season is a learning period, with younger fishers honing their technique and understanding the nuances of the riverbed—skills that will serve them when the peak fishing season arrives. Spring dipnetting also allows fishers to observe the initial phases of whitefish migration. Villagers often gather along the riverbank, chatting and exchanging news about the catch of the day, while preparing for the transition to shift dipnetting, a more intensive, organized phase of fishing that starts in late July or early August when whitefish are in abundance.

Shift dipnetting, a highly organized practice managed by the local fishing association, begins when whitefish migration is in full swing. This rigorous schedule requires fishers to work in shifts, with 24-hour rotations and breaks only for meals and short rests. Starting each evening at 6 pm, the fishing shifts involve a set number of fishers actively dipnetting, either from the family or as hired laborers. The catch from each 24-hour shift is shared among the community's families, and the sharing of whitefish becomes a daily ceremony that persists until mid-September. During this intense fishing season, the community follows the rhythm of the river and the whitefish migration, with every family or farm taking its turn at fishing. This system reflects the deep-rooted values of cooperation and mutual reliance within the community, as the ownership and responsibility of fishing rights are passed down through generations.

4. Architecture and Urban Factors

The architectural and urban dimensions focus on the physical composition of the villages, including building forms, spatial organization, and their uses. Historical mapping, cartographic analysis, field surveys, and archival research are used to examine changes in architectural layout and urban fabric. This approach identifies key elements such as settlement patterns, material usage, and building functions that have evolved in response to social, economic, and environmental pressures.

Kukkola is a settlement that spans both the Finnish and Swedish sides of the river, featuring a similar urban layout on each side (Figure 10). On the Swedish side, the settlement is arranged in a linear composition and includes 18 traditional cultural structures. A main road runs parallel to the river, providing direct access to the buildings, with only two located close to the riverbank—the 'Vattenkrafwerk' Hydroelectric Power Station and the 'Nedre kvarnen' Lower Flour Mill—both taking advantage of the river's energy. To the south of the village, a group of 10 huts is closely linked to fishing activities. These structures consist of five 'fiskebodar' (fishing huts), the 'Iskällare' (icehouse), the 'Stekbodan' (grill hut), the 'Kontoret' (office), and other shared facilities such as "kalapuohi" (a fish storage building), a

Figure 10: The village of Kukkola and the traditional bridge. Source: S. Tornieri.

"paistokota" (cooking and sleeping hut), and equipment storage warehouses. The central placement of the ice house emphasizes the importance of preserving the catch, while the grill hut and office, positioned near the river, ensure oversight and safety. These huts are near the hydroelectric station and watermill, both within ten meters of the river. The Flour Mill, constructed in 1860 in the center of the village, was a communal asset but under state control, requiring a portion of grain as tax, earning it the nickname 'tax-mill.'

The absence of private spaces around the buildings underscores the community's strong dedication to fishing activities. The design of the outdoor areas is intentionally minimal, prioritizing green spaces and the connectivity between structures over purely functional infrastructure. This approach not only enhances the visibility of fishing practices but also creates an open environment conducive to this purpose. A dominant feature in the outdoor space is the traditional wooden pier, known as the Krenkku, extending into the Torne River to access the rapids. The origins of the Krenkku can be traced to the early 20th century, when U.T. Sirelius (1906) documented stationary salmon weirs in Norrbotten, known as mockapata. These weirs consist of trestles connected by ridgepoles and laths, forming a solid barrier. Among the various types of weirs, the most notable is the Hevos-pata in Kukkola Village, which is built after the salmon season ends and whitefish season begins. This temporary, human-constructed structure is assembled annually by experienced builders using local wood. The construction techniques of these fishing structures remain entirely natural, evolving in harmony with the environment and reflecting ancient traditions. Although different piers and weirs exist, all share a functionalist design. Each component serves a specific purpose, with the primary structure featuring an 'A-shaped' frame inserted into the water, supported by horizontal beams and a boardwalk.

Due to the force of the river currents, a secondary, reverse-V structure is added to reinforce the pier. The design of these piers is also influenced by the volume and flow of the water, as well as the empirical knowledge passed down through generations of fishermen. The type and shape of the pier vary according to water conditions and the behavior of fish, which changes in response to the currents. The Finnish side features a similar urban and architectural arrangement. A row of nine huts is aligned along the riverbank, with their facades facing the water in a rational, orderly sequence. At the end of the street stands a larger, multifunctional community building. The kitchen is the only structure located on the opposite side of the street, directly facing the huts and positioned close to the water. This side appears more modest, with smaller structures supporting traditional fishing activities. The Krenkku here is shorter and serves to connect the riverbank to a small "island" in the river stream. Despite these differences, the construction technique remains the same.

The village of Korpikylä features a simpler urban layout compared to Kukkola, with only three huts dedicated to fishing, all positioned as close to the water as possible. Among these structures is a cooking hut designed for preparing traditional dishes like skewer-roasted whitefish and whitefish soup. Additionally, the ice storage hut, located near the wooden ramp and stairs, provides easy access to fishing areas. Several small modifications to the riverbank vegetation cater to fishing activities, including the placement of an ice storage hut, a fire hut for cooking, outdoor benches, and wooden stairs leading to the water. The ice hut, constructed with stone pieces and built partially underground using a natural dip in the riverbank, allows for optimal fish preservation.

In recent years, several transformations have taken place in the Kukkola village with the aim of preserving and revitalizing its cultural heritage. Here we are referring to the Swedish side as the Finnish side, is less developed from the touristic point of view and the old village is preserved as memory of the place without a specific re-design of functions. One of the most significant restoration projects is the establishment of a fish museum in the upper part of the Swedish village. This museum was created in a large hut that was previously used for storage and has now been repurposed to house a two-floor permanent exhibition. The museum provides a comprehensive look into the traditional life of the village through a range of mediums, including photographs, videos, object displays, and scale models. Adjacent to the fish museum is the village's historic mill, which, while no longer operational in its original capacity, now serves as an exhibit space showcasing traditional fishing tools and techniques. In the central part of the village, modern amenities have been introduced, including a restaurant, café, and visitor centre, all housed within new wooden hut. These facilities offer tourists the opportunity to sample local cuisine through curated tasting menus, while also providing views of the Krenkku, the traditional fishing pier that extends into the river. Further south, a new area dedicated entirely to tourism has been developed. This section of the village includes seven double-houses, an L-shaped hotel, 12 small bungalows, and designated spaces for camper lots. This area, built more recently, caters to visitors seeking a deeper connection with the village's natural surroundings while providing modern comforts. Close to the water, two original saunas stand as a testament to the village's enduring lifestyle. These saunas, constructed from wood and featuring green earth roofs, offer an authentic glimpse into the traditional way of life still present in the village today. They are not only functional but also symbolic of the village's efforts to maintain its cultural identity amidst modern developments.

5. Community Engagement as Intangible Heritage

UNESCO and the International Council on Monuments and Sites (ICOMOS) have recently emphasized the interconnectedness of tangible and intangible cultural heritage (ICH). In recent years, this dialogue has expanded beyond traditional focus areas—such as artifacts, monuments, architectural structures, and archaeological sites—to include a broader spectrum of historically significant elements. Today, the concept of cultural heritage encompasses a diverse array of cultural expressions, including oral traditions, music, languages, religious and cultural festivals, ceremonies, and traditional events. These intangible aspects of heritage are increasingly recognized for their deep connections to material culture, such as tools, architecture, and infrastructure. Brulotte and Di Giovine (2016) highlighted the challenges posed by "colonialist" and "hierarchical" perspectives that historically undervalued intangible cultural elements. In response, UNESCO has established criteria for recognizing practices as part of Intangible Cultural Heritage. According to UNESCO (Lenzerini 2011), a cultural practice must be: (a) traditional yet "living," with continued relevance, (b) inclusive, (c) representative, and (d) rooted in the community. The principles of inclusivity and social cohesion are essential in valuing shared activities, which have been practiced for centuries in these communities and remain vibrant today, fostering a strong sense of community and continuity across generations.

Figure 11: The fishermen's daily catch-sharing ritual. Stefano Tornieri.

Fishing in these villages isn't just about catching fish, it's a community-driven practice. During the fishing season, locals engage in various related activities, including the construction and maintenance of piers, repairs of traditional buildings, organizing fishing rounds, and evening gatherings, where the day's catch is shared among villagers (Figure 11). These social gatherings reinforce community bonds, preserving both the fishing culture and the knowledge of sustainable, environmentally conscious techniques that have shaped the region's identity. While shift dipnetting is on, the catch of the day is shared between the fishing right owners by an age-old system. About the same size whitefish are shared in piles which are then raffled to their owner farms. Each farm further shares the pile in smaller parts among the member houses according to the historical tax value, "mantal". This numeric practice makes the sharing equal and easy to execute without weighing the catch.

The sharing happens on the rapids field in front of the old fish shed when the dipnetting shifts change at 6 pm. Villagers and visitors alike gather to small talk and hear stories about the happening of the day. Seeing the fishermen throwing the whitefish on several piles according to some non-verbal rule feels like a visit to the history which it actually is, the tradition having existed for centuries. Dipnet fishing culture has been nominated by Sweden and Finland as a candidate to the UNESCO's representative list of Intangible Cultural Heritage of Humanity (Interreg Nord 2012).

6. Conclusion

Our analyses show the importance of heritage for shaping the community's identity of the fishing villages in northern Sweden and to promote an alternative, to mining and extractive industries - a form of development that is more long term and grounded in local qualities. This knowledge is particularly important in light of the rapid transformations occurring in Arctic regions; climate change, industrial investments for the decarbonization of the extraction industry, impeding mass tourism (Soroudi et al. 2024, Ma and Rizzo 2024). The interconnections between fishing elements, techniques, and practices in these river communities are subtle and not immediately apparent, especially as fishing seasons vary and only a portion of these activities are accessible to tourists. Each fish species—whether burbot, lamprey, whitefish, or salmon—requires unique techniques, yet the tools share notable similarities, particularly in their use of materials like wood and metal. Simple tools, such as the gaff, are basic and utilitarian, consisting of a single wooden handle with nails for grip, while more complex traps for lamprey and burbot rely on fishermen's knowledge to position them effectively within the river's current for maximum efficiency. The relationships between objects, labor, and seasonal activities unfold at several levels which can be summarized into three statements:

Fish-Technique-Season Interdependence: Each fish species is associated with specific seasonal conditions and fishing techniques. For example, as winter freezes the river, the fishing season for burbot begins, with traps carefully placed beneath the ice in optimal spots. In summer, the arrival of whitefish and salmon shifts practices to techniques like dipnetting. The controlled catch rate across these seasons helps maintain the river's ecological balance, allowing these species to thrive in cycles linked to both seasonal shifts and community knowledge.

Objects-Infrastructure-Buildings Relationship: The interaction between fishing tools, infrastructure, and buildings is essential. For instance, the "Krenkku", a key structural element of piers, supports the framework for lamprey traps, which are suspended from the main pier structure. This setup means that the trap size must align with the dimensions of the bridge or pier it's connected to. Once the fish are caught, they are stored in dedicated buildings where, at the end of the day, they are divided among the community near simple, functional urban facilities - such as tables and benches—that are thoughtfully positioned with views of the river and bridge, emphasizing the community's connection to the water.

Infrastructure-Buildings-Activities Interplay: A further layer of interconnection appears in the relationship between structures, communal buildings, and activities. Specific areas, like the cooking hut, grilling spaces, and outdoor seating areas, are reserved for food preparation and consumption. Cooking methods, such as grilling and smoking fish, are centuries-old traditions, preserved through communal gatherings and seasonal festivals that celebrate and maintain these practices.

This knowledge should serve as a resource for local actors and policymakers, aiding them in the preservation of the Torne River's delicate ecosystem and supporting efforts to expand our mapped inventory of regional elements. These efforts could foster local opportunities, allowing residents to remain in place and potentially reversing the trend of population decline in small villages. A recent success story, reported by Swedish national news (Svt 2024), demonstrates this trend: a community-led crowdfunding campaign will soon fund an art gallery dedicated to displaying works inspired by the Torne Valley, strengthening cultural and economic resilience in the area.

Acknowledgements

The authors would like to acknowledge funding received from the Arctic Five in the Arctic five chair in architecture and planning project (Agatino Rizzo, 2022–2024) and the Lerici Foundation Research Fellowship (Stefano Tornieri 2022). All the paragraphs are e attributed to S.Tornieri except the "introduction" and the "conclusion" paragraphs, which are attributed to S. Tornieri and A.Rizzo.

References

AA.VV. 1965. Hunting and Fishing. Nordic Symposium on Life in a Traditional Hunting and Fishing Milieu in Prehistoric Times and Up to Present Day. Edited by H. Hvarfner. Luleå: Norrbotten Museum.

Allen, K. A., Kern, M. L., Rozek, C. S., McInerney, D. M., and Slavich, G. M. 2021. Belonging: a review of conceptual issues, an integrative framework, and directions for future research. Australian Journal of Psychology, 73(1): 87–102. https://doi.org/10.1080/00049530.2021.1883409.

Azzopardi, E., Kenter, J. O., Young, J., Leakey, C., O'Connor, S., Martino, S. et al. 2023. What are heritage values? Integrating natural and cultural heritage into environmental valuation. People and Nature, 5: 368–383. https://doi.org/10.1002/pan3.10386.

Belanche, D., Casaló, L. V., and Rubio, M. Á. 2021. 'Local place identity: A comparison between residents of rural and urban communities', Journal of Rural Studies, 82: 242–252. https://doi.org/https://doi.org/10.1016/j.jrurstud.2021.01.003.

Berkes, F. 2012. Sacred Ecology. Routledge.

Nilsson, Bo. (2010). Ideology, environment and forced relocation: Kiruna - a town on the move. European Urban and Regional Studies - Eur Urban Reg Stud. 17: 433-442. https://doi.org/10.1177/0969776410369045.

Brulotte, R. L. and Di Giovine, M. A. 2016. Edible Identities: Food as Cultural Heritage. London: Routledge.

Bullo, G. 1940. Le valli salse da pesca. *In:* Brunelli, G., Magrini, G., Miliani, L., Orsi, P. (Eds), La Laguna di Venezia, III (6) C. Ferrari, Venezia: 27–48.

Demajorovic, J., Xavier, A., Pimenta, A. C., and Barreto, R. (2022). Social aspects in the process of mine closure: evolution and avenues for future research agenda. *Proceedings of the International Conference on Mine Closure*. https://doi.org/10.36487/acg_repo/2215_10.

Donoghue, 2017. Beyond Lights-Out: Future Data Centers Will Be Human-Free. In *Data Center Knowledge*. https://www.datacenterknowledge.com/build-design/beyond-lights-out-future-data-centers-will-be-human-free.

Gayday, R. R. (2024). *Automation of processes in aquaculture based on artificial intelligence*. 77–81. https://doi.org/10.23947/aquaculture.2024.77–81.

Hedlund, M., Carson, D., Eimermann, M., and Lundmark, L. 2017. Repopulating and Revitalising Rural Sweden? Re-Examining Immigration as a Solution to Rural Decline. The Geographical Journal 183 (4): 400–413. https://doi.org/10.1111/geoj.12227.

Interreg Nord. 2012. Summer, Whitefish and Torne River: The Whitefish and the Living Fishing Culture of Torne River – the Border River Between Finland and Sweden. Accessed December 8: 2023. https://summerwhitefish.wordpress.com/.

Kaslegard, A. 2011 Climate Change and Cultural Heritage in the Nordic Countries, TemaNord NV - 2010:599. Copenhagen: Nordic Council of Ministers. https://doi.org/10.6027/TN2010-599.

Kent, N. 2014. The Sámi Peoples of the North. A Social and Cultural History. Hurst.

Lenzerini, F. 2011. Intangible Cultural Heritage: The Living Culture of Peoples, European Journal of International Law, Volume 22, Issue 1: 101–120. https://doi.org/10.1093/ejil/chr006.

Ma, J., and Rizzo, A. 2024. 'Arctic-tecture': Teaching Sustainable Urban Planning and Architecture for Ordinary Arctic Cities. Urban Planning, 9.

Mármol, C., and Vaccaro, I. (2020). New extractivism in European rural areas: How twentieth first century mining returned to disturb the rural transition. *Geoforum*, 116: 42–49. https://doi.org/10.1016/J.GEOFORUM.2020.07.012.

Morata, B., Cavalieri, C., Rizzo, A., and Luciani, A. 2020. Territories of Extraction: Mapping Palimpsests of Appropriation. Urban Planning, 5(2): 132–151.

Paredes, D., and Fleming-Muñoz, D. A. (2021). Automation and robotics in mining: Jobs, income and inequality implications. *The Extractive Industries and Society*, 8(1): 189–193. https://doi.org/10.1016/J.EXIS.2021.01.004.

Ransley, J. 2011. Maritime communities and traditions. *In:* A. Catsambis, B. Ford, D. Hamilton (Eds.), Oxford Handbook of Maritime Archaeology, Oxford University Press.

Ravagnan, G. 1980. Principi di progettazione di allevamenti ittici in acque salmastre. Nova Thalassia, 4 (suppl.): 203–218.

Rizzo, A., and Sordi, J. 2020. Resources and urbanization in the global periphery : Perspectives from urban and landscape studies. Cities, 100.

Rizzo, A., Sjöholm, J., and Luciani, A. (2023). Smart(en)ing the Arctic city? The cases of Kiruna and Malmberget in Sweden. *European Planning Studies*, 32(1): 59–77. https://doi.org/10.1080/09654313.2023.2217850.

Senanayake, S. G. J. N. 2006. Indigenous knowledge as a key to sustainable development, Journal of Agricultural Sciences, 2. https://doi.org/10.4038/jas.v2i1.8117.

Shamsi, S., Williams, M., and Mansourian. Y. 2020. An Introduction to Aboriginal Fishing Cultures and Legacies in Seafood Sustainability. Sustainability 12, no. 22: 9724. https://doi.org/10.3390/su12229724.

Singhal, A., Kaur, H., Duggal, A. S., Malik, P. K., and Gehlot, A. (2023). *Emerging Technologies in the Fish Farming by Deploying Artificial Intelligence*. https://doi.org/10.1109/icesc57686.2023.10193206.

Gideon Fink Shapiro, Kiruna, Forever Changing, *Places Journal*, August 2020. Accessed 27 Apr 2025. https://doi.org/10.22269/20080.

Sirelius, U. 1906. Suomalaisten kalastus. Suomalaisen Kirjallisuuden Seura (Kansatieteellisiä tutkimuksia, 1), p. 3 v. file://catalog.hathitrust.org/Record/008866354.

Soini, K., Vaarala, H., and Pouta, E. 2012. 'Residents' Sense of Place and Landscape Perceptions at the Rural-Urban Interface', Landscape and Urban Planning - Landscape Urban Plan, 104. https://doi.org/10.1016/j.landurbplan.2011.10.002.

Soroudi, A., Rizzo, A., and Ma, J. 2024. Urban Sustainability in Arctic Cities: Challenges and Opportunities of Implementing the Sustainable Development Goals. Urban Planning, 9.

Suopajärvi, L., Ejdemo, T., Klyuchnikova, E., Korchak, E., Nygaard, V., and Poelzer, G. A. 2017. Social impacts of the "glocal" mining business: case studies from Northern Europe. *Mineral Economics*, 30(1): 31–39. https://doi.org/10.1007/S13563-016-0092-5.

Svt Nyheter, 2024. Konsthall i Tornedalen kan bli av – tack vare crowdfunding. Konsthall i Tornedalen kan bli av – tack vare crowdfunding. Accessed on 29th April 2025.

Tavares, A. 2024. Architecture Follows Fish. An Amphibious History of the North Atlantic. The Mit Press, Chicago.

Tepecik Diş, A., and Karimnia, E. (2021). Reframing Kiruna's Relocation—Spatial Production or a Sustainable Transformation? *Sustainability*, 13(7): 3811. https://doi.org/10.3390/su13073811.

Tornieri, S., Ma, J., and Rizzo, A. 2024. Which urban and landscape qualities make Arctic villages attractive? The Torne River villages in Sweden. European Planning Studies, 32(8): 1731–1751.

Tornieri, S. 2022. Produrre pesce, produrre paesaggi. Stato dell'arte e opportunità degli impianti vallivi nell'Alto Adriatico, in Currà E., Docci M., Menichelli C., Russo M., Severi L. (Eds.). Stati Generali del Patrimonio Industriale, Ricerche Marsilio, Venezia.

UNESCO. 2003. Convention for the Safeguarding of the Intangible Cultural Heritage.

Walter, N. 2023. 'Lost in Translation: Tangible and Non-Tangible in Conservation', Architecture, 3(3): 578–592. https://doi.org/10.3390/architecture3030031.

Warhurst, A., Macfarlane, M., and Wood, G. (2024). Issues in the Management of the Socioeconomic Impacts of Mine Closure: A Review of Challenges and Constraints. 81–100. https://doi.org/10.1201/9780203756966-5.

5

The City of Montreal and the St. Lawrence River Shores

A Recent History of Rediscovery and Reconnection

Anouk Bélanger,[1] *Maxim Bonin*[2,*] and *Xavier Massicotte*[3]

1. Introduction

Waterfront revitalization is part of a broader phenomenon of urban transformation observed over the last few decades affecting large and medium-sized cities alike. The natural element of water has become a major attraction for quality of life, for citizens, tourists and investors. It has also become a real asset and challenge for urbanists, designers and developers. Visual contact with the water, pedestrian promenades along the banks and the introduction of river transport and water sports, where possible, add value to these areas and make them more attractive from many points of view (residents, symbolic capital, tourists and promoters). Like many 'Cities on water' around the world, Montreal has been reconnecting with its blue spaces over the last two decades. For Montreal, regenerating its waterfronts, and reconnecting with the river could be the starting point for a broader revitalization of the city, and contribute to its repositioning on the international scene. Since the 1980s, a series of nature parks were created based on the acquisition of land and notably on the banks of the Rivière des Prairies on the northwest side of the city (Ile Bizard, Cap-St. Jacques).

[1] University of Quebec in Montreal (UQAM), Atelier de chronotopies urbaines (UQAM), Montréal, Canada.
Email: belanger.anouk@uqam.ca.

[2] Université du Québec en Abitibi-Témiscamingue (UQAT), confluences.lab (UQAT), Rouyn-Noranda, Canada.

[3] Université du Québec en Abitibi-Témiscamingue (UQAT), and Coop Le Comité, Montréal, Canada.
Email: xavier@lecomitemtl.com.

* Corresponding author: maxim.bonin@uqat.ca.

DOI: 10.1201/9781032711737-5

Albeit these parks are not easily accessible with public transports. Other nature parks with a visual connection to the rivers continued to be developed by the city in the 1990s. This coincided with a process of reconnection between the city and the St. Lawrence River through revitalization projects of the industrial heritage around the old port. More recently, other initiatives of reconnection were designed integrating a rising environmental consciousness and care for water quality.

In this chapter, we focus on what we consider as two waves of reconnection with waterfronts; the reconversion of industrial heritage sites (docks, quays, warehouses, railway relics) into spectacular site of consumption, and the design of linear park running along the water and based on human scale diversified public uses. We develop a sociocultural analysis of the appropriation of waterfronts through formal and informal uses of two urban blue spaces, each expressing one of these development phases. Our analysis constitutes the initial phase of a joint research project interested in the impact of design on urban and territorial development with the objective of reflecting on the role of design practices in the current context. This project follows the manifesto "Reconstruire doucement" (Le Comité 2022), which explores different forms of public space utilization in a post-pandemic context, with a specific section dedicated to reclaiming riverbanks. As such, "À nous les rives!" (The Shores are Ours!), is the research project aiming specifically at rediscovering uses of riverbanks in urban environments. Concretely, this chapter aims at understanding the historical uses of two riverside areas, the Old Port of Montreal and the South (Figure 1). After contextualising the city's relation to the river, we present the complex cultural and environmental strata of these two areas, to construct a typology of urban blue spaces for Montreal. Following this, and from a perspective combining cultural studies, urban studies and political ecology we analyse the areas based on their design, cultural uses and governance.

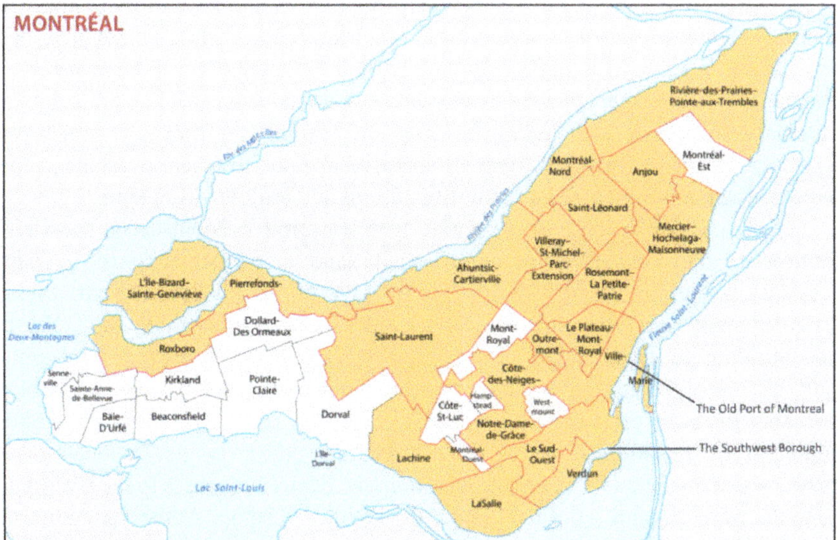

Figure 1: Map of Montreal Boroughs. Authors: Maxim Bonin (2024).

2. The River and the Metropolis

Like canalized water, fresh water has been part of Montreal's urbanization process. It is in fact a mutual process and it seems important to consider the richly conflicting and contradictory role of blue spaces in the city. Already, at the time of European colonization, the presence of water appeared as both an asset and a problem:

"As communication routes linking the city to the mainland and the rest of the world, the rivers encircling the island were also seen as obstacles when the metropolitan area grew, and bridges had to be built to cross them. These tributaries have constituted a formidable resource for human consumption while being used as garbage cans and dumping grounds for waste, to the point of becoming a threat to health. The lure of the riverbanks has always generated its share of pleasures and worries."

(Dagenais 2011, p. 11 [translated by the authors]).

Water is a multifaceted element, simultaneously representing a source of enjoyment, a potential hazard, and a crucial resource for development and tourism. Over time, affects and collective imagination in relation to the rivers surrounding Montreal have also been both an inspiration and a challenge for decision-makers.

2.1 Montreal and its Historical Relation to Water and Riverside

In the 19th and 20th centuries, continental industrialization influenced Montreal's urbanization and significantly impacted its relationship with the riverbanks. The city is strategically positioned with a river connection for international trade and a railway connection for continental transportation, making it a pivotal hub for commerce and human mobility in North America. The city's connection to the St. Lawrence River carries a distinct emphasis on manufacturing and industrial efficiency. Some of the city's waterways have been buried, while others have been canalized to preserve Montreal's original fortified city boundaries (Dagenais 2011). The island's southern shores were extensively developed from east to west to support industrial port activities, including factories and warehouses, and to establish road and rail connections. This development occurred at a period where every city relied on technological advancements to modify natural and built environments.

Green spaces and large urban parks, first appeared in Montreal in the 1870s. The need for protected and sometimes partially landscaped natural spaces to enable sociability, leisure and recreation marked the face of the city in a context where a particular society of rights and democratization policies was asserting itself. Yet this trend in urban development implied turning its back from the river and designing small artificial blue spaces in large parks. Yet, since the 1960s, there has been an increasing pressure on public authorities to clean up rivers and lakes and restore people's access to natural spaces and green areas along the riverside (Dagenais 2011). Slowly, and with the city's deindustrialization, the desire for access to the riverbanks and to the water became a political issue and a matter of rights. At the same time, the city was reimagining itself to compensate for the relocation of industrial activities to southern countries (Bélanger 2005, Gibson 2011). During this time, the city shifted from being an industrial hub to being an attraction for recreation and tourism. Montreal was trying to carve out a place for itself in the interurban competition. It was important to position itself among the many North American cities, like Seattle, San Francisco, Philadelphia and Boston, which were banking on revitalizing their port facilities (Marsan 1994, Gibson 2004).

In many post-industrial cities at the time, wharves, warehouses, and railway lines were the focus of urban development projects to attract developers, tourists and consumers. A Federal corporation (The Old Port of Montreal Corporation) was funded to undertake a major redevelopment project on the city's industrial wharves. In 1992, the flagship destination of the *Old Port of Montreal* was officially inaugurated on the city's 350th anniversary, restoring public access to these areas. This opened new visual access to the river, unveiling a fresh perspective on the river. In the same period, though a few years later, in 1990, the then mayor suggested removing a highway (Bonaventure) southwest of downtown to create more access points to the St. Lawrence River. However, this proposal faced intense criticism and opposition, leading to the mayor's defeat in the municipal elections and the subsequent abandonment of the project.

The mutual development of the city and water banks reveals two key elements more distinctly in this context. On the one hand, "*The municipalization of the territory of the city of Montreal unfolds through water services*" (Dagenais 2011, p.13, [translated by the authors]). On the other hand, the various urban developments have the effect of "*… transforming the island's hydrology by integrating secondary watercourses, along with part of the St. Lawrence River and Rivière des Prairies, into Montreal's drainage plan, and generating environmental inequalities.*" (ibid., p.14, [translated by the authors]). Water, like landscapes, is political, and it is fascinating to see how urban blue spaces allow us to think about the city in the age of global urbanization. It bears political, design, cultural, environmental dimensions along with questions of urban density and urban sprawl.

During the 2002 *Sommet de Montreal*, the city's future priorities were drawn up, including the proposal to protect and improve natural heritage by creating a comprehensive plan for water management. Related actions included increasing the contact points with the water and protecting/naturalizing riverbanks. Following this and since 2007, the *Réseau bleu* (Blue Network)[1] has been committed to preserving the integrity of the aquatic environments of the city and recognizing its invaluable benefits. These initiatives positioned water as a key element to enhance Montreal's brand image. Water (rivers, riverbanks, artificial lakes, docks, etc.) can be a very effective promotional tool. The recent landscaping work at the Grand Quai of the Old Port of Montreal, the Yacht Club, river shuttles and the docking of international cruises is a convincing example of Montreal's brand image as a blue metropolis.

---

[1] *The Réseau bleu* (Blue Network) refers to all water bodies, riverbanks and shoreline areas, including the urban development frontage of Montreal. Through the use of appropriate technologies, the development and enhancement of the *Réseau bleu* (Blue Network) harmonize the protection of the environment, the improvement of the quality of life of the citizens and the development of sustainable financial models. All of this is made possible through consultation with the partners and by listening to the needs and expectations of residents and users (Ville de Montréal 2006).

Today, water culture plays a significant role in discussions about access to riverbanks, prevalent in political documentation, development projects, and frequent media references to aquatic activities. People are increasingly focusing on utilizing, safeguarding, and improving the recreational value of the St. Lawrence River. In this context, the utilization of riverbanks, water activities, and landscaped spaces with views of the river has slowly but successfully integrated blue urban ambiances into the collective imagination of Montrealers. Several development projects are underway to enhance the presence of water in the public life of the city. In the meantime, and in relation to the city's increasing density, informal uses of the riverbanks have opened new pathways and connections with the riverbanks, some of which have been officialised and landscaped for public uses (fishing, kayaking, swimming). Recently, a series of citizen associations and communities are reclaiming the right to be involved in the development and preservation of Montreal's urban blue spaces (Fondation rivière, Surf Grand Montréal, Les Amies du courant, Grand Splash).

2.2 Montreal and Shoreline Access

The city of Montreal is situated on an island surrounded by 266 km of shoreline. Despite this, the island's population faces challenges in accessing the waterfront. The 2024 report from the Jacques-Cartier ZIP Committee[2] highlighted that industrial presence and the privatization of waterfront land have restricted access to riverbanks in many areas. As a result, only 137 kilometers of the shoreline is accessible to the public through 114 riverside parks, limiting the various potential uses of the riverbanks. For example, the by-law concerning parks and public places (Ville de Montréal 2015) stipulates that access to these spaces is prohibited between 11 p.m. and 7 a.m., thus limiting the possibilities for nighttime use of riverside spaces. Port infrastructures prevent access to a good part of the riverbanks (completely denaturalized), despite the requalification of some of these infrastructures near Old Montreal to enhance tourism development and festival activities. The most recent data from the City of Montreal (2024) reveals that this area boasts the highest population density, with multiple sectors averaging around 11,000 people per square kilometer. Additionally, many of these sectors are home to individuals living in more vulnerable conditions.

Several significant park development projects are underway on the island's eastern and western sides. One notable project is the Grand Parc de l'Est, which aims to "safeguard and improve the remaining natural habitats in Montréal's east end" (Ville de Montréal 2024, [translated by the authors]). Other sites, notably the Parc naturel du Cap Saint-Jacques, have provided access to the shoreline for swimming and water sports requiring a boat for several decades, from paddleboarding to jet skiing. However, these resort-like spots, located in the far west and east of the

[2] The Jacques-Cartier ZIP Committee is a non-profit organization whose mission is to use its expertise as a facilitator, its knowledge and its network of partners to promote the harmonious cohabitation of human activities that affect the health of the Montreal island's aquatic and riparian ecosystems (Comité ZIP Jacques-Cartier 2024).

city, are not easily accessible by public transit. For instance, someone living in the Mercier-Hochelaga-Maisonneuve borough would require over two hours of public transit. Unlike the public bus system, the metro does not serve these areas of the island, making travel times much longer. Therefore, both residents and city officials have been wanting to open up various points of access to the river in more central areas and boroughs.

This is taking place in the context of recent public awareness of the necessity of fresh water as well as environmental protection and justice. The "2050 Land Use and Mobility Plan Timeline" (Ville de Montréal 2024[3]) expresses concerns about the degradation of certain riverbanks due to erosion and the loss of wildlife habitats. It highlights the importance of maintaining and enhancing diverse educational, community, cultural, sports, and recreational activities in collaboration with the community to improve natural and aquatic environments (ibid.). While these are commendable goals of the current administration of the City of Montréal, few details are provided about the considerations regarding access to the riverbanks from downtown Montréal areas.

It is interesting to mention the seasonal character of the city, although it is not at the forefront of our following discussion. Montreal is a Nordic city and is characterized by drastic temperature changes which affects what bleu spaces, and water offers in terms of assets and challenges. Historical images indicate that the river around the island of Montreal used to freeze partially and create space for Carnivals, activities and transportation. The need for industrial effectiveness made it so that the ice is now broken every winter and more recently climate change has affected the state of the river during the winter season.

An increasing number of festivals and events is occurring in the city during winter, and it would be interesting to observe the specific uses of banks and rivers in this context. Many of these events (*Nuit Blanche à Montréal* and *Montréal en Lumière*) are hosted in the *Quartier des Spectacles*, others, such as the Igloofest electronic music festival, occur near the waterfront. While some promoters tend to move closer to the river, winter activities presented directly on the banks are rarer. Informal uses rooted in Quebec heritage, such as ice fishing, are uncommon on the outskirts of Montreal. The City of Montreal prohibits certain water sports between October and April, yet some will brave the cold to surf the famous "Vague à Guy" (Guy's wave), an eternal wave on the St. Lawrence River.

These factors linked to the privatization of the riverbanks by the industrial sector and combined with the City of Montreal's legislation (uses of the banks), complicate the governance of these spaces along with the possibilities for community-based or citizen driven projects, or any type of bottom-up development. This is linked to the present movement reclaiming urban blue spaces in Montreal, which is mainly organized through punctual mobilizations responding to crying needs in terms of shoreline access and whitewater swimming opportunities. The claims made by

---

[3] https://montreal.ca/en/articles/2050-land-use-and-mobility-plan-timeline-15575 (consulted, October 2024).

organizations such as the Société des gens de baignade[4] and the Fondation Rivières[5] point in this direction. It is in this context that we have been exploring, with the coop Le Comité, various types of riverside uses, particularly those located on the south side of the Island of Montreal, bordered by the St. Lawrence River and the Prairie Basin.

3. Analyzing Urban Blue Spaces Ambiances in Montreal

Moving our analysis onto the ground (and onto the water), our intention was then to understand the riverside spaces from a cultural and urban studies perspective. We also borrowed from a political ecology approach, as outlined by Keil (2020). As Gandy (2021) mentioned, the urban political ecology perspective has attracted many researchers, although there needs to be more emphasis on cultural uses, design, and affect. Drawing on this hybrid perspective, we can adopt what Latour calls "*a new way to handle all the objects of human and non-human collective life*" (Latour 1998: p. 221, qtd. in Keil).

We designed an analytical grid based on a literature review and previous field observations. Our grid includes the following themes: 1) degree of citizen engagement; 2) land ownership; 3) politics; 4) heritage (industrial and natural); 5) urban planning; 6) uses and 7) culture. While aspects relating to more factual criteria were pursued through a preliminary documentary analysis, our photographic survey focused on uses and cultural aspects. This approach allowed us to include affect and be affected by the ambience of urban landscapes both from the shore and from the water. In addition, our cultural approach bonified by political ecology helped understanding the articulation of human and non-human objects that make up the observed spaces and contribute to both uses and ambiances. Hence, walking and biking on the riverbanks and kayaking on the St. Lawrence River over a period of four months, we observed and documented two distinct types of urban blue spaces: large integrated ambiance for entertainment and commercial purposes on the one hand, and linear assemblage of human-scale and mainly public micro-ambiances.

3.1 The Old Port of Montreal as a Large Integrated Ambiance

As mentioned above, the Old Port (Figure 2), with its industrial docks and grain silos, was completely transformed for the celebrations of the city's 350th anniversary

[4] The Société des Gens de Baignade is a non-profit organization founded in 1996. It is open to all individuals, associations and administrations interested in developing, improving, maintaining and enhancing public access to bodies of water, especially but not exclusively in urban areas (Société des gens de baignade 2024).

[5] For over 20 years, the Fondation Rivières has been working to preserve, restore and highlight the natural character of rivers. It contributes to ensure water quality and access to water for the people of Quebec, while taking into account social, environmental and economic dimensions (Fondation Rivières 2024).

in 1992. Consultations leading up to this requalification began in the early 1980s when specific infrastructures were demolished to open the docks and provide a visual window onto the river. Since the renovation, the docks have been used for events throughout the year. During the summer, Cirque du Soleil erects its tents to showcase its last international spectacle next to major cultural and entertainment permanent venues (IMAX, Science Center, Ferris Wheel). Some water-level-controlled basins are used for activities such as pedal boating in summer and skating in winter. Plage de l'Horloge, an artificial beach very similar to Paris Beaches, is also accessible in Summer although swimming is prohibited. While the photos of Plage de l'Horloge offer a seaside ambiance with a view of the St. Lawrence River and the Jacques-Cartier Bridge, access is limited by the beach's capacity, opening hours, regulations and fees especially so when event promoters occupy the site. The incentive to develop riverbank access for tourism has faced significant opposition from citizen groups like Les AmiEs du Courant Sainte-Marie[6]. In response and with the intention of improving general public access, a large beach was created in the Old Port, which could accommodate up to 7,000 sunbathers on hot summer days. Sitting on this beach, residents and tourists are set in a hybrid urban resort environment with a view on the water, yet a 6-meter-high cement wall separates the beach from the water.

The Old Port of Montréal creates a resort-like ambiance geared more towards tourists and consumers. The area is enclosed between the water; a main street and old warehouses recycled in entertainment venues. Many elements signal the entrance into a specific area as one walks in. A similar situation can be observed at Parc Jean-Drapeau, situated on Île Sainte-Hélène just south of the Old Port, in the middle of the St. Lawrence River. This island and its neighbor, Île Notre-Dame, are partly artificial and were built from residues left over from digging the tracks for the Montréal metro. The islands hosted the 1967 *Terre des Hommes* World's Fair and some of the competitions at the 1976 Summer Olympics. Over the past few decades, the islands where Parc Jean-Drapeau is situated have experienced significant growth, primarily due to the hosting of major concerts and festivals such as the Osheaga Music Festival and îleSoniq. These large-scale events, which draw tens of thousands of attendees

Figure 2: The Old Port of Montreal. Photos by the authors (2024).

[6] Les AmiEs du Courant Sainte-Marie is a citizens' group whose main objective is to improve access to the river in the Centre-Sud area of Montreal (Les AmiEs du Courant Sainte-Marie 2022).

annually, have profoundly impacted the new development initiatives launched in 2017 as part of the commemorative projects for Montréal's 375th anniversary. These examples illustrate how festival promoters wield significant influence over local urban development, often resulting in biased planning favoring larger events over smaller ones (Broudehoux and Bonin 2021).

In the case of the Old Port the development is rooted in the principle of large waterfront promenades developed in the context of a western wave of spectacular post-industrial urban developments (in reference to the works of Zukin, Harvey and others). The vestiges of the industrial era are very much in evidence. White-water swimming is virtually impossible and is occasionally permitted during major events such as the Grand Splash,[7] organized annually by the Fondation Rivières to assert the right to swim in this sector of the Old Port of Montreal. Only boat owners or water sports concessions have access to the water from the docks and marinas, or the yacht Club, listed in this sector of the city. The promenade's-built environment is based on broad architectural principles.

These integrated environments are conducive to the development of tourist and commercial activities, making this sector of the city one of the major tourist destinations on a par with such major urban centers like The Navy Pier in Chicago and Fisherman's Wharf in San Francisco. And in a way, very similar to them. The waterfront strip is virtually non-existent, since the Old Port of Montreal promenade is mainly made up of quays connected by automobile and pedestrian traffic lanes running from the river towards Old Montreal. Here, the waterfront strip could simply represent the twenty feet or so separating the top of the quay from the St. Lawrence River. It is in the verticality of the riverside space (or lack of it) that the urban political ecology of large integrated ambiances can be observed. These spaces offer limited access to water. Here, urban blue spaces serve as a backdrop to a commercial and tourist offer, leaving little or no room for other? forms of reclaiming urban blue spaces.

3.2 The Southwest Borough as an Assemblage of Micro-ambiances

In the southwestern part of (Figure 3), there is a significant area where the river's natural flow was disrupted over a century ago to create a major hydroelectric power station. Historically, important rapids at this point of the river were challenging to navigate and forced ships to stop for portage, creating a place of trade and cultural exchange of all kinds.

The "rocky beach" turned this area into a popular recreational spot, whereas it remained unused when the river itself was primarily used for transporting timber. Downstream, the river was eventually modified to aid in the advancement of both commerce and urbanization. Hence, the river's strong current led to the construction

[7] The Grand Splash is a series of festive events held on the banks of various waterways in Quebec. Citizens are invited to jump into the water at safe locations that do not have public water access but have the potential to become so (Fondation Rivières 2024).

Figure 3: Ambiances along the banks in Montreal's Southwest Borough. Photos by the authors, (2024).

of a hydroelectric power station and dam in the late 19th century, along with a series of mills. These installations were partly demolished to make way, in 1948, for a public green space. Named *Parc des Rapides*, a part of the *Réseau des grands parcs Montréalais*, this space faces the St. Lawrence River and include seven small islands protected as a natural habitat for migratory birds and other species. A linear park, connecting with Parc des Rapides, is 22 km long and is organised alongside bicycle and walking paths running along the riverbanks. From there, the river opens to reveal seven small islands, creating an ambiance far removed from the city's hustle and bustle. The park has preserved some urban history, including several water mills and a hydroelectric dam.

The linear park is organically connected to both the water on one side of the trail, and to residential areas on the other side. The paths seem to balance minimal infrastructure on the residential side (which include volleyball fields, and a public pool) and a series of points of entry (paths, quays, etc.) amidst protected natural spaces on the side of the river. There is a progressive water and nature ambiance as we approach the water. This is radically different from the enclaved entertainment space of the Old Port. These riverside spaces are characterized by their underdesigned appearance, creating freedom of use. Small recreation and tourism businesses have sprung up on both sides of the park, allowing visitors to take advantage of the water's resources by renting boards, surfing, and rafting expeditions. Since the 2000s, more informal and subcultural uses have developed in this hybrid public/private space, shaping or destined to shape the dominant uses of these banks and this part of the river. The "Vague à Guy," an eternal wave in this part of the St. Lawrence River, was discovered about twenty years ago, like the wave at Habitat 67 (See surfgrandmo. It has quietly become a popular spot for river surfing and freestyle kayak subculture. Even though it is not officially part of *Parc des Rapides*, the city has recognized and developed it due to its popularity. In 2021, the city inaugurated improvements to direct user traffic and protect the weakened riverbank. They stabilized and naturalized nearly 150 m of the bank to prevent erosion. This marks the city's first tangible step towards recognizing its river surfing culture, which aligns with the urban blue spaces trend for leisure, outdoor activities, and sports.

The 30ha *Parc des Rapides* connected to the 22 km linear park or bank trail embodies and modernizes the area's heritage, serving as a place dedicated to environmental preservation and nature observation. With a view of the river, it also contains recreational areas and informal micro-spaces for various uses. The ambiance here differs significantly from Montréal's Old Port area, providing a complete riverfront experience geared towards a more diverse public experience. Although this area displays the design principles of a large linear riverside park, the riparian strip is distinguished by what we describe as a series of flexible assemblage of micro-ambiences. These ambiances are both constituted by official installations of the City parks' services, but also by more informal, citizen uses and appropriation. The remnants of the landfilling of this waterfront area (ran in the 1970s) are still clearly visible, and in the spirit of the current urban wasteland, nature seems to claim these places back. In this waterfront strip, we observe a different articulation of the concept of urban political ecology. In fact, public signs are there to inform about nature (fauna, flora, and water) as opposed to directing towards commercial venues of industrial heritage buildings.

The landscape of micro-ambiences between the large linear park and access to water presents itself as underdetermined or not fixed, but rather open and flexible. In this riparian strip, power relations are articulated as well as the complexity of the jurisdiction surrounding the space and the desire of citizens to appropriate these spaces in informal ways which complement or confront the planning principles of large urban parks. The space tends more towards a form of horizontality (democratic). Citizens' uses add to the range of prescribed uses of large parks, allowing informal uses sometimes defying municipal authorities, notably regarding opening hours, the possibility of making fires and even swimming in white water, even if it is contrary to the City of Montreal's regulations. Regarding this, the riversurfing subculture and its impact on short term urban planning of this area is a convincing example.

4. Conclusion: Shoreline Cultures and Eco Transition of Urban Blue Spaces

The analysis of these two urban blue space types allows us to highlight their articulation with the post-industrial city context (Giovinazzi and Moretti 2009). The culture of these riverside spaces serves as convincing indicators of the central role of water in the transformation of cities. Shoreline cultures, whether it be commercial or civic, determined or free, play an important role in defining what could be the expression of a distinct wave of post-industrial cities' development: from spectacular to vernacular, from economic to ecologic, from vertical to horizontal.

The recent wave of waterfront development represents a development that is no longer anchored solely in a logic of tourism development and city branding. While large integrated ambiances have been prioritized in Montreal's urban development projects in recent decades in favor of tourism and mega-events, other projects on a human scale and connecting residential areas to the water are emerging. Rooted in more vernacular practices of riverside space, these uses are part of everyday life for Montrealers, just like fishing, water sports, family pic-nics or simply the art of *flânage*. If crowds gather on the docks of the Old Port of Montreal during major festivals, citizens turn to off centered riverside spaces to reclaim the right of access to white water or even just an access to the river front in an urban environment.

From a political-economic to a political-ecology logic of the development of riverbanks, we can see how a first incentive to open a window to the river has intensified to a complete and horizontal access to the water. Riparian spaces, which play a key role in our adaptation to climate change (Gouvernement du Québec, 2022), can be considered barometers of our willingness and need to protect these urban ecosystems. It is also interesting to reflect upon the expressed desire for underdetermined or underdesigned spaces as they offer a sense of freedom and imagination.

From vertical to more horizontal principles of governance, tolerance and permission for informal uses of these spaces enables the expression of citizen diversified preferences. While large integrated ambiances leave little room for free uses or expression, assemblage of green and blue micro-ambiances can offer greater modularity in the practice of public space. It also provides a more direct and complete connection with the river, as various paths and quays lead directly onto or into the water. The visual access provided in the Old Port, even from the beach area, comes with a 20 feet high vertical disconnection with the river.

These contrasting development waves, and their consequential spaces lead us to reflect upon further opportunities of urban blue spaces design and cultures. Can collaborative urban design practices adapt more closely to the desired practices of riverside spaces? Can we imagine mixed urban blue spaces rather than univocational destinations? The city's densification has pushed us to the shores of the rivers. How will we want to connect, design and protect it?

Our objective in this chapter was to analyze power dynamics, stakeholder priorities, the role of design in contrasting post-industrial urban blue spaces, and its potential connection with the political economy and ecology of the city. These observations serve as a steppingstone for the authors aiming at a critical approach to urban riverbank development. The overall objective is to contribute to the socio-ecological transition of cities with civic planning projects considering the intersection of formal and informal uses; the collective imagination and engagement; the challenge of diversity and inclusion; and above all an increasingly profound connection with water, knowing how it will always constitute both an asset and a challenge.

Acknowledgements

This research project brings together researchers from confluences.lab, at University of Quebec in Abitibi-Témiscamingue, from l'Atelier de chronotopies urbaines at University of Quebec in Montreal, and from Le Comité, a local urban design cooperative based in Montreal (Quebec).

References

Bell, S., Wilczyńska, A., and Balicka, J. 2021. Docklands, harbours and post-industrial sites. pp. 372–405. *In: Urban Blue Spaces*. Routledge.

Bélanger, A. 2005. Montréal vernaculaire/Montréal spectaculaire: dialectique de l'imaginaire urbain. *Sociologie et sociétés*, 37(1): 13–34.

Bouchard, S., and Naud, L. 2001. Un fleuve inaccessible. *Téoros*, 20(1): 9–12. https://doi.org/10.7202/1071904ar.

Brenner, N., and Keil, R. 2014. From Global Cities to Globalized Urbanization. The City Reader. Seventh Edition. London: Routledge.

Broudehoux, A.-M., and Bonin, M. 2020. Beyond the Scene Rethinking Event Assessment at Montreal's Music Festival. *In: Planning and Managing Smaller Events: Downsizing the Urban Spectacle*, London: Routledge.

Comité ZIP Jacques-Cartier. 2024. Notre organisme. https://www.zipjc.org/fr/.

Dagenais, M. 2011. *Montréal et l'eau.* Montréal: Boréal.

Dubinina, A., Wawrzyńska, A., and Krośnicka, K. A. 2022. Permeability of Waterfronts—Contemporary Approach in Designing Urban Blue Spaces. *Sustainability*, 14(15): 9357.

Durán Vian, F., Pons Izquierdo, J.-J., and Serrano Martínez, M. 2021. River-city recreational interaction: A classification of urban riverfront parks and walks, Urban Forestry and Urban Greening, Volume 59. https://doi.org/10.1016/j.ufug.2021.127042.

Fondation Rivières 2024a. Grand Splash. https://fondationrivieres.org/nos-actions/grand-splash/.

Fondation Rivières 2024b. Notre mission. https://fondationrivieres.org/.

Forgaci, C. 2018. *Integrated Urban River Corridors: SpaHal Design for Social-ecological Resilience in Bucharest and Beyond*, Doctoral dissertayon, University of Technology, Delz *Securing the Spectacular City: The Politics of Revitalization and Homelessness in Downtown Seattle.* Lanham: Lexington Books.

Gandy, M. 2022. Urban political ecology: A critical reconfiguration. *Progress in Human Geography*, 46(1): 21–43.

Gibson, T. A. 2003. The trope of the organic city: Discourses of decay and rebirth in downtown Seattle. *Space and Culture*, 6(4): 429–448.

Giovinazzi, O., and Moretti, M. 2009. Port cities and urban waterfront: Transformations and opportunities. *TeMA-Journal of Land Use, Mobility and Environment, 2.*

Goodling, E. 2021. Urban political ecology from below: Producing a "peoples' history" of the Portland Harbor. *Antipode*, 53(3): 745–769.

Gouvernement du Québec. 2022. Comité consultatif sur les changements climatiques Climat et biodiversité: redéfinir notre rapport à la nature. https://www.quebec.ca/gouvernement/ministeres-organismes/comite-consultatif-changements-climatiques/publications.

Hochhalter, A. 2013. *Waterfront spectacular* (Doctoral dissertation, University of Illinois at Urbana-Champaign).

Hussein, H. 2006. Urban recreational riverfronts: successful revitalization elements. *J. Des. Built Environ.*, 2(1).

Keil, R. 2020. An urban political ecology for a world of cities. *Urban Studies*, 57(11): 2357– 2370.

Le Comité. 2022. Manifeste Reconstruire Doucement. https://www.lecomitemtl.com/manifeste/.

Le Comité. 2024. La coopérative. https://www.lecomitemtl.com/la-cooperative.

Les AmiEs du Courant Sainte. 2022. 2012-2022: Un désir devient une mobilisation pour le changement. https://www.facebook.com/profile.php?id=100064267616613.

Mishra, H. S., Bell, S., Roberts, B. R., and White, M. P. 2023. Theory-based design for promoting positive behaviours in an urban blue space: Pre-and-post observations of a community co-created intervention in Plymouth, United Kingdom. *Landscape and Urban Planning*, 233: 104708.

Shaw, K. 2013. Docklands dreamings: Illusions of sustainability in the Melbourne docks redevelopment. *Urban Studies*, 50(11): 2158–2177.

Schubert, D. 2008. Transformation processes on waterfronts in seaport cities–causes and trends between divergence and convergence. *Port Cities as Areas of Transition. Ethnographic Perspectives*, 25–46.

Ville de Montréal. 2006. Le réseau bleu. Montreal vers une culture de l'eau. https://ville.montreal.qc.ca/pls/portal/docs/page/commissions_perm_v2_fr/media/documents/Document_reseau_bleu_20061101.pdf.

6

Implementing Nature-Based Solutions in Urban River Landscapes

Opportunities and Constraints in the Vistula River Valley, Warsaw

Piotr Sikorski[1] and *Kinga Kimic*[2,*]

1. Introduction

Nature-based Solutions (NBS) represent a set of strategies and interventions inspired and powered by nature to address complex environmental and social challenges. These range from protecting or restoring existing ecosystems to the creation of new ones (Sutton-Grier et al. 2018, Albert et al. 2021, Sikorska et al. 2021, Sowińska-Świerkosz and Garcia 2022). By enhancing natural functions, NBS simultaneously deliver social and economic benefits (Gómez Martín et al. 2021, Djukic et al. 2025). Their effectiveness, however, depends on the degree to which they support biodiversity, which underpins ecosystem resilience and the sustained provision of ecosystem services. NBS that prioritize native species diversity and ecosystem restoration are significantly more effective for climate change mitigation and adaptation than those based on monocultures (Seddon et al. 2020). As such, NBS are increasingly recognized as foundational tools for sustainable urban governance

[1] Warsaw University of Life Sciences – SGGW, Department of Remote Sensing and Environmental Assessment, Institute of Environmental Engineering, Warsaw, Poland.
Email: piotr_sikorski@sggw.edu.pl, 0000-0002-5611-4731

[2] Warsaw University of Life Sciences – SGGW, Department of Landscape Architecture, Institute of Environmental Engineering, Warsaw, Poland.

* Corresponding author: kinga_kimic@sggw.edu.pl, 0000-0001-8336-347X

DOI: 10.1201/9781032711737-6

and resilience-building (Lafortezza et al. 2018). Water resources are central to this integrated approach, requiring holistic and system-based planning (Voulvoulis et al. 2017). In riverine contexts, NBS can be implemented across multiple spatial scales – from localized Blue and Green Infrastructure (BGI) elements in dense urban settings (Kimic and Ostrysz 2021, Kimic and Otręba 2022) to broader catchment-level actions in peri-urban or rural landscapes (Sauliotis and Voulvoulis 2022). Full-scale interventions, although potentially more impactful, remain more difficult to realize (Fastenrath et al. 2020). While many NBS initiatives are assessed at local levels, a growing awareness of global challenges calls for more systemic, large-scale solutions. In urban river landscapes, these may include slope greening, river and stream renaturation, floodplain restoration, sandy shore revitalization, the creation of natural wetlands, and the integration of gardens, parks, green corridors, and bioretention areas into the urban fabric (World Bank 2021).

The effectiveness of NBS varies widely depending on context and design. Among the most impactful are river renaturation efforts that promote natural succession and restore ecosystem dynamics (Lundy and Wade 2011, Cohen-Shacham et al. 2016). Key to such interventions is enhancing biodiversity and ensuring the sustainable use of natural resources (Seddon et al. 2020, Gómez Martín et al. 2021, Martín Muñoz et al. 2024). Especially in urban environments, where ecological conditions are heavily altered, ecosystem functions initiated or supported by humans can generate long-term benefits. These interventions are often cost-effective, transferring the responsibility for ecosystem regulation to natural processes, and resulting in self-sustaining patches of urban green space (Aznarez et al. 2022). To ensure resilience and functionality, such areas must be designed as high-ecological-quality ecosystems, where biodiversity is the central driver of system efficiency and durability.

Equally important is the integration of ecological and social objectives. Riverine NBS that restore ecosystems also provide valuable spaces for recreation and human–nature interaction (Cohen-Shacham et al. 2016, Kabisch et al. 2017, Nesshöver et al. 2017). These interactions enhance public health and well-being (White et al. 2020, Smith et al. 2022, Calapez et al. 2023, Trzeciak and Sikorska 2023, Grzyb 2024), while promoting awareness of environmental values. For such outcomes to be sustainable, the involvement of local communities is essential, both through participatory planning and educational outreach. Social participation enhances the acceptance and success of climate adaptation strategies grounded in NBS (Richardson and Razzaque 2006, Reid et al. 2018).

The Vistula River in Warsaw, the capital of Poland, is a vital water resource influencing residents' health, well-being, and access to high-quality urban nature. The green and open areas along its course provide attractive recreational spaces (Warsaw City Council 2020). The river is notable for its relatively low level of anthropogenic transformation; in particular, the right bank remains free of regulatory infrastructure along nearly 4 km of its course. The natural heritage – comprising riparian forests, swamps, wet meadows, and peat bogs – acts as ecological nodes requiring protection and is considered unique in a European urban context.

The natural riverbeds and surrounding green banks play a key role in preserving biodiversity within both Warsaw and the broader Mazovia region. The Warsaw section of the river is recognized as a biodiversity hotspot, known as the *Central*

Vistula Valley, and forms part of the city's green infrastructure system. This network includes ecological linkages to adjacent areas, such as meadow-woodland mosaics in modified embankment zones, and is integrated into the European Natura 2000 network. Key nature reserves along the valley include Kępa Zawadowska and Ławice Kiełpińskie (Warsaw City Council 2020).

Biodiversity hotspots extend along the entire length of the Vistula River Valley within the city (Warsaw City Council 2018b). Particularly valuable areas are found in the Wilanów district, which includes the historic Wilanów Park. Here, natural resources are reinforced through multiple greenery and landscape types that enhance habitat diversity. Supporting biodiversity is one of the operational goals of the *#Warsaw2030 Strategy*, which aims to shape a high-quality natural environment by 2030 (Warsaw City Council 2018a). This strategy also highlights the importance of recreational accessibility along riverfront areas. These principles have guided the *Study of Conditions and Directions of Spatial Development of the Capital City of Warsaw*, which directly informs local spatial development plans (Warsaw City Council 2018c). Despite these ambitions, two centuries of urbanization have significantly altered the natural conditions of the river in the city's central section (Wierzbicki et al. 2021). Today, urban expansion pressures continue to threaten remaining open spaces, while protection efforts for ecologically valuable areas – particularly those forming the Warsaw Nature System – remain insufficient (Warsaw City Council 2020).

This chapter aims to assess the current state of development in the Vistula River Valley and to identify NBS that contribute to the provision of ecosystem services in two key dimensions: (1) the natural, focused on enhancing biodiversity and supporting high-quality ecosystems; and (2) the social, related to opportunities for recreation in the river's natural surroundings. The analysis is structured across three spatial scales: the entire Vistula River Valley in Warsaw, the Wilanów district as a representative local context, and a single green space – Warsaw's Park Akcji 'Burza' methodological approach combines a qualitative review of strategic urban planning documents, spatial analysis of ecological and recreational conditions, and quantitative data from social surveys conducted among Warsaw residents in 2021–2022. Together, these cases and methods help to illustrate the limitations and opportunities for NBS implementation in a complex urban river landscape.

2. Materials and Methods

2.1 Case Study

The study was conducted in Warsaw (52°13′56″N 21°00′30″E), the capital of Poland, which covers an area of 517.2 km^2 and has a population exceeding 1.79 million residents (Central Statistical Office 2024). The Vistula River, Warsaw's main watercourse, flows through the city for approximately 31 km, forming the largest green corridor within the urban landscape. Due to its variable geomorphology, Warsaw's green spaces are located either on the Vistula floodplain or on the upland terraces, resulting in distinct ecological and spatial conditions. These differences influence the planning, design, and implementation of Nature-based Solutions (NBS) at multiple scales.

The case study focuses on the identification and classification of NBS across three spatial levels: (1) city-wide (the entire Vistula River Valley), (2) district-level (Wilanów), and (3) site-level (Park Akcji 'Burza'). At each scale, NBS were characterized in terms of their type (e.g., conservation of natural ecosystems, design of novel ecosystems, restoration of degraded areas), governance context (e.g., local government, planning offices, or municipal institutions such as the Warsaw Greenery Authority), and intended outcomes (e.g., biodiversity enhancement, recreational value, landscape connectivity). The analysis distinguishes between implemented, planned, and proposed NBS, evaluating their capacity to support high-quality ecosystems and ecosystem services.

2.2 Methodology

The identification and assessment of Nature-based Solutions (NBS) along the Vistula River in Warsaw were based on a multi-scalar, mixed-methods approach combining qualitative document analysis, spatial data interpretation, and the integration of social survey results. The research covered the period from 2018 to 2024 and drew upon key strategic and planning documents, including the *#Warsaw2030 Strategy* (Warsaw City Council. 2018a), the *Environmental Protection Programs for the Capital City of Warsaw* for 2021–2024 (Warsaw City Council 2020) and 2025–2030 (Warsaw City Council 2024a), and local spatial development plans (Warsaw City Council 2024b). These were complemented by an analysis of implementation activities led by the Warsaw Greenery Authority (2024). The analytical framework focused on three spatial scales (Cortinovis et al. 2022, Figure 1):

1. **City scale** – assessing the Vistula River Valley in Warsaw in terms of protected areas, ecological connectivity, and designated biodiversity hotspots.
2. **District scale** – examining the Wilanów district due to its proximity to the river and the presence of historic and protected landscapes, with reference to local planning documents and landscape protection measures.
3. **Site scale** – analyzing the Park Akcji 'Burza' as a designed green space built on degraded land, representing a novel urban ecosystem based on NBS principles.

At each scale, two dimensions of NBS were examined:

- Natural aspect – referring to biodiversity support, habitat quality, and ecosystem resilience. This was assessed through the spatial occurrence of biodiversity hotspots, rare and protected species, and ecological continuity, based on data and classifications from the Study of Conditions and Directions of Spatial Development of the Capital City of Warsaw (elaborated until 2023).
- Social aspect – referring to opportunities for nature-based recreation and contact with urban nature. This was evaluated using data on population density and residents' preferences for nature contact, derived from quantitative social research conducted in 2021 and 2022 on a representative sample of 401 Warsaw residents (Melon et al. 2024).

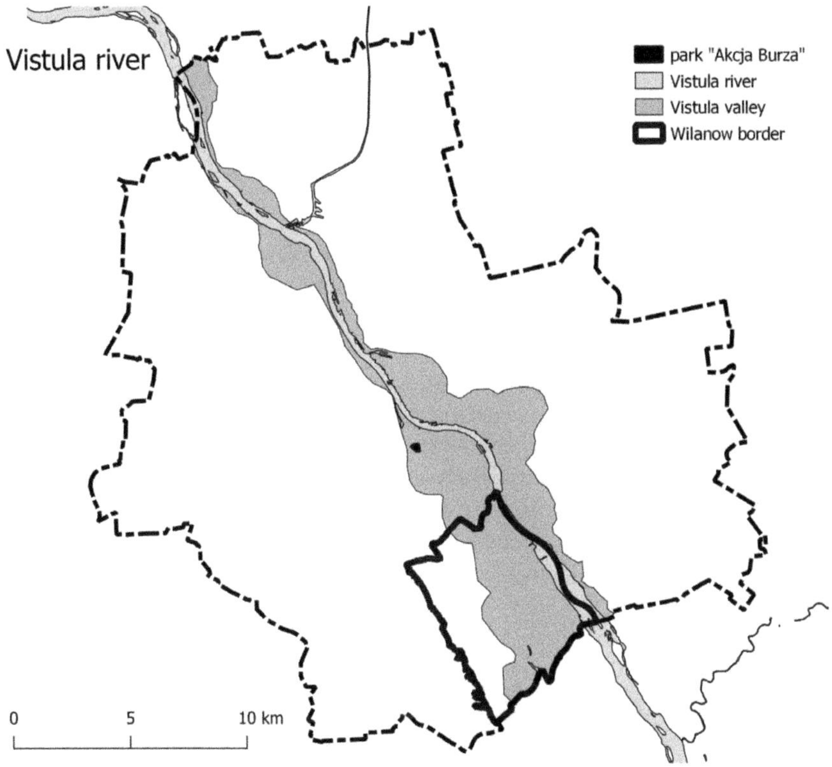

Figure 1: Areas of Nature-based Solutions' implementation within the Vistula River analyzed at three scales: Vistula River (light grey), Vistula River Valley in Warsaw (dark grey), Wilanów District (black border line), Park Akcji 'Burza' (black area). Source: Authors' elaboration.

To link the three scales with the two analytical dimensions, a matrix approach was used. Each case (city, district, site) was evaluated according to the presence, type, and quality of NBS interventions, and cross-referenced with both ecological indicators (e.g., hotspot designation, vegetation types) and social indicators (e.g., perceived access to green spaces, proximity to residential areas). This allowed for a comparative understanding of how different scales of intervention contribute to the delivery of ecosystem services and the realization of NBS principles in practice.

3. Results

3.1 The Vistula River Valley in Warsaw

The Vistula River Valley runs longitudinally through the city of Warsaw, forming a prominent natural corridor. In the northern and southern outskirts of the city, the river retains semi-natural characteristics, as it remains unregulated in these sections. The riverbed width varies from 700 to 1,000 meters and can extend up to 1.4 kilometers during high water conditions. However, the central section of the river has undergone significant modification since the 1960s, when a system of hydraulic engineering structures—primarily longitudinal dikes and transverse spurs (groynes) – were introduced to stabilize the channel. These interventions narrowed the riverbed to approximately 340 meters and straightened its course. Along the left bank, the river is bounded by an artificial embankment built of concrete, which limits natural processes. In contrast, the right bank is characterized by a set-back dike, allowing for the development of a relatively undisturbed riparian zone between the river and the flood protection infrastructure (Wierzbicki et al. 2021).

Natural aspect: The Vistula River Valley in Warsaw includes numerous river islands, oxbow lakes, and floodplain habitats, supporting exceptional biodiversity. Within a 500-meter buffer along the river, several high-value natural and semi-natural habitats are present, including riparian forests (e.g., in Tarchomin district), urban wetlands (e.g., Wawerskie Corner), aquatic ecosystems (e.g., Czerniakowskie Lake), and parkland forests (e.g., Bielański Forest) (Sikorski et al. 2021b, Wierzbicki et al. 2021). These areas collectively form the Warsaw Nature System, an ecological network supporting both landscape connectivity and rare species. Figure 2 illustrates the spatial distribution of biodiversity hotspots and rare species occurrences within the valley. Preserving and enhancing these hotspots is a strategic priority for Warsaw's sustainable development, as stated in the #Warsaw2030 Strategy (Warsaw

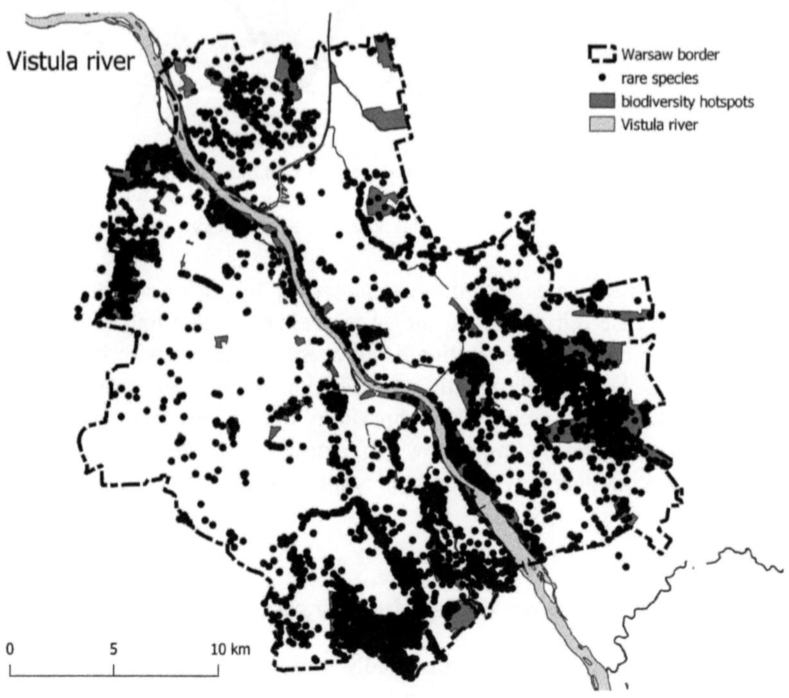

Figure 2: Natural conditions of Warsaw, including the Vistula River Valley area: biodiversity hotspots (marked in dark grey), distribution of rare species (marked by black dots). Source: Authors' elaboration based on Sikorski et al. (2021b).

City Council 2018a) and the Environmental Protection Program for 2025–2030 (Warsaw City Council 2024a).

As part of the 2019–2023 elaboration of the *Study of Conditions and Directions of Spatial Development of the Capital City of Warsaw* (Warsaw City Council 2018b), over 70 biodiversity hotspots were delineated. Group A hotspots include areas where rare or protected species occur, or where valuable habitats sustain natural ecological processes; these zones are recognized for both their ecological and recreational potential. Group B hotspots refer to areas with less valuable natural vegetation or anthropogenic plant communities, yet offering significant recreational and landscape values (Architecture and Spatial Planning Office of the Capital City of Warsaw 2023). Many of these hotspots overlap with the Vistula River Valley. However, proposals to legally protect them through local spatial plans have faced criticism from some urban planning stakeholders, who argue that such designations would restrict urban development (Society of Polish Urban Planners - Warsaw Branch 2023).

Social aspect: The Vistula River Valley is under increasing social and infrastructural pressure due to urban expansion along the river corridor. Engineering interventions are needed to service new residential developments, particularly in the embankment zone. Within a 1 km radius of the riverbanks, over 100 000 people currently reside, and this figure is projected to rise significantly by 2050, potentially reaching close to 1 million. Given this trajectory, the demand for green and accessible spaces along the river will grow, particularly in areas where population density is high (Melon et al. 2024). Figure 3 presents a spatial overlay of population density (500 × 500 m grid) and locations identified by residents as key places for contact with nature.

The city has responded with the gradual implementation of NBS aimed at improving recreational infrastructure and ecological connectivity. On the highly urbanized left bank, revitalized green areas and urban boulevards dominate, while the right bank, which retains higher ecological value, is being equipped with nature-compatible facilities such as pedestrian and cycling paths, informal beaches, scenic viewpoints, and picnic areas. These interventions aim to provide multifunctional spaces that support both ecological and recreational needs. According to recent surveys, this approach is well received by residents and visitors, though source validation of public opinion is needed (Melon et al. 2024).

3.2 Wilanów District

Wilanów is one of Warsaw's southern districts, located on the city's left bank of the Vistula River. A portion of its territory lies within the Vistula River Valley. The district covers an area of 36.73 km² and is inhabited by approximately 50 363 residents (Warsaw City Council 2024c). The hydrographic system of Wilanów includes the Vistula River and its tributaries – the Wilanówka River and Służewiec Stream – as well as remnant floodplain features such as Powsinkowskie Lake and Wilanowski Pond. These elements constitute important aquatic ecosystems but are currently under considerable anthropogenic pressure (Hejduk et al. 2022).

Natural aspect: The vegetation of the Wilanów district reflects the characteristic flora of the Vistula River Valley within the Masovian Lowland, with the presence

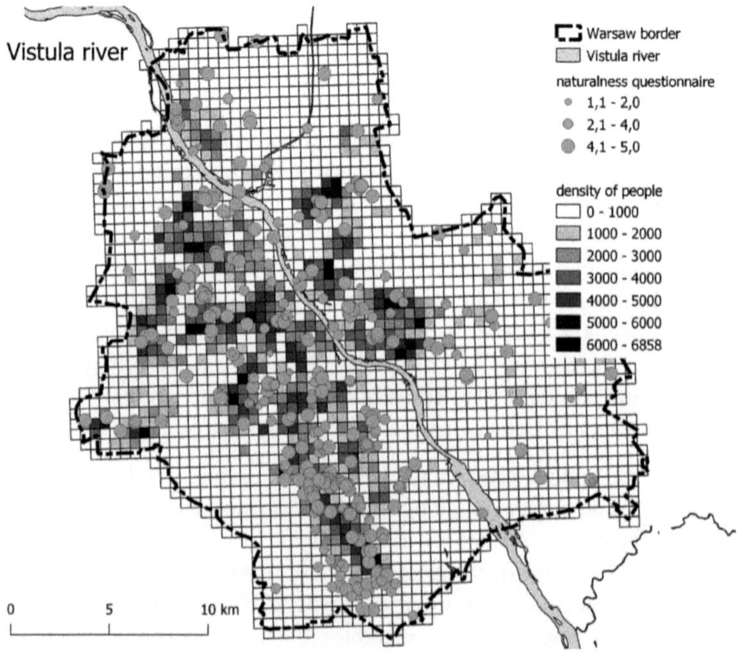

Figure 3: Social conditions for the development of Nature-based Solutions in Warsaw, population density in a square of 500 x 500 m (marked with a white-grey-black scale), places indicated by residents as enabling contact with nature (marked by gray dots). Source: Authors' elaboration based on Melon et al. (2024).

of riparian forests, oak-hornbeam woodlands, wet meadows, and floodplains. This ecological diversity is considered of national significance, with a high richness of vascular plant and lichen species (Sikorski et al. 2022). As a result, various legal forms of nature protection are in place (Wilanów District Office 2011). The district is intersected by a Natura 2000 area and includes several designated nature reserves: the Zawadowskie Islands (comprising riparian habitats and sandbanks), the Natoliński Forest, and the Ursynów Escarpment. Additionally, two ecological areas ('Powsin' and 'Powsinek') and numerous natural monuments further contribute to biodiversity conservation. A particularly notable site is the Morysin Nature Reserve (Regional Directorate for Environmental Protection in Warsaw 2010), which forms part of the Wilanów Palace Museum landscape and has been protected since 1996. It encompasses riparian forests, monumental trees, and a rich array of flora and fauna. Managed by the Wilanów Palace Museum (Museum of King Jan III's Palace at Wilanów 2024), the reserve lies within the Warsaw Protected Landscape Area (Mazowieckie Voivodeship Office in Warsaw. 2007). These areas are essential for maintaining biodiversity hotspots and serve as ecological buffers in the face of expanding urban development (Figure 4).

The identification of biodiversity hotspots within the Wilanów district was an integral component of the *Study of Conditions and Directions of Land Development of the Capital City of Warsaw* (Warsaw City Council 2021). The study proposed

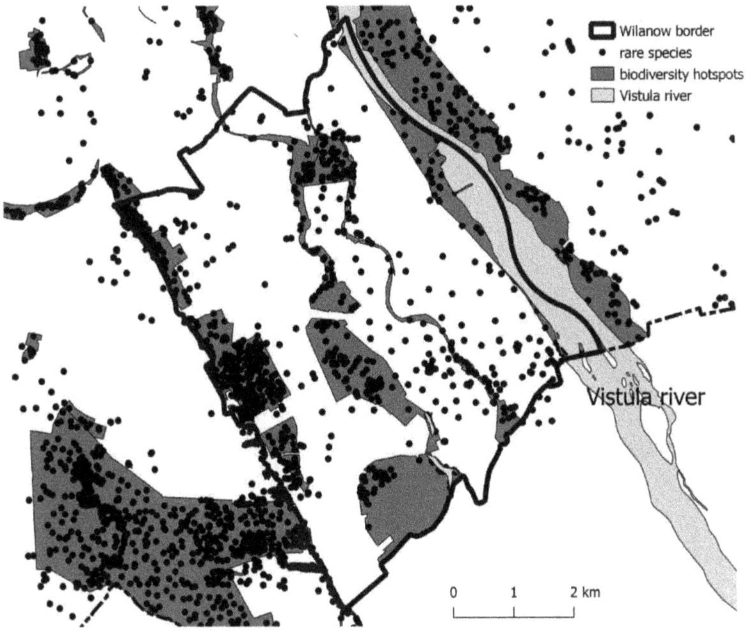

Figure 4: Natural values in the Wilanów district: biodiversity hotspots (areas marked in dark grey), distribution of rare species in the district and its buffer (marked by black dots). Source: Authors' elaboration based on Sikorski et al. (2021b).

safeguarding key sites along the riverfront by expanding the protected area network. However, as in other parts of the city, these measures faced significant opposition from urban planners and residents, who feared that such designations would constrain future development (Society of Polish Urban Planners - Warsaw Branch 2023). Although the formal approval of the study has been suspended, the ecological assessments continue to inform local spatial planning decisions.

Social aspect: Wilanów is one of Warsaw's most dynamically developing districts, shaped by several decades of intensive residential and infrastructural investment. Many former agricultural or open spaces remain undeveloped, presenting considerable potential for urban expansion. This dynamic growth, combined with the district's natural amenities, has made Wilanów a highly attractive area for new residents. At the same time, the rapid urbanization process generates strong social pressure to preserve and improve access to natural areas for recreation and contact with nature. As Figure 5 illustrates, areas with high ecological value, such as the Vistula River corridor and adjacent reserves, are frequently used for outdoor activities, which intensifies the functional role of these ecosystems. Population projections and ongoing development suggest that social demand for green space will continue to rise, further increasing the need for strategic implementation of NBS in urban planning.

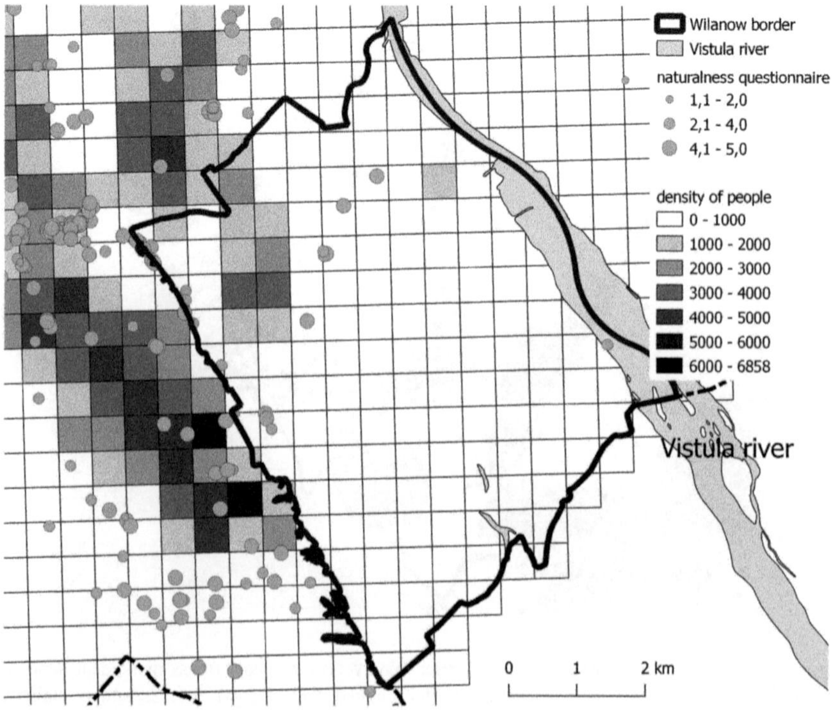

Figure 5: Social conditions related to the development of Nature-based Solutions in the Wilanów district: population density in a square of 500 x 500 m (marked with a white-gray-black scale), places indicated by residents as enabling contact with nature in the district (marked by grey dots). Source: Authors' elaboration based on Melon et al. (2024).

3.3 Park Akcji 'Burza'

The Warsaw Uprising Mound and its surroundings constitute a highly symbolic, yet historically degraded site in central Warsaw. Originally created between the late 1940s and mid-1960s as a rubble landfill from the destruction of Warsaw during World War II, the artificial mound rises approximately 35 meters above the flat, marshy terrain of the Vistula River floodplain. In the decades that followed, the area gradually became overgrown with spontaneously developing vegetation, while the surrounding landscape was increasingly urbanized. The summit of the hill now hosts a memorial monument commemorating the 1944 Warsaw Uprising, reinforcing the site's cultural and historical significance.

In 2019, a landscape architecture competition led to the selection of a design proposal by topoScape and Archigrest studios, aiming to transform the site into a multifunctional public green space that would combine recreational use, ecological function, and heritage values (Celiński 2024). The resulting Park Akcji 'Burza', completed in 2023, has received widespread acclaim, including national and international awards such as the Landezine International Landscape Award (LILA

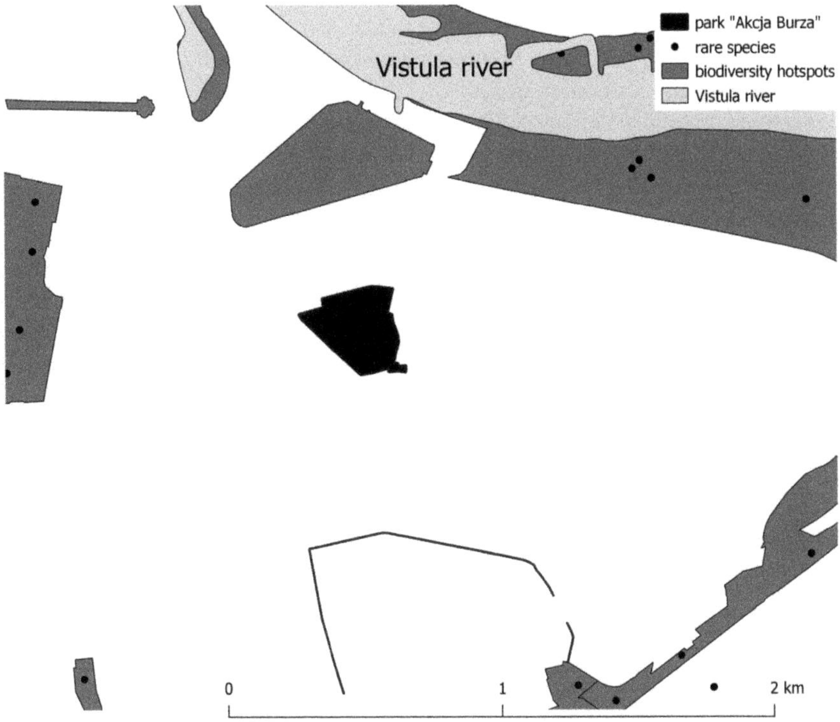

Figure 6: Natural conditions of the site before the park implementation: biodiversity hotspots (marked in dark grey) in the vicinity of the planned park, distribution of rare species in the vicinity of the planned park (marked by black dots). Source: Authors' elaboration based on Sikorski et al. (2021).

2024) and the European Prize for Urban Public Space (2024). The park is highly rated by both residents and visitors.

Natural aspect: The park was established on a brownfield site, which at the time of intervention did not contain any recognized biodiversity hotspots or occurrences of rare species (Figure 6). However, its location in the broader ecological corridor of the Vistula River Valley presented opportunities for enhancing urban biodiversity through the implementation of Nature-based Solutions (NBS). The design approach was grounded in the concept of *novel urban ecosystems*, defined as emergent ecological assemblages shaped by both natural succession and anthropogenic disturbance (Higgs 2017, Cooper et al. 2024).

Rather than replacing spontaneous vegetation, the strategy embraced natural processes already underway, including colonization by pioneer and even invasive species. These were selectively managed to encourage successional development, increase ecological complexity, and create microhabitats. Biodiversity was further supported through targeted plantings, enhancement of soil formation processes, and water retention interventions such as organic matter conservation and surface runoff management. The design also included wilderness zones with limited human access, intended to function as refuge habitats for fauna.

A section of the park, located within the former floodplain of the Vistula River, involved the partial restoration of a historical riverbed, further reinforcing ecological connectivity. In total, 450 trees, 8,500 shrubs, and herbaceous plants were introduced, including over 50,000 specimens of native species. These actions contributed to the park's integration with nearby natural areas, extending the green infrastructure network within the urban core.

Social aspect: Prior to its redevelopment, the area designated for the park had remained largely inaccessible and devoid of recreational or ecological function for several decades. This spatial underutilization was also reflected in social perception: residents of Warsaw—including those from nearby newly developed housing estates—did not identify the site as a space offering contact with nature (Figure 7). The redevelopment initiative aimed to reclaim this degraded brownfield for public use, combining recreational functions with ecological enhancement.

Since its completion in 2023, Park Akcji 'Burza' has become an important space for recreation and nature experience within the urban core, offering amenities and access that were previously absent. The park effectively addresses the needs of

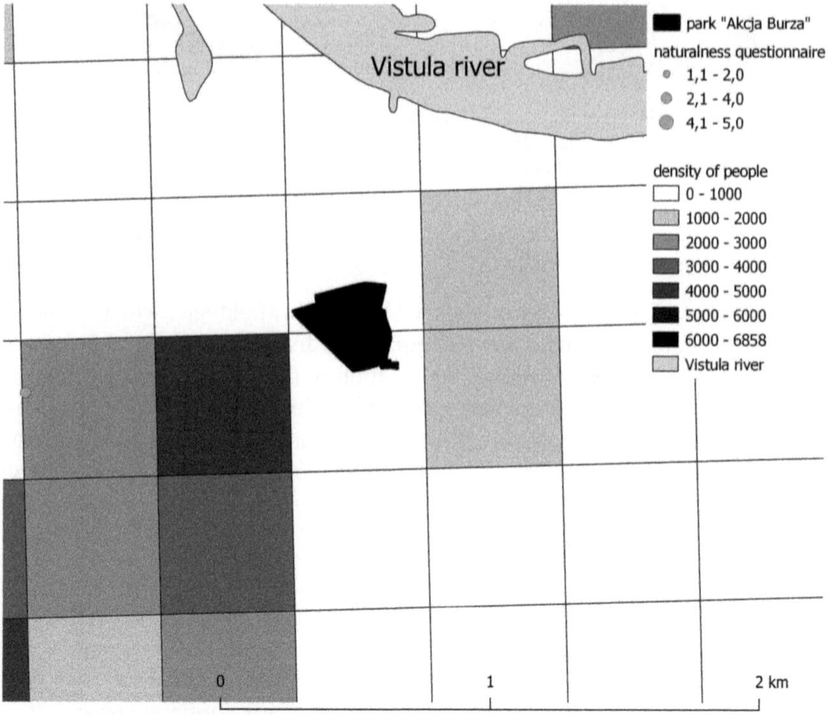

Figure 7: Social conditions in the vicinity of the planned park before its construction: population density in a square of 500 x 500 meters (marked with a white-grey-black scale). Source: Authors' elaboration based on Melon et al. (2024).

residents from surrounding housing developments and complies with the principles of universal design. Two distinct pathways lead to the summit of the Warsaw Uprising Mound and its commemorative monument: a staircase with integrated rest areas and an alternative route using a system of gently sloping ravines, whose retaining walls are constructed from site-specific rubble concrete, preserving historical continuity.

The park's nature trail meanders through newly established tree cover and features educational signage that introduces visitors to local flora and fauna, with a particular emphasis on ruderal vegetation and pioneer species typical of novel ecosystems. In addition, the recreational program includes a children's activity area equipped with climbing structures, a zip-line, swings, and slides—all integrated into the terrain and constructed primarily from natural materials. A designated dog recreation zone has also been provided, reflecting the multifunctional design and inclusivity of the space.

4. Discussion

Implementing Nature-based Solutions (NBS) in urban environments requires more than isolated interventions. Especially in large and densely populated cities, such as Warsaw, achieving comprehensive and measurable outcomes depends on integrating NBS into strategic planning, spatial policy, and long-term governance frameworks (Dorst et al. 2019). While creating green spaces – such as a new park – is a tangible and visible act, the effectiveness of NBS must be evaluated in relation to their ability to improve ecological and social functions across different scales. Superficial planning efforts risk becoming a form of greenwashing (Gałecka-Drozda et al. 2021), where the establishment of a natural structure is not followed by measurable improvements in biodiversity or accessibility. Therefore, successful NBS implementation should be understood as a process involving planning, integration into spatial development strategies, and effective realization (Dorst et al. 2019, Corgo et al. 2024). This requires ongoing cooperation between policymakers, urban planners, the scientific community, and local stakeholders (Kauark-Fontes et al. 2023).

The case of the Vistula River Valley in Warsaw illustrates the complexity of integrating ecological and social dimensions of NBS in a strategic urban context. Although this area includes protected natural habitats with high ecological value, the lack of cohesive planning and limited implementation tools constrain the ability to scale up NBS across the river corridor. Most actions remain fragmented, and planning instruments often fall short of supporting ecological connectivity or long-term ecosystem enhancement. The designation of new biodiversity hotspots, for instance, is met with resistance at both the planning and community levels, limiting the impact of such efforts.

The Vistula River Valley, as a linear natural corridor, holds strategic significance for Warsaw's urban resilience. Current forms of nature protection help reduce urban pressure in ecologically sensitive zones, yet the integration of recreational infrastructure with ecological integrity remains challenging. Recent strategic documents such as the #Warsaw2030 Strategy and environmental protection programs acknowledge this ambition, but implementation often prioritizes recreation over biodiversity (Warsaw City Council 2018a, 2020, 2024a). Notably, biodiversity hotspot proposals in new urban plans were contested by local actors, including urban planners and developers, despite public support for preserving natural areas (Society of Polish Urban Planners - Warsaw Branch 2023).

The Wilanów district represents a microcosm of these tensions. As one of the most rapidly developing residential areas of Warsaw, Wilanów faces intense urbanization pressures, while simultaneously hosting historically and ecologically valuable ecosystems. These include wet meadows, riparian forests, and multiple protected areas (Sikorski et al. 2022). In this context, the main role of NBS lies in protecting existing green infrastructure, enhancing recreational accessibility, and maintaining the ecological function of remnant natural patches. However, fragmented land ownership and speculative real estate investments limit the ability to designate new green zones or ecological corridors. The potential of historical meadows or post-agricultural land is thus constrained, even when ecological assessments highlight their importance.

In contrast, site-scale projects like Park Akcji 'Burza' demonstrate how integrated NBS approaches can succeed when natural potential, social needs, and symbolic meaning align. The park was established on a brownfield site with no prior biodiversity value and was transformed using novel ecosystem principles (Higgs 2017, Cooper et al. 2024). Successional processes and minimal interventions allowed for the emergence of biodiverse plant communities, supported by the selective planting of native species and the creation of microhabitats. The design also accounted for ecological connectivity with the Vistula River corridor and introduced spatial zoning to balance human access and habitat development. This site-based implementation shows that small-scale interventions, if well-designed, can create multifunctional spaces with both ecological and cultural significance.

Socially, the transformation of the former rubble site into a multifunctional park responds directly to residents' expectations. Prior to its redevelopment, the site was unused and perceived as inaccessible. Post-implementation studies confirm that the park now functions as a vital recreational and educational space, supporting nature contact for surrounding communities. Its universal design features and informal nature education elements (e.g., information on ruderal species and succession) highlight how NBS can be aligned with broader goals of inclusion and urban livability. The project's recognition in national and international competitions further confirms its value as a model of urban green space creation that respects historical memory while enhancing ecosystem services.

5. Conclusion and Policy Implications

The experience of Warsaw's Vistula River Valley shows that while protected areas form a strong foundation for urban ecological resilience, a broader and more cohesive NBS framework is needed. Fragmented planning and weak integration of biodiversity hotspots into local development plans hinder the systemic implementation of NBS. Legal protection measures (e.g., nature reserves or cultural heritage sites) are more effective than planning-based 'soft' designations, which often remain unrealized due to stakeholder opposition. In areas like Wilanów, this leads to isolated patches of biodiversity, disconnected from the wider ecological network, and vulnerable to future development.

Site-scale NBS, especially when they combine ecological rehabilitation with public accessibility, tend to gain higher levels of social acceptance. As shown in the Park Akcji 'Burza', such projects are more feasible to implement and can act as demonstrators for larger-scale strategies. However, their ecological impact is limited unless embedded within a wider network of natural and semi-natural areas.

A key limitation of the current approach is the lack of binding mechanisms to safeguard proposed biodiversity hotspots or to scale up NBS across fragmented ownership structures. This also limits monitoring and evaluation, as social and ecological indicators are not systematically applied. Future efforts should focus on developing a multi-scalar NBS strategy that spans from citywide ecological corridors to district-level planning and localized park interventions. This requires aligning planning provisions, participatory processes, and investment mechanisms.

Sustainable management of the Vistula River Valley thus depends on recognizing NBS as a complex, process-based approach. Ensuring ecological continuity, enhancing recreational value, and addressing climate resilience must be integrated at all stages – from metropolitan strategy to local implementation. Ultimately, long-term success depends on understanding the diverse roles of ecosystems in shaping urban well-being and ensuring that NBS are embedded in both planning logic and implementation practice.

References

Albert, C., Brillinger, M., Guerrero, P., Gottwald, S., Henze, J., Schmidt, S. et al. 2021. Planning nature-based solutions: Principles, steps, and insights. Ambio 50: 1446–1461. https://doi.org/10.1007/s13280-020-01365-1

Architecture and Spatial Planning Office of the Capital City of Warsaw. 2023. Raporty z planowania Warszawy (Reports on Warsaw planning). Warsaw. https://pawilonzodiak.pl/wp-content/uploads/2023/07/Jakie-nowe-Studium_internet.pdf (access: 15.10.2024).

Aznarez, C., Svenninh, J.-Ch., Taveira, G., Baró, F., and Pascual, U. 2022. Wildness and habitat quality drive spatial patterns of urban biodiversity. Landscape and Urban Planning 228: 104570. https://doi.org/10.1016/j.landurbplan.2022.104570.

Calapez, A. R., Serra, S. R. Q., Mortágua, A., Almeida, S. F. P., and Feio, M. J. 2023. Unveiling relationships between ecosystem services and aquatic communities in urban streams. Ecological Indicators 153: 110433. https://doi.org/10.1016/j.ecolind.2023.110433.

Celiński, A. 2024. Park Akcji Burza to najlepsza przestrzeń publiczna w Europie! Właśnie ogłoszono wyniki konkursu The European Prize for Urban Public Space. https://architektura.muratorplus.pl/konkursy/park-akcji-burza-to-najlepsza-przestrzen-publiczna-w-europie-wlasnie-ogloszono-wyniki-konkursu-the-european-prize-for-urban-public-space-aa-1fBg-h3Fn-1pcd.html (access: 29.10.2024).

Central Statistical Office (2024). Warszawa w liczbach (Warsaw in numbers). Warsaw. https://um.warszawa.pl/waw/warszawa-w-liczbach (access: 18.10.2024).

Cohen-Shacham, E., Janzen, C., Maginnis, S., and Walters, G. 2016. Nature-based solutions to address global societal challenges. IUCN International Union for Conservation of Nature. https://doi.org/10.2305/IUCN.CH.2016.13.en.

Cooper, C., Collier, M. J., Pineda-Pinto, M., Rodriguez Castañeda, N., O'Donnell, M., and Nulty, F. 2024. Urban novel ecosystems as affective landscapes. Sustainability Science 19: 1921–1933. https://doi.org/10.1007/s11625-024-01539-w.

Corgo, J., Cruz, S. S., and Conceição, P. 2024. Nature-based solutions in spatial planning and policies for climate change adaptation: A literature review. Ambio 53: 1599–1617. https://doi.org/10.1007/s13280-024-02052-1.

Cortinovis, C., Olsson, P., Boke-Olén, N., and Hedlund, K. 2022. Scaling up nature-based solutions for climate-change adaptation: Potential and benefits in three European cities. Urban Forestry & Urban Greening 67: 127450. https://doi.org/10.1016/j.ufug.2021.127450.

Djukic, A., Maric, J., Šuklje Erjavec, I., and Kimic, K. 2025. Towards Sustainable Urban Green Spaces: Increasing Health and Well-Being. pp. 215–244. *In:* Delgado-Jiménez, A., Ruchinskaya, T., Palmese, C., Smaniotto Costa, C., Kirdar, G., and Horan, C. (Eds.). Placemaking in Practice Volume 3. The Future of Placemaking and Digitization. Emerging Challenges and Research Agenda (Vol. 3). Brill, Leiden, The Netherlands. https://brill.com/display/book/9789004691926/BP000016.xml.

Dorst, H., van der Jagt, A., Raven, R., and Runhaar, H. 2019. Urban greening through nature-based solutions – Key characteristics of an emerging concept. Sustainable Cities and Society 49: 101620. https://doi.org/10.1016/j.scs.2019.101620.

Fastenrath, S., Bush, J., and Coenen, L. 2020. Scaling-up nature-based solutions. Lessons from the living Melbourne strategy. Geoforum 116: 63–72. 10.1016/j.geoforum.2020.07.011.

Gałecka-Drozda, A., Wilkaniec, A., Szczepańska, M., and Świerk, D. 2021. Potential nature-based solutions and greenwashing to generate green spaces: Developers' claims versus reality in new housing offers. Urban Forestry & Urban Greening 65: 127345. https://doi.org/10.1016/j.ufug.2021.127345.

Gómez Martín, E., Máñez Costa, M., Egerer, S., and Schneider, U. A. 2021. Assessing the long-term effectiveness of Nature-Based Solutions under different climate change scenarios. Science of The Total Environment 794: 148515. https://doi.org/10.1016/j.scitotenv.2021.148515.

Grzyb, T. 2024. Mapping cultural ecosystem services of the urban riverscapes: the case of the Vistula River in Warsaw, Poland. Ecosystem Services 65: 101584. https://doi.org/10.1016/j.ecoser.2023.101584.

Hejduk, L., Krajewski, A., Kaznowska, E., and Okruszko, T. 2022. Wody Wilanowa w kontekście zmian w XXI wieku. pp. 87–100. *In:* Sikorski, P. (ed.). Wartości przyrodniczo-kulturowe Wilanowa w kontekście postępującej urbanizacji i zmian klimatycznych w XXI wieku. Wydawnictwo SGGW, Warszawa.

Higgs, E. 2017. Novel and designed ecosystems. Restoration Ecology 25(1): 8–13. https://doi.org/10.1111/rec.12410.

Kabisch, N., van den Bosch, M., and Lafortezza, L. 2017. The health benefits of nature-based solutions to urbanization challenges for children and the elderly – A systematic review. Environmental Research 159: 362–373. https://doi.org/10.1016/j.envres.2017.08.004.

Kauark-Fontes, B., Marchetti, L., and Salbitano, F. 2023. Integration of nature-based solutions (NBS) in local policy and planning toward transformative change. Evidence from Barcelona, Lisbon, and Turin. Ecology and Society 28(2): 25. https://doi.org/10.5751/ES-14182-280225.

Kimic, K., and Ostrysz, K. 2021. Assessment of Blue and Green Infrastructure Solutions in Shaping Urban Public Spaces - Spatial and Functional, Environmental, and Social Aspects. Sustainability 13: 19. https://doi.org/10.3390/su131911041.

Kimic, K., and Otręba, M. 2022. Assessment of blue and green infrastructure solutions used in housing estates – spatial and functional aspects. Desalination and Water Treatment 275: 122–130. https://doi.org/10.5004/dwt.2022.28857.

Lafortezza, R., Chen, J., van den Bosch, C. K., and Randrup, T. B. 2018. Nature-based solutions for resilient landscapes and cities. Environmental Research 165: 431-441. 10.1016/j.envres.2017.11.038.

Lundy, L., and Wade, R. 2011. Integrating sciences to sustain urban ecosystem services. Progress in Physical Geography: Earth and Environment 35(5): 653–669. https://doi.org/10.1177/0309133311422464.

Martín Muñoz, S., Schoelynck, J., Tetzlaff, D., Debbaut, R., Warter, M., and Staes, J. 2024. Assessing biodiversity and regulatory ecosystem services in urban water bodies which serve as aqua-Nature-based Solutions. Frontiers in Environmental Science 11: 1304347. https://doi.org/10.3389/fenvs.2023.1304347.

Mazowieckie Voivodeship Office in Warsaw (2007). Rozporządzenie Nr 3 Wojewody Mazowieckiego z dnia 13 lutego 2007 r. w sprawie Warszawskiego Obszaru Chronionego Krajobrazu (Dz. Urz. Woj. Maz. Nr 42/14.02.2007 r., poz. 870) (Regulation No. 3 of the Mazovian Voivode of 13 February 2007 on the Warsaw Protected Landscape Area (Journal of Laws of the Mazovian Voivodeship No. 42/14.02.2007, item 870)). https://siskom.waw.pl/komunikacja/kolej/LK7/lk7_ros_2011-05_zal8_10_pozostale.pdf (access: 22.10.2024).

Melon, M., Sikorski, P., Archiciński, P., Łaszkiewicz, E., Hoppa, A., Zaniewski, P. et al. 2024. Nature on our doorstep: How do residents perceive urban parks vs. biodiverse areas? Landscape and Urban Planning 247: 105059. https://doi.org/10.1016/j.landurbplan.2024.105059.

Museum of King Jan III's Palace at Wilanów. 2024. https://www.wilanow-palac.pl/o_muzeum (access: 19.10.2024).

Nesshöver, C., Assmuth, T., Irvine, K. N., Rusch, G. M., Waylen, K. A., Delbaere, B. et al. (2017). The science, policy and practice of nature-based solutions: An interdisciplinary perspective. Science of the Total Environment 579: 1215–1227. https://doi.org/10.1016/j.scitotenv.2016.11.106.

Regional Directorate for Environmental Protection in Warsaw. 2010. Zarządzenie Regionalnego Dyrektora Ochrony Środowiska w Warszawie z 17 czerwca 2010 r. w sprawie rezerwatu przyrody Morysin (Order of the Regional Director for Environmental Protection in Warsaw of 17 June 2010 regarding the Morysin nature reserve). Warsaw. https://edziennik.mazowieckie.pl/WDU_W/2010/155/3826/akt.pdf (access: 2.10.2024).

Reid, H., Bourne, A., Muller, H., Podvin, K., Scorgie, S., and Orindi, V. 2018. A framework for assessing the effectiveness of ecosystem-based approaches to adaptation. pp. 207–216. *In:* Zommers, Z., and Alverson, K. (eds.). Resilience. Elsevier. https://doi.org/10.1016/B978-0-12-811891-7.00016-5.

Richardson, B. J., and Razzaque, J. 2006. Public participation in environmental decision making. Environmental Law for Sustainability, 165–194. Available at SSRN: https://ssrn.com/abstract=1794203.

Sauliotis, I., and Voulvoulis, N. 2022. Operationalising nature-based solutions for the design of water management interventions. Nature-Based Solutions 2: 100015. https://doi.org/10.1016/j.nbsj.2022.100015.

Seddon, N., Chausson, A., Berry, P., Girardin, C. A. J., Smith, A., and Turner, B. 2020. Understanding the value and limits of nature-based solutions to climate change and other global challenges. Philosophical transactions of the Royal Society of London. Series B, Biological Sciences 375(1794): 20190120. https://doi.org/10.1098/rstb.2019.0120.

Semeraro, T., Gatto, E., De Bellis, L., Luvisi, A., Emmanuel, R., and Buccolieri, R. 2023. A decision-making framework for promoting the optimum design and planning of Nature-based Solutions at local scale. Urban Forestry & Urban Greening 84: 127945. https://doi.org/10.1016/j.ufug.2023.127945.

Sikorska, D., Ciężkowski, W., Babańczyk, P., Chormański, J., and Sikorski, P. 2021. Intended wilderness as a Nature-based Solution: Status, identification and management of urban spontaneous vegetation in cities. Urban Forestry & Urban Greening 62: 127155. https://doi.org/10.1016/j.ufug.2021.127155.

Sikorski, P., Archiciński, P., Sikorska, D., Zaniewski, P., and Zaniewska, E. 2022. Szata roślinna Wilanowa w kontekście zmian w XXI wieku. pp. 109–136. *In:* Sikorski, P. (ed.). Wartości przyrodniczo-kulturowe Wilanowa w kontekście postępującej urbanizacji i zmian klimatycznych w XXI wieku. Wydawnictwo SGGW, Warszawa.

Sikorski P., Gawryszewska, B., Sikorska, D., Chormański, J., Schwerk, A., Jojczyk, A. et al. 2021a. The value of doing nothing – How informal green spaces can provide comparable ecosystem services to cultivated urban parks. Ecosystem Services 50: 101339. https://doi.org/10.1016/j.ecoser.2021.101339.

Sikorski, P., Sudnik-Wójcikowska, B., Zaniewska, E., Zaniewski, P., Kowalska, A., and Wrzosek, M. 2021b. Charakterystyka gatunkowa i przestrzenna ostoi bioróżnorodności – roślin naczyniowych, grzybów i porostów na potrzeby projektu Studium uwarunkowań i kierunków zagospodarowania przestrzennego Warszawy. Biuro Planowania i Architektury Warszawy, Warszawa.

Smith, N., Georgiou, M., King, A. C., Tieges, Z., and Chastin, S. 2022. Factors influencing usage of urban blue spaces: A systems-based approach to identify leverage points. Health Place 73: 102735. 10,1016/j.healthplace.2021.102735.

Society of Polish Urban Planners - Warsaw Branch. 2023. Uwagi do projektu Studium uwarunkowań i kierunków zagospodarowania przestrzennego m.st. Warszawy złożone w trybie Art. 11 pkt, 11 ustawy o planowaniu i zagospodarowaniu przestrzennym (Comments on the draft Study of conditions and directions of spatial development of the capital city of Warsaw submitted in accordance with Art. 11 item 11 of the Act on spatial planning and development). Warsaw. https://mazowiecka.iarp.pl/wp-content/uploads/2023/08/OW-TUP-uwagi-do-projektu-Studium-W-wy-2.pdf (access: 19.10.2024).

Sowińska-Świerkosz, B., and García, J. 2022. What are nature-based solutions (NBS)? Setting core ideas for concept clarification. Nature Based Solutions 2: 100009. https://doi.org/10.1016/j.nbsj.2022.100009.

Sutton-Grier, A., Gittman, R. K., Arkema, K. K., Bennett, R. O., Benoit, J., Blitch, S. et al. 2018. Investing in natural and nature-based infrastructure: building better along our coasts. Sustainability 10: 523. https://doi.org/10.3390/su10020523.

Trzeciak, M., and Sikorska, D. 2023. Application of UAV and ground measurements for urban vegetation cooling benefits assessment, Wilanów Palace case study. Scientific Review Engineering & Environmental Sciences 33: 1–14. 10.22630/srees.5619.

Voulvoulis, N., Arpon, K. D., and Giakoumis, T. 2017. The EU Water Framework Directive: From great expectations to problems with implementation. Science of the Total Environment 575: 358–366. https://doi.org/10.1016/j.scitotenv.2016.09.228.

Warsaw City Council. 2018a. #Warsaw2030 Strategy. https://um.warszawa.pl/documents/56602/38746844/%23Warsaw2030+Strategy.pdf/8845940c-99df-aca4-1a03-524adafe4f63?t=1639549741334 (access: 11.10.2024).

Warsaw City Council. 2018b. Atlas ekofizjograficzny m.st. Warszawy (Ecophysiographic Atlas of the Capital City of Warsaw). https://architektura.um.warszawa.pl/documents/12025039/26223617/atlas_ekofizjograficzny.pdf/94816176-87cd-2d9b-5bf8-79e37a12dea0?t=1634497935335 (access: 14.11.2024)

Warsaw City Council. 2018c. Załącznik 4. do Strategii #Warszawa2030. Wytyczne dla Studium Uwarunkowań i Kierunków Zagospodarowania Przestrzennego, Warszawa (Annex 4 to the Strategy #Warsaw2030. 2018. Guidelines for the Study of Conditions and Directions of Spatial Development, Warsaw) https://um.warszawa.pl/documents/56602/38746844/a%C5%82%C4%85cznik+4+do+strategii.+Wytyczne+dla+studium+uwarunkowa%C5%84+i+kierunk%C3%B3w+zagospodarowania.pdf/fa93512d-3ef1-d250-6258-49c8d42c3798?t=1639549946344 (access: 11.10.2024)

Warsaw City Council. 2020. Program ochrony środowiska dla m.st. Warszawy na lata 2021-2024 (Environmental Protection Program for the Capital City of Warsaw for 2021–2024). Warsaw. https://konsultacje.um.warszawa.pl/uploads/decidim/attachment/file/4801/projektprogramuochronysrodowiskadlamstwarszawy20212024.pdf (access: 11.10.2024).

Warsaw City Council. 2021. Studium uwarunkowań i kierunków zagospodarowania przestrzennego Warszawy (Study of Conditions and Directions of Land Development of the Capital City of Warsaw). https://bip.warszawa.pl/documents/53790/4419672/1611_uch_zal_1_tekst_studi.pdf/0eae17b0-49ab-262d-943b-97f537bea0d1?t=1728491468737 (access: 11.10.2024).

Warsaw City Council. 2024a. Program ochrony środowiska dla m.st. Warszawy na lata 2025-2030 (Environmental Protection Program for the Capital City of Warsaw for 2025–2030). Warsaw. https://konsultacje.um.warszawa.pl/uploads/decidim/attachment/file/6589/Pos%CC%81_Warszawa_2025-2030_08.24.pdf (access: 19.10.2024)

Warsaw City Council. 2024b. Miejscowe plany zagospodarowania przestrzennego (Local spatial development plans). Biuro Strategii i Analiz. Warsaw. https://architektura.um.warszawa.pl/informacje-o-planach

Warsaw City Council. 2024c. Raport z badania telemetrycznego liczby mieszkańców Warszawy (Report on a telemetric study of the number of Warsaw inhabitants). Warsaw. https://um.warszawa.pl/documents/55043703/0/Raport+ORANGE+2024+Ludno%C5%9B%C4%87+WAW+v7_dostepna.pdf/60691334-5a1b-7bc3-3ee6-8414fb724312?t=1724334407886 (access: 1.10.2024)

Warsaw Greenery Authority. 2024. Projekty (Projects). https://zzw.waw.pl/projekty/ (access: 29.10.2024)

White, M. P., Elliott, L. R., Gascon, M., Roberts, B., and Fleming, L. E. 2020. Blue space, health and well-being: A narrative overview and synthesis of potential benefits. Environmental Research 191: 110169. https://doi.org/10.1016/j.envres.2020.110169.

Wierzbicki, G., Ostrowski, P., Bartold, P., Bujakowski, F., Falkowski, T., and Osiński, P. (2021). Urban geomorphology of the Vistula River valley in Warsaw. Journal of Maps 17: 170–185. https://www.tandfonline.com/doi/full/10.1080/17445647.2020.1866698.

Wilanów District Office. 2011. Formy ochrony przyrody (Forms of nature protection). Urząd Dzielnicy Wilanów m.st. Warszawy. https://wilanow.um.warszawa.pl/-/formy-ochrony-przyrody (access: 12.10.2024).

World Bank. 2021. A Catalogue of Nature-Based Solutions for Urban Resilience. Washington, DC. http://hdl.handle.net/10986/36507.

7

Integrating Rivers into the Urban Fabric of Kanpur, India

Regenerating Ganga River to Meet Community Needs and Mitigate Environmental Challenges

Sagarika Nayak,[1] *Nazish Abid*,[2,*] *Somi Sareen*[3] and *Sameen Fatma*[4]

1. Introduction

The degradation of the Ganga River is not an isolated event but rather a part of a more general global disaster. According to the World Water Development Report (UN Water 2023), 2.2 billion people globally lack access to safely controlled drinking water; meanwhile, urban rivers are being used more and more as dumping sites for garbage. It also suggests that by 2025, absolute water scarcity will affect about 1.8 billion people. In developing countries, where increasing urbanisation, poor environmental laws, and industrial growth place unsustainable pressures on freshwater supplies, this problem is exacerbated. Simon and Joshi (2022) argue that similar industrial contamination and excessive extraction are also occurring in river systems around the world, including the Yangtze, Nile, and Mississippi.

[1] Independent Researcher, Ontario, Canada.
Email: ar.sagarika87@gmail.com, 0009-0002-7409-6204

[2] Department, of Architecture and Interior Design, University of Bahrain, Sakhir, Bahrain.

[3] Independent Researcher, New Delhi, India.
Email: khansareen8@gmail.com, 0000-0002-9360-7033

[4] Department of Architecture and Planning, National Institute of Technology, Patna, Bihar, India.
Email: fatmasameen12@gmail.com, 0000-0003-3212-0712

* Corresponding author: Nabid@uob.edu.bh, 0000-0002-8875-9614

DOI: 10.1201/9781032711737-7

The Ganga River is of utmost importance, both naturally and culturally. Not only is it revered as a holy river in Hinduism, but it is also a vital ecological resource that supports the way of life for millions of people living near its banks. The first Prime Minister of India, Pt. Jawaharlal Nehru said about the river, "The Ganga to me is the symbol of India's memorable past, *which has been flowing into the present and continues to flow towards the ocean of the future*." The Ganga River is the longest and most significant river in India. It originates from the Gangotri Glacier in the Himalayas in Uttarakhand. Then, it passes through Uttar Pradesh, covering a significant portion of the state. It continues to Bihar, briefly through Jharkhand and finally culminates in the Bay of Bengal in West Bengal. Commonly known by Indians as Ganga, it is absolutely vital in Hinduism and is regarded as the holiest river.

Moreover, India's spiritual and social character has always benefited much from this river. Rising urbanisation and industry, which led to notable environmental degradation, compromised this dual role more than ever. Once thought of as a purifying force, the Ganga River is among the most contaminated rivers in the world, therefore highlighting a global issue with urban sustainability and water management. The Ganga represents a heavenly energy fundamental to Hindu rituals regarding life and death, transcending its mere river character. Important sites for religious rites include the immersion of ashes post-cremation, and daily offerings include the ghats and stairs leading to the riverfront. Historically, these ghats have been gathering spots that show how symbiotically the river and people interact. Still, the Ganga river's environmental contamination - particularly in cities like Kanpur - has seriously compromised its ability to be both a holy site and a supply of food. For its dependents, the pollution from untreated sewage, industrial discharges, and solid waste has rendered the river increasingly dangerous, therefore compromising public health and cultural legacy.

The Ganga River stretches 2,510 km, comprising an 861,404 square kilometre basin region (Kumar 2017) Ganga River. Currently, over 10% of the world's population lives in the Ganga River basin; by 2030, this figure is expected to rise to one billion (NIUA and NMCG 2021). With over 52 cities, 48 towns, and several thousand villages spread over the Ganga River basin, the river clearly influences the economic life in northern India (NIUA and NMCG 2021). The fact that almost all of these villages' waste is dumped into the Ganga River is crucial. Together with an extra 260 million litres of industrial waste, runoff from 6 million tonnes of fertilisers, and 9,000 tonnes of insecticides used in the basin, this waste amounts to 1.3 billion litres every day (Batabyal 2023).

Second most populous city in Uttar Pradesh and twelfth most populated city in India (Troch 2018), Kanpur is 90 km southwest of Lucknow, the capital of Uttar Pradesh; Kanpur is among the first industrial cities in Northern India and sits on the banks of the Ganga River. It is limited northward by the Ganga River and southward by the Pandu River. According to the 2011 census, the city boasts a population of 2,656,348. According to official Kanpur Nagar Nigam figures, there were 522,242 homes in the fiscal year 2019–20, which translates to an expected population of 3.2 million (32,50,000) (Ahmad and Chaurasia 2019). This represents a significant estimated population increase for the past eight to nine years - 19%. Given the large

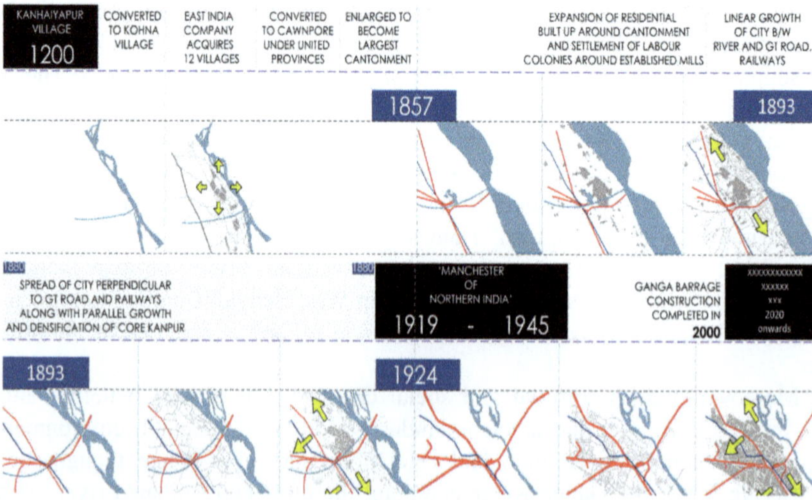

Figure 1: Evolution of Kanpur. Source: Authors.

population increase, it is estimated that 25 to 30% of people live in slums (NIUA and NMCG 2021).

The historical evolution of Kanpur can be traced back to 1200 AD when a village named Kanhaiyapur was established, and today, Kanpur is located there. It was developed as Cawnpore under British rule in the United Province, and it later went on to become the largest cantonment town (Troch 2018). In the following years, after the establishment of mills, a labour colony started to develop around it. Afterwards, it densified, and the settlement expanded with enormous industrial development, so much so that the city started to be known as Manchester of Northern India from 1919 to 1945 (as shown in Figure 1, Paarcha et al. 2023). Currently, industrial development has slowed down due to various factors. Still, the city continues to grow as the largest city in the state and a major institutional hub in the region (Batabyal 2022).

Industrial toxins mostly contribute to the Ganga River's environmental degradation in Kanpur. Being an industrial centre, Kanpur hosts chemical businesses releasing large amounts of untreated waste straight into the river, as well as tanneries and textile mills. According to (Ahmad and Chaurasia 2019b), these industries significantly contribute to heavy metal contamination of the river, therefore compromising water quality and harming aquatic life as well as human health. With dissolved oxygen levels often below permissible limits, i.e., less than 3mg per Litre, the Central Pollution Control Board (CPCB) has named Kanpur as one of the most contaminated areas of the Ganga River (Shakti 2021). The riverine ecosystem suffers greatly from this pollution, as do the people who depend on it for religious ceremonies, agriculture, and drinkable water. Initiated in the 1980s, the Ganga Action Plan sought to reduce pollution. Still, it failed because of insufficient local participation and poor regulatory enforcement (Shukla et al. 2020). Initiated in 2014, the Namami Gange Mission is a rejuvenated project requiring localised plans

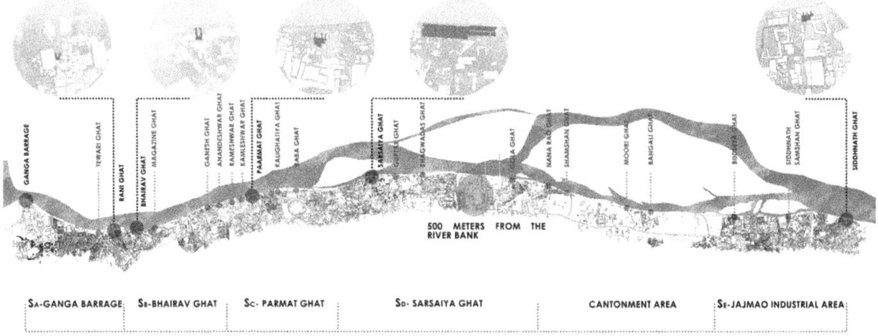

Figure 2: Mapping the locations of ghats along the stretch of the Ganga River in Kanpur. Source: Authors.

to balance economic activities with river restoration. To properly lower pollution and restore the Ganga River's cultural and biological integrity, national projects must combine localised, community-oriented techniques.

Urban river redevelopment is not only urgent but also complex. Apart from the environmental problem, it is essential to solve the socioeconomic dynamics of river regeneration. Along the Ganga River, many people depend on the river; they engage in religious tourism, farming, and fishing, among other activities. The challenge is balancing the river's ecological recovery with its preservation of cultural and religious values. Urban areas function as dynamic testing grounds where various strategies, decisions, and innovations are continuously implemented, assessed, and refined - often experiencing setbacks alongside achievements and efforts at urban renewal. They have to take the several needs of the populations involved into account. In Kanpur, the ghats serve as social and economic hubs as well as religious and cultural venues. Restoring the river calls for a collaborative approach involving nearby populations, religious leaders, and business and governmental partners. This is consistent with the idea of "river-sensitive urban planning", which emphasises the need to include ecological, cultural, and social issues in urban design (Chaudhary and Walker 2019).

Mostly from urban and industrial sources, the World Bank (2018) finds that over 80% of India's surface water is contaminated (Troch 2018). The situation in Kanpur mirrors the challenges faced by other communities along the Ganga River and internationally in riverine areas, where fast urbanisation and poor environmental governance have led to the unsustainable use of water resources (Das and Tamminga 2012a). The tension between development and sustainability is more intense as urban areas grow, which drives necessary research on the cohabitation of urbanisation and the protection of natural resources. Thus, the study is essentially required as it is paramount to the intersection of three systems: industrial development, environmental degradation and cultural preservation, which govern the everyday life of the riverine communities in Kanpur. It also fits in the global need for interventions aimed at balancing ecological sustainability, industrial growth and cultural heritage preservation, highlighting the need for integrated solutions to demonstrate how urban rivers can be sustainably managed and culturally respected.

One of the most important strategies for addressing this worldwide problem now is urban river regeneration. Globally, river-based regeneration projects have come under fire for stressing aesthetic and financial improvements at the price of ecological repair and community needs. For instance, the rehabilitation of the Cheonggyecheon River in Seoul has attracted praise from all around for turning a once-polluted river into an urban green area. Yet, academics argue that the project displaced informal communities and poorly addressed more general environmental problems (Mukherjee 2022). Likewise, although mostly successful in improving water quality, the Thames River clean-up in London has struggled to balance industrial needs with natural river restoration (Mould et al. 2020). These events highlight the complexity of urban river rejuvenation and the need to balance ecological restoration, economic growth, and cultural preservation.

This chapter intends to investigate the dynamics of meeting the spatial needs of riverine communities in Kanpur and simultaneously solve the environmental vulnerabilities coming from urbanisation and industry. The study will analyse the issues at the river edge on a city scale. It will demonstrate focused discussion and interventions, covering the Ganga Barrage and the Jajmao Industrial Area, as shown in Figure 2. The chosen stretch constitutes the most important elements of man-river interaction in the city, such as Bhairav Ghat, Parmat Ghat, and Sarsaiya Ghat. It also covers distinct morphological neighbourhoods along the river, such as ample green and planned space of Cantonment, the major pollutant of the river, i.e., Jajmau Industrial Area and Ganga Barrage. The chapter especially intends to (i) investigate the socio-cultural interactions between Kanpur residents and the Ganga River, (ii) identify the environmental challenges related to the river, and (iii) propose sustainable urban design and policy measures that help river regeneration and redevelopment along the ghats while preserving the community's cultural and spatial requirements.

For sustainable urban planning in Kanpur there is a need of thorough plan for river restoration projects from the global projects knowledge. Projects like green infrastructure, nature based solutions and community driven methods offer ways for improving the water quality for restoring the cultural and biological relevance of the rivers. Kanpur still struggles and faces challenges in fitting these strategies to socio- cultural and environmental context of the city because of the poor governance, industrial population and ingrained traditions of Ganga River. Urbanisation has escalated environmental challenges due to the socio cultural link between the Ganga River and the local communities. So the sustainable design ideas may help for river regeneration in Kanpur as a balanced approach.

2. The Socio-Cultural Significance of the Ganga in Kanpur Cultural and Religious Ties

The Ganga River is regarded as sacred and serves as a symbol of India's ancient civilisation and cultural heritage while holding a distinctive status as it is regarded as sacred by the Hindu population in India while being unmatched in economic, social, and cultural parameters for its inhabitants at banks of Ganga River (Chopra et al. 2004). The city of Kanpur offers an incredibly rich history and cultural heritage. One of the most prominent centres in Kanpur are the Ganga ghats that touch the city, and the living traditions, crafts, festivals, and rituals are celebrated at this site. The ghat is a primary pilgrimage centre for Hindus and is set against the backdrop of the mighty Ganga. The river is referred to as Maa Ganga, and it attracts millions of individuals who partake in ritualistic dips in its waters daily. Additionally, there is a belief that the water possesses healing properties that can cleanse individuals of sin and that cremation along the banks of the Ganga River guarantees a more favourable afterlife (Chopra et al. 2004, Singh 2001).

Bathing, prayer, and festivities are held on the Ganga River Front Ghats, which lead down to the river. Ghats were originally built for bathing and religious rites but later became important social and cultural centres. The Ghats bring local populations and draw tourists from around the world. Millions of people visit India's Ghats, a cultural treasure. Tourism in the Ghats has boosted the local economy and created jobs. However, the cultural identity of riverine people is being undermined by the spiritual dissonance—that is, the physical harm done to a holy river. Sometimes, local officials and government health recommendations warn against direct water contact, so religious rites have had to be changed or cut in frequency. Further contamination from non-biodegradable materials such as POP and other chemical paints results from the immersion of idols and ritual items during celebrations (Goswami and Mazumdar 2016). Ganga River is now recognised for spreading diseases while originally it was considered as purifier hence contradicting the thought and tradition of healing and cleaning the health of people. This spiritual contradiction emphasises both the erosion of these people's sense of cultural identity and the great breakdown in the human-river interaction.

- COMMUNITY INTERACTIONS:
  - DAILY ENGAGEMENT IN KANPUR INHABITANTS ASSOCIATED WITH THE GANGA RIVER:
    - *RELIGIOUS PRACTICES:* The Ganga River is of considerable religious importance to the inhabitants of Kanpur. Numerous individuals participate in daily practices, including immersing themselves in its waters, which is thought to cleanse the soul and eliminate sins. This practice illustrates the profound spiritual relationship that the community maintains with the river, which is regarded as sacred within the context of Hinduism (Khatoon et al. 2013). These sacred sites are thought to purify and free the soul from reincarnation (Swaroop et al. 2023.) Cremation, known as Antim Samskara, represents the final stage among the sixteen samskaras that a Hindu is believed to experience upon death. The procedure involves a systematic approach in which the physical remains of a person are subjected to combustion. The remains are subjected to cremation or incineration at the riverbank(Indian Institutes of Technology 2013). The process of cremation is subsequently accompanied by the immersion or disposal of the ashes and remains into the river.

- *DAILY CHORES:* The river functions as an essential resource for a range of daily activities. The local population frequently utilises the water for laundering garments, cleaning dishes, and performing various clean-up tasks. This practical engagement underscores the river's significance as a crucial water source in daily life, establishing it as an essential component of their routines.
- *CULTURAL CELEBRATIONS:* During various celebrations, such as Ganga Dussehra, Ganga Jayanti, Nag Panchami, Ganga Mahotsav, and Ganga Mela, residents congregate along the riverbanks to engage in rituals and commemorate their cultural heritage. The occurrence of these events serves to enhance communal relationships while simultaneously underscoring the river's importance within the framework of their cultural identity. During the Ganga Mela, it undergoes a significant transformation into a spiritual centre, characterised by numerous ghats, which are steps leading to the river that are embellished with vibrant decorations. Individuals congregate at this location to engage in sacred rituals, present prayers, and partake in ritualistic bathing in the waters of the Ganga River.
- *RECREATIONAL ACTIVITIES:* The river functions not only for religious and practical purposes but also as a venue for recreational activities. Individuals residing in the area have the opportunity to participate in various activities, including picnicking, fishing, or appreciating the aesthetic qualities of the river's natural environment(Khatoon *et al.* 2013). The recreational dimension plays a significant role in shaping the community's overall interaction with the Ganga River, establishing it as a site of both leisure and practical use.
- *THE HISTORICAL AND CULTURAL CONNECTIONS BETWEEN THE COMMUNITY AND THE RIVER:* Hindus revere the Ganga River as a mother figure, making it India's holiest river. Many myths surround the river, emphasising its holiness. Hindus believe Prince Bhagirath brought the river from heaven to earth to purify his siblings' souls after Kapil burned them. The final rites after death are performed in the Ganga River because it has been seen as a crucial crossing between the terrestrial and the celestial (Prakash Kala 2018). The body is cremated on its banks, and the ashes are scattered into its sacred waters. A custom of placing a few drops of water in the deceased's mouth before cremation is considered to aid the afterlife trip. Pilgrims bring the water home to use in rituals. The Ganga's religious and social connotations have frequently overshadowed its economic and utilitarian uses. British engineers built the Indian canal system, which enabled agriculture, manufacturing, pisciculture, navigation, wastewater disposal, drinking water supply, and recreation. The Ganga River is almost synonymous with Hinduism and culture. People use its waters for religious bathing, drinking, morning ablutions, and cooking (Bhargava 1987).

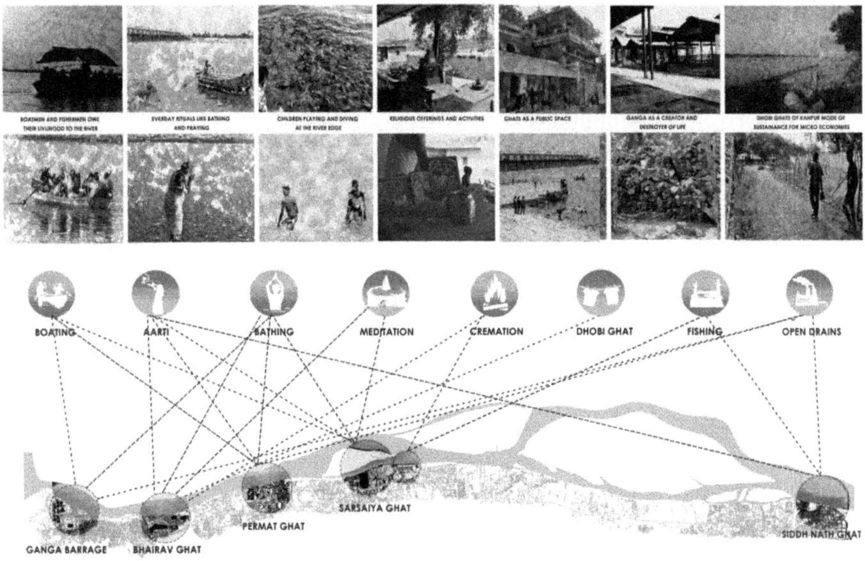

Figure 3: Various ghats along the Ganga River in Kanpur showcasing religious, social, and cultural activities. Source: Authors.

- **SPATIAL NEEDS OF RIVERINE COMMUNITIES:**

Kanpur residents depend on the Ganga River for food and income, suggesting a close correlation between their spatial needs and daily routines. Clean water is an initial spatial need. The Ganga River supplies most domestic cooking, washing, and drinking water. Kanpur's rich history, diverse ethnicity, and continuous cultural activities affect riverine populations' spatial needs, as shown in Figure 3.

  - **SIGNIFICANT PRACTICES:**

    - *RIVER WORSHIP:* The residents of Kanpur attribute significant spiritual importance to the Ganga River . In riverine villages, it is customary to observe morning and evening prayers, make offerings, and engage in immersion. These activities necessitate the presence of accessible, safe, and well-maintained ghats.

    - *CULTURAL FESTIVAL THE KANPUR MAHOTSAV:* This event is significant for its presentation of classical dance, music, and local cuisine. The festival necessitates expansive areas to accommodate concerts, food stalls, and various commercial activities, in addition to the requisite transportation and Wi-Fi infrastructure (Zingraff-Hamed et al. 2022)

    - *ECONOMIC ACTIVITIES:* The inhabitants of the riverine areas in Kanpur engage in fishing and boat-making as primary sources of livelihood. The activities necessitate the docking along riverbanks and the clean fishing areas of equipment for the sale of the fishing catch (Das and Tamminga 2012).

- *COMMUNITY SPACES AND PUBLIC GARDENS:* They are utilised by riverine communities for gathering and relaxation, particularly in shaded locations near riverbanks. This approach facilitates the enhancement of public gardens and shaded communal spaces, contributing to the spatial development along riversides.
- *INDUSTRIAL BASE:* The historical context of Kanpur, along with its designation as a significant industrial hub, often referred to as the Manchester of the East, presents complexities in the realm of water management (Khatoon et al. 2013). Industrial operations consume significant quantities of water and have the potential to affect river ecosystems adversely. The management of water resources plays a critical role in the preservation of cultural heritage as well as in supporting the industrial and economic activities of the city. The city's main industry is tanning, the nation's largest exporter of finished and semi-finished leather goods. Revenue from this sector exceeds 2.75 billion rupees or $61 million US dollars. Tanneries dominate Jajmau, the city's industrial area downstream. Munitions and metalworks are also found inland behind the core business centre and military cantonment. On the southern bank of the river, close to civil lines, are textile factories. Leather production is centred on Jajmau, with 305 tanneries. Hazardous effluent from hides stripping and curing plants is processed at shared Common Effluent Treatment Plants (CETPs). It contains trace minerals, organic matter, chromium, sulphide, ammonium, and salts(Kedzior 2015).

3. Environmental Challenges and Urbanisation Impact in Kanpur

The Ganga River's environmental condition in Kanpur shows a degradation brought about by fast urbanisation and industry. Deeply affecting communities depending on the river, this deterioration is not just an ecological but also a social and financial problem (Gondwal and Verma 2023). Untreated waste, industrial discharges, and urban growth have gradually degraded the health of the river. Already one of the most polluted areas along the river, the Ganga River in Kanpur is under threat that compromises not only human but also environmental survival. This part tries to explain the challenges Kanpur face in the Ganga River by investigating the main sources of pollution, the function of industrialisation and urban growth, and the greater susceptibility of riverine communities.

Especially in Kanpur, pollution has grown to be one of the most important hazards to the Ganga River. The Holy Ganga has been turned into a cesspool in Kanpur, where more industrial and household waste flows than river water (Kumar 2016). The industrial growth of the city, population increase, and antiquated infrastructure have

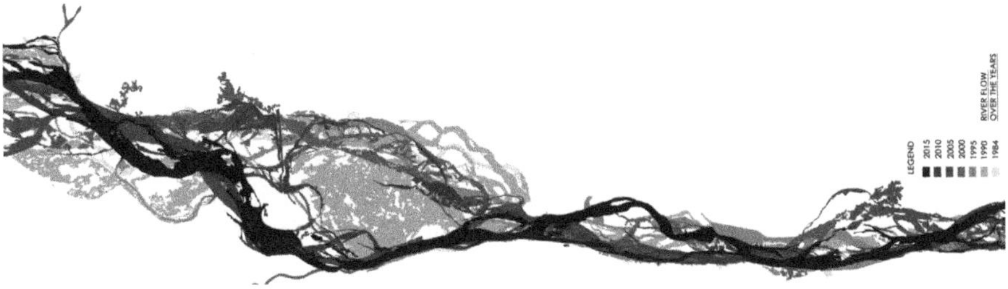

Figure 4: Temporal Mapping of Ganga River Channel Shifts in Kanpur. Source: Authors.

seriously degraded water quality. With measurements of 9-10 mg/L, the CPCB noted that Biological Oxygen Demand (BOD) levels in Kanpur's section of the Ganga River routinely exceed the permissible threshold of 3mg/L (Ahmad and Chaurasia 2019). Such high BOD levels show that the river's ability to break down organic waste naturally is seriously hampered, which has negative ecological implications. The urban activities around the river edges and other physiographic factors have resulted in the change in the course of the Ganga River, which is illustrated in Figure 4.

Though extreme, Kanpur's environmental problems fit a larger trend seen in other Ganga River cities. For example, Varanasi struggles similarly, especially concerning untreated sewage and solid waste contamination (Mukherjee 2022). However, Kanpur stands out because of the degree of industrial pollution, especially from the tannery segment (Batabyal 2023). Although Varanasi's pollution mostly results from religious and household waste, Kanpur's heavy metal poisoning poses more serious environmental and health hazards (Ahmad and Chaurasia 2019). Kolkata suffers from pollution issues related to its port operations and businesses further downstream. However, the wide delta area makes some dilution and mitigating effects possible. As the river is narrower and its flow is more limited in this area, Kanpur lacks this geographic advantage (Mukherjee 2022). Moreover, cities like Kolkata have been more successful in implementing pollution control policies, especially in terms of controlling industrial waste, thereby contrasting with the rather poor governance systems of Kanpur.

4. Riverine Community Vulnerability to Environmental Hazards

The vulnerability of riverine populations to environmental risks in Kanpur is a complex problem with cultural, financial, and medical aspects as well. According to the World Health Organisation (WHO), consuming contaminated Ganga River water directly causes over 60% of water-borne diseases in the area (Goswami and Mazumdar 2016). River pollution is also seriously affecting the agricultural industry, which depends mostly on the Ganga River for irrigation. High amounts of heavy metals building up in the soil have farmers noting notable drops in crop production. Five times more than the safe limits, the study found chromium levels in Ganga River floodplain crops close to Kanpur (Gaurav *et al.* 2018). Consequently, the livelihoods of nearby farmers are in jeopardy, and long-term food instability looms big. The financial ramifications are considerable; many farmers experience crop failure while others battle lower market prices brought on by concern about contamination.

The tannery sector in Kanpur is a major contributor to pollution impacting the Ganga River, as its swift growth surpasses the establishment of essential wastewater treatment facilities. It is reported that only 88 of the 110 chrome tanning facilities mandated to establish chrome recovery plants (CRPs) have complied with the regulations, and numerous CRPs are underutilised owing to their elevated operational expenses. (Batabyal 2023) Consequently, untreated or insufficiently treated wastewater containing hazardous compounds such as chromium and sulphides persists in discharging into the river, adversely affecting water quality and aquatic habitats (Batabyal 2022). The deficiencies of the Ganga Action Plan (GAP) in Kanpur have significantly affected adjacent populations, especially farmers in Jajmau, where the majority of tanneries are situated. The farmers are currently unable to utilise wastewater for irrigation, exacerbating environmental degradation and economic difficulties (NIUA and NMCG 2021).

Likewise, the fishing villages in Kanpur have seen a terrible drop in fish numbers since the pollution of the river has sharply lowered dissolved oxygen levels, therefore harming aquatic life. Once abundant in the Ganga River, fish species are quickly declining in response to hazardous compounds, including ammonia and heavy metals, which upset aquatic ecosystems (Sinha 2014). Data from the Central Inland Fisheries Research Institute (CIFRI) shows that over the past two decades, fish capture in the Ganga River's Kanpur section has dropped by 50% (Nelms et al. 2021). Many fishermen now lack a viable source of income.

Moreover, for millions of people, the Ganga River represents a spiritual and cultural lifeline as well as a means of economic survival. Hindus revere the ghats of the river as significant places of worship in Kanpur, where they carry out daily ablutions and ashes immersion. But the growing pollution makes these activities more and more dangerous. Many religious celebrations centre on ritual bathing in the Ganga River since believers think it cleans the soul and atones for sins. Still, the severely polluted waters expose followers to dangerous microorganisms that cause skin infections, gastrointestinal disorders, and other health issues (Shukla 2020).

The religious and cultural burden on Kanpur's riverine people goes beyond mere ceremonial customs. For many groups, including washermen, the Ganga River is essential for daily existence. Those who usually make their living from the river still feel that the growing population pressure has seriously disrupted the connection (NIUA and NMCG 2021). For millennia, these people have cleaned clothing beside riverbanks using the natural resource of running water. Apart from making their work unclean, contaminated water causes skin illnesses and respiratory problems by means of industrial chemicals and sewage filling the river (Chaudhary and Walker 2019).

Likewise, previously vibrant riverfront leisure activities on the banks of the Ganga River have suffered greatly. Historically, the riverbanks were venues for leisure, picnics, and small-scale business for both locals and visitors, including tea shops, snack vendors, and boat operators. But as pollution levels have surged, the one lively riverbank environment has vanished with much less foot traffic in these places. Small company owners that depend on everyday visitors for income have thus suffered financial losses (Troch 2018). Furthermore, among the most badly impacted by the river's environmental degradation are the colonies and slums close to river margins; often, low-income families have little to no access to other sources of water and must thus rely on the contaminated Ganga River for their daily needs. With water-borne diseases, including cholera, typhoid, and dysentery, becoming regular in contaminated water, public health crises have been sparked (UN Water 2023). Moreover, untreated waste and industrial discharges have aggravated the already poor air quality close to riverbanks, which has led to more respiratory ailments among the people, so increasing their vulnerability (Kumar 2016).

5. Mapping Government Initiatives

From the Ganga Action Plan (GAP) in 1985 to the Namami Gange Mission in 2014, the projects aiming at cleansing and rehabilitation of the Ganga River highlight their great ecological, cultural, and financial importance. Still, the Ganga River remains much contaminated despite many efforts over several decades, especially in industrial areas like Kanpur. This underlines continuous implementation shortcomings and the difficulties in balancing urban growth with environmental protection. Although the Ganga Action Plan stressed industrial waste management and sewage treatment facilities (STPs), poor local participation and inadequate execution hampered it (Das and Tamminga 2012). Many Sewage Treatment Plants (STPs) were underused or ineffective, and local authorities' cooperation was poor. Lack of community involvement made the method difficult to properly enforce and monitor compliance (Chaudhary and Walker 2019).

The Namami Gange Mission developed a thorough plan in 2014 that prioritised infrastructure development and community involvement. It also heavily worked on improving the abatement of industrial pollution. Although many areas have improved their sewage treatment, Kanpur's industrial effluent—especially from tanneries—

Table 1: Initiatives focused on the Ganga River.

Year	Initiative	Focus Areas	Achievements & Challenges
1985	Ganga Action Plan (GAP)	STPs, industrial effluent control	Poor execution, technical failures, lack of local engagement
2014	Namami Gange Mission	Community involvement, stricter pollution control, afforestation, STPs	Some success in capacity building, but industrial pollution continues in areas like Kanpur
2018	Urban River Management Plan (NIUA)	Sustainable urban planning, green infrastructure, river-sensitive development	Limited implementation, especially in smaller cities
2020 onward	River-Sensitive Master Plans (NIUA)	Zoning laws, floodplain protection, eco-friendly development	Strong theoretical foundation but minimal on-ground implementation

still impedes significant development (Batabyal 2023). Furthermore, the openness of pollution monitoring has come under fire as recorded statistics contradict real situations. Promoting green infrastructure and sustainable development, the NIUA's Urban River Management Plan (2018) marks a change towards the integration of urban planning with river health (NIUA and NMCG 2021). Nonetheless, its influence is constrained in smaller municipalities, where local administrations frequently lack resources. Similarly, River-Sensitive Master Plans offer a strong theoretical basis for zoning and floodplain preservation. Yet, practical implementation is constrained since sometimes economic motivations supersede environmental concerns (Gondwal and Verma 2023).

6. Urban Design Strategies for Sustainable River Regeneration

The regeneration of urban rivers represents a critical challenge in contemporary urban design, particularly in rapidly urbanising regions like Kanpur, India. Urban rivers, once the lifeblood of cities, are now often synonymous with pollution, neglect, and environmental degradation. Sustainable urban design offers a pathway to reverse this trend, restoring rivers to their natural state while supporting the needs of urban populations. This section will focus on sustainable urban design principles and strategies for river regeneration, with a specific focus on the Ganga River in Kanpur. This section will also explore strategies for improving water quality, the role of green infrastructure and nature-based solutions, and the challenges of balancing cultural preservation with environmental sustainability.

6.1 Sustainable Urban Design in Riverine Environments

Sustainable urban design in riverine environments necessitates an integrated approach that considers ecological, social, and economic factors. The primary objective is to restore the ecological health of rivers while ensuring that urban populations benefit from improved quality of life. In the context of Kanpur, this involves addressing the city's historical relationship with the Ganga River.

Key principles for sustainable urban design in river regeneration include:

- *ECOLOGICAL RESTORATION:* This involves reestablishing natural habitats along riverbanks, restoring native vegetation, and improving water quality. The Ganga River, for instance, has suffered from the loss of natural habitats due to urban encroachment and industrial activity. Reintroducing native plant species can help stabilise riverbanks, reduce erosion, and provide habitats for wildlife.
- *PUBLIC ACCESS AND CONNECTIVITY:* Ensuring that urban populations have access to riverfronts is essential for fostering a connection between people and their environment. In Kanpur, where the Ganga River has been largely cut off from the public by industrial zones and informal settlements, creating accessible riverfront parks and walkways can promote community engagement and environmental stewardship.

- *MULTI-FUNCTIONAL SPACES:* Urban riverfronts should serve multiple functions, including recreation, flood management, and habitat creation. Designing riverfronts that can accommodate seasonal flooding while providing public spaces for leisure and cultural activities is crucial in cities like Kanpur, where monsoons can cause significant river swelling.
- *POLLUTION MITIGATION:* Reducing pollution at the source is critical for sustainable river regeneration. In Kanpur, this means implementing stricter regulations on industrial effluents and improving waste management infrastructure. Additionally, promoting the use of non-toxic, biodegradable materials in urban development can reduce the overall pollution load on the river.
- *COMMUNITY ENGAGEMENT AND EDUCATION:* Sustainable river regeneration requires the active participation of local communities. In Kanpur, public awareness campaigns, community clean-up drives, and educational programs about the cultural and ecological significance of the Ganga River can help foster a sense of ownership and responsibility towards the river.

6.2 Strategies for Improving Water Quality and Reducing Pollution

Improving water quality and reducing pollution in urban rivers like the Ganga is a multifaceted challenge that requires coordinated efforts at multiple levels, including government policy, industrial regulation, and community action. The following strategies are essential for achieving these goals:

- *UPGRADING WASTEWATER TREATMENT INFRASTRUCTURE:* Untreated wastewater, due to Kanpur's outdated treatment infrastructure, is a major cause of river pollution. To address this, substantial investments in modern wastewater treatment plants (WWTPs) are required. These plants should be equipped with advanced technologies capable of removing both organic and inorganic pollutants, as well as heavy metals, which are prevalent in industrial effluents. In addition to upgrading existing facilities, decentralising wastewater treatment systems can also prove effective. Smaller, localised treatment plants can be installed near industrial areas or dense urban settlements, reducing the load on central treatment facilities and ensuring that wastewater is treated before it enters the river.
- *INDUSTRIAL POLLUTION CONTROL:* Implementing stringent regulations on industrial discharges is essential to curb this pollution. This can be achieved through a combination of government enforcement and industry incentives. For example, industries could be mandated to install effluent treatment plants (ETPs), which are subject to regular inspections to ensure compliance. In addition to regulatory measures, adopting cleaner production techniques can help reduce the environmental impact of industrial activities. This includes using less harmful chemicals in the tanning process, recycling wastewater

within industrial facilities, and transitioning to more sustainable energy sources to power industrial operations.

- *PROMOTING GREEN INFRASTRUCTURE FOR STORMWATER MANAGEMENT:* Urban runoff is another major contributor to river pollution, as it carries with it a variety of contaminants, including oils, heavy metals, and sediments. Green infrastructure, such as bioswales, green roofs, and permeable pavements, can help manage stormwater more sustainably. By allowing water to infiltrate the ground rather than running off into the river, these systems reduce the volume of polluted stormwater entering the Ganga River. In Kanpur, integrating green infrastructure into urban planning can mitigate the impact of stormwater on the river. For example, constructing bioswales along roadsides and in public parks can capture and filter runoff before it reaches the river. Similarly, promoting the use of permeable pavements in new developments can help reduce surface runoff.
- *WASTE MANAGEMENT REFORMS:* Solid waste management is another critical issue affecting water quality in Kanpur. The improper disposal of solid waste, particularly plastic waste, often ends up in the Ganga River, contributing to both visual pollution and harm to aquatic life. To address this, Kanpur needs to implement comprehensive waste management reforms that prioritise waste reduction, recycling, and proper disposal. Initiatives such as community-based waste segregation programs, the promotion of recycling industries, and the enforcement of anti-littering laws can significantly reduce the amount of solid waste entering the river. Additionally, establishing more efficient and frequent waste collection services, particularly in informal settlements, can help prevent the accumulation of waste in areas prone to runoff.

6.3 The Role of Green Infrastructure and Nature-Based Solutions

Green infrastructure and nature-based solutions play a pivotal role in the sustainable regeneration of urban rivers. These approaches work with natural processes to address environmental challenges, providing multiple co-benefits such as improved water quality, enhanced biodiversity, and increased resilience to climate change. In the context of Kanpur, where urbanisation has put immense pressure on the natural environment, integrating green infrastructure and nature-based solutions into urban design is essential for the successful regeneration of the Ganga River.

- *RESTORING RIPARIAN BUFFERS:* Riparian buffers—strips of vegetation along riverbanks—are crucial for maintaining the health of river ecosystems. They act as natural filters, trapping pollutants and sediments before they enter the water and provide habitats for wildlife. In Kanpur, restoring degraded riparian buffers along the Ganga River can significantly improve water quality and enhance biodiversity. This can be achieved through reforestation efforts, where native tree and shrub species are planted along the riverbanks. In addition to stabilising the soil and reducing erosion, these buffers can serve as green corridors that connect fragmented habitats, supporting the movement of species across the urban landscape.

- *CREATING WETLANDS FOR WATER TREATMENT:* Constructed wetlands are another effective nature-based solution for improving water quality. These engineered systems mimic the functions of natural wetlands, using plants, soil, and microorganisms to remove pollutants from wastewater. In Kanpur, constructed wetlands could be used to treat wastewater from both domestic and industrial sources before it is discharged into the Ganga. For example, treating wastewater from smaller settlements or industrial clusters through constructed wetlands could reduce the load on centralised treatment plants while providing additional environmental benefits, such as habitat creation and carbon sequestration. These wetlands can also be integrated into public parks and open spaces, providing recreational and educational opportunities for local communities.
- *ENHANCING URBAN GREEN SPACES:* Urban green spaces, such as parks, gardens, and street trees, contribute to the overall sustainability of a city by improving air quality, reducing the urban heat island effect, and providing habitats for urban wildlife. In Kanpur, enhancing urban green spaces, particularly those adjacent to the Ganga, can support the river's regeneration by reducing runoff, filtering pollutants, and providing shade that helps regulate water temperatures. Designing these spaces to incorporate water-sensitive urban design (WSUD) principles, such as rain gardens and swales, can further enhance their role in managing stormwater and protecting the river. Moreover, green spaces can serve as gathering points for community activities and environmental education, fostering a sense of connection between residents and the Ganga.
- *CULTURAL PRESERVATION AND ENVIRONMENTAL SUSTAINABILITY*: Ganga River regeneration must balance cultural heritage with environmental goals. In Kanpur, the Ganga River is not just an ecological resource but a sacred river, so its restoration must respect and incorporate its cultural significance.
- *INTEGRATING CULTURAL PRACTICES INTO ENVIRONMENTAL MANAGEMENT:* To harmonise urban development with cultural heritage, it is essential to integrate traditional practices and knowledge into environmental management strategies. For example, many rituals associated with the Ganga River, such as offering flowers or ashes, contribute to the pollution of the river. Instead of banning these practices, sustainable alternatives could be promoted. This might include providing biodegradable materials for offerings or designated areas for ritual activities that are equipped with facilities to manage waste. Additionally, engaging religious and cultural leaders in river regeneration efforts can help garner community support and ensure that environmental initiatives align with cultural values. In Kanpur, involving local temples and religious organisations in clean-up drives and awareness campaigns could strengthen the connection between cultural practices and environmental stewardship.
- *PRESERVING CULTURAL LANDSCAPES:* Cultural landscapes along the river, such as ghats (steps leading down to the river) and historic buildings, are integral

to the identity of the Ganga River in Kanpur. Preserving these landscapes while regenerating the river presents a complex challenge. On the one hand, these structures need to be protected from the impacts of flooding and pollution; on the other hand, they should not hinder the river's natural processes. A possible strategy is to adopt adaptive reuse and conservation techniques that respect the historical significance of these landscapes while making them more resilient to environmental changes. For example, reinforcing the foundations of historic ghats to withstand flooding or integrating green infrastructure around cultural sites can help protect these assets while contributing to the overall health of the river.

- *PROMOTING ECO-TOURISM AND CULTURAL EDUCATION:* Eco-tourism and cultural education can play a vital role in supporting both environmental sustainability and cultural preservation. By promoting the Ganga River as a destination for responsible tourism, Kanpur can generate economic benefits that support river regeneration efforts. Eco-tourism initiatives could include guided tours of restored riverfronts, educational programs about the ecological and cultural significance of the Ganga River, and community-based tourism projects that empower residents. Incorporating cultural education into school curricula and public awareness campaigns can also foster a deeper understanding of the connection between the Ganga River and Kanpur's cultural heritage. This approach can help cultivate a sense of pride and responsibility among residents, encouraging them to participate actively in river regeneration efforts.

The sustainable regeneration of urban rivers like the Ganga River in Kanpur requires a holistic approach that integrates ecological restoration, pollution control, green infrastructure, and cultural preservation. By adopting sustainable urban design principles, investing in modern wastewater treatment infrastructure, promoting green infrastructure, and respecting the cultural significance of the river, Kanpur can restore the health of the Ganga River while supporting the needs of its urban population.

The challenges are immense, but so are the opportunities. The successful regeneration of the Ganga River would not only improve the environmental quality of Kanpur but also serve as a model for other cities facing similar challenges. It would demonstrate that with the right strategies, it is possible to reconcile urban development with the protection and restoration of our natural and cultural heritage.

7. Recommendations for Policy and Implementation

Urban river regeneration is a critical component of sustainable urban design, especially in rapidly developing cities like Kanpur. Effective policies and implementation strategies are essential to transforming neglected riverfronts into vibrant, eco-friendly spaces that benefit both the environment and the community. This essay outlines specific policy recommendations for urban river regeneration, suggests strategies for community engagement and participation, and discusses the potential benefits and challenges associated with these measures.

7.1 Policy Recommendations

- *INTEGRATED RIVER BASIN MANAGEMENT (IRBM):* Integrated River Basin Management (IRBM) should be the cornerstone of urban river regeneration policies. IRBM approaches emphasise the coordination of land use, water management, and environmental protection across entire river basins (Schramm 2020). Policies should mandate the development of comprehensive river management plans that include provisions for pollution control, habitat restoration, and flood management. For example, the European Union's Water Framework Directive provides a model for integrating water quality standards with ecological health objectives (EU 2000). Kanpur could benefit from adopting similar frameworks to address its riverine challenges holistically.
- *GREEN INFRASTRUCTURE INCENTIVES:* Policies should promote green infrastructure solutions such as riparian buffers, wetlands, and urban parks. These features help manage stormwater runoff, enhance biodiversity, and improve water quality. Incentives for private developers to incorporate green roofs, permeable pavements, and rain gardens into their projects can accelerate the adoption of sustainable practices (Tzoulas et al. 2007). In Kanpur, zoning laws could be updated to include requirements for green infrastructure in new developments along the riverfront.
- *REGULATION OF INDUSTRIAL DISCHARGES:* Tightening regulations on industrial discharges into rivers is crucial for improving water quality. Policies should set stringent limits on pollutants and enforce regular monitoring and reporting requirements for industries (Gibbs 2015). Establishing a river health monitoring system similar to those in developed countries can help track improvements and ensure compliance. In Kanpur, partnerships with local industries and environmental agencies can facilitate the implementation of these regulations.
- *ECONOMIC AND FINANCIAL SUPPORT:* Economic and financial support mechanisms are necessary to fund river regeneration projects. Public-private partnerships (PPPs) can mobilise investment for large-scale projects. At the same time, government grants and subsidies can support smaller initiatives (Elliott et al. 2013). In Kanpur, creating a dedicated river regeneration fund could provide financial resources for ongoing maintenance and enhancement efforts.

7.2 Strategies for Community Engagement and Participation

- *PUBLIC AWARENESS CAMPAIGNS:* Public awareness is a vital part of river regeneration, especially in Kanpur, where aligning with the Namami Gange Mission's principles—Aviral Dhara, Nirmal Dhara, and Jan Bhagidari—can enhance impact. Efforts such as school-based education, media collaborations for documentaries, and street theatre can engage a wide audience. Involving Ganga Praharis—trained local volunteers—adds credibility and reach to these campaigns. At the same time, participatory planning ensures that regeneration reflects local needs. Establishing River Management Committees, hosting community workshops, and partnering with institutions like IIT Kanpur can encourage meaningful citizen input. These inclusive, locally grounded approaches help build trust and shared responsibility in the riverfront's transformation.

- *PARTICIPATORY PLANNING PROCESSES:* Engaging community members in the planning and decision-making processes ensures that regeneration efforts reflect local needs and preferences. Participatory approaches, such as public consultations, community advisory boards, and workshops, allow residents to contribute their ideas and concerns (Arnstein 1969). In Kanpur, creating a community advisory committee for river regeneration projects can facilitate meaningful dialogue and enhance project outcomes. Participatory planning ensures that riverfront regeneration reflects community needs and values. The Namami Gange Mission promotes decentralised planning by involving local governance and community representatives. In Kanpur, this can be achieved through local River Management Committees, design workshops, and public exhibitions that allow citizens to share their ideas. Collaborating with institutions like IIT Kanpur can provide technical support for these efforts. This approach bridges the gap between top-down governance and grassroots action, fostering shared ownership and legitimacy.
- *VOLUNTEER PROGRAMS:* Volunteer programs can mobilise community members to take an active role in river restoration activities. In Kanpur, targeted programs like citizen science initiatives, where volunteers monitor water quality and biodiversity, and Adopt-a-Stretch programs, where communities take responsibility for river sections, can make a big impact. Annual Ganga Festivals and cleanliness drives can foster environmental pride while bringing people together. Incorporating *Ganga Doots*—local volunteers as environmental ambassadors—can also connect these efforts to the community's cultural and spiritual values.
- *INCLUSIVE ENGAGEMENT FOR NON-VOLUNTEERS:* For those who can't participate in physical volunteer activities, it's essential to create other ways for everyone—especially women, the elderly, children, and marginalised groups—to get involved. This could include interactive storytelling sessions that highlight the river's cultural importance, mobile kiosks and pop-up exhibitions in public places to share information and gather feedback, or digital platforms like apps and surveys for easier participation. Creative activities like art competitions and photo exhibitions can also inspire people to contribute in their unique way. By offering these options, river regeneration becomes a community effort where everyone can play a part.

7.3 Benefits and Challenges of Implementing the Proposed Measures

- *BENEFITS:*
  - *ENVIRONMENTAL IMPROVEMENTS:* Implementing the recommended policies and strategies can lead to significant environmental benefits, including improved water quality, enhanced biodiversity, and reduced flood risks. Green infrastructure, for instance, can mitigate the effects of urban runoff and support the health of aquatic ecosystems (Tzoulas et al. 2007).
  - *ECONOMIC GROWTH:* Revitalising riverfronts can stimulate economic growth by attracting tourism, increasing property values, and creating job opportunities. Investment in river regeneration can also lead to long-term cost savings by reducing the need for expensive flood control and water treatment measures (Elliott et al. 2013).
  - *ENHANCED QUALITY OF LIFE:* Well-designed riverfronts contribute to the quality of life by providing recreational spaces, improving aesthetics, and fostering community pride. Urban parks and green spaces can enhance residents' physical and mental well-being, promoting healthier and more vibrant communities (McDonald et al. 2017).
- *CHALLENGES:*
  - *FINANCIAL CONSTRAINTS:* One of the primary challenges in implementing river regeneration measures is securing adequate funding. Large-scale projects require substantial investment, and financial constraints can limit the scope and effectiveness of regeneration efforts. Developing sustainable financing mechanisms and leveraging public-private partnerships are crucial to overcoming this challenge (Elliott et al. 2013).
  - *COORDINATION AND GOVERNANCE:* Effective river regeneration requires coordination among various stakeholders, including government agencies, private developers, and community organisations. Ensuring that all parties work together towards common goals can be challenging, particularly in complex urban environments. Establishing clear governance structures and communication channels can help address this issue (Schramm 2020).

- *COMMUNITY RESISTANCE:* While community engagement is essential, residents or stakeholders with vested interests may resist change. Overcoming opposition requires transparent communication, addressing concerns, and demonstrating the tangible benefits of regeneration efforts (Arnstein 1969). Engaging stakeholders early and continuously can mitigate resistance and build support for the project.

Urban river regeneration in Kanpur presents a unique opportunity to enhance environmental sustainability, economic growth, and community well-being. By implementing policies such as Integrated River Basin Management, promoting green infrastructure, regulating industrial discharges, and providing economic support, Kanpur can set a precedent for effective river regeneration. Strategies for community engagement, including public awareness campaigns, participatory planning, and volunteer programs, are crucial for ensuring the success and sustainability of these efforts. Despite challenges such as financial constraints, coordination issues, and community resistance, the potential benefits of river regeneration are substantial, making it a worthwhile endeavour for the city's future.

8. Implications for Global Urban River Regeneration

Kanpur's journey towards reviving the Ganga River offers powerful insights for cities worldwide that are struggling with the decline of their urban rivers. As cities continue to grow, their rivers often become dumping grounds for industrial effluents, sewage, and unchecked construction—straining both the environment and local traditions. The challenges faced in Kanpur are not unique; they mirror a global trend where fast-paced development, lack of accountability, and weak enforcement have led to rivers becoming dangerously polluted.

One of the most valuable takeaways from Kanpur's efforts is the importance of combining eco-friendly infrastructure with meaningful community involvement. Solutions rooted in nature—such as planting vegetation along riverbanks, creating rain gardens, and restoring wetlands—have the potential to be applied in cities around the world. But the real key to lasting success is involving the people who live alongside these rivers. Kanpur, like other cities such as Seoul with its Cheonggyecheon River restoration, highlights that when local communities are not part of the restoration process, efforts fail. By bringing them into clean-up initiatives, awareness programs, and planning, we ensure the river's future respects the environment and the local culture. Globally, the biggest obstacle to progress often lies in poor governance and weak policies. Stronger laws and community-driven approaches are the foundations for bringing life back to our rivers.

9. Future Directions

Future efforts should focus on creating smarter, more adaptable ways to treat wastewater, especially in cities with heavy industrial activity like Kanpur. Natural solutions like wetlands, buffer zones, and porous surfaces could hold the key to the long-term restoration of our rivers' health. Research could also explore alternatives, like using biodegradable materials for offerings or creating special spaces where these traditions can function sustainably. Collaboration between urban designers, ecologists, engineers and sociologists should be encouraged to co-create context-specific solutions and improve implementation frameworks. Additionally, eco-tourism and educational initiatives that highlight the cultural and ecological value of rivers should be promoted to elevate the relationship between city dwellers and the natural world around them.

10. Conclusion

Kanpur's example highlights how restoring an urban river is much more than just cleaning up water—it's where nature, culture, livelihoods, and city growth all intersect. The deterioration of the Ganga River, along with the diverse efforts to bring it back to life, reflects the broader water crises that many cities around the world are facing today. While technologies like sewage treatment and pollution control are important, what truly makes a difference is rebuilding the connection between people and the rivers they depend on. Bringing the Ganga back to life in Kanpur isn't just about fixing an environmental problem; it's about reconnecting with heritage, tradition, and everyday life. This journey teaches us that the most effective and lasting way to regenerate rivers lies in approaches that honour local culture, embrace natural solutions, and involve the communities who live along their banks—not just in India but everywhere.

References

Ahmad, I., and Chaurasia, S. 2019. Water quality index of Ganga River at Kanpur (UP). *Thematics J. Geogr*, 8(11): 66–77.

Batabyal, A. A. 2022. Tanneries in Kanpur and Pollution in the Ganga River: A Theoretical Analysis. *SSRN Electronic Journal*. 115032: 1–23.

Batabyal, A. A. 2023. Tanneries in Kanpur and pollution in the Ganga River: A theoretical analysis. *Reg. Sci. Policy Pract.*15(5): 1114–1123.

Bhargava, D. S. 1987. Nature and the Ganga. *Environ. Conserv*. 14(4): 307–318.

Chaudhary, M., and Walker, T. R. 2019. Ganga River pollution: Causes and failed management plans. *Environ. Int.*126(02): 202–206.

Chopra, A. K., Prasad, G., and Khanna, D. R. 2004. Ganga River water pollution and its management through religion. *Environ. Conserv. J.* 5(1–3): 35–40.

Das, P., and Tamminga, K. R. 2012. The Ganga River and the GAP: An assessment of efforts to clean a sacred river. *Sustainability*. 4(8): 1647–1668.

Gaurav, V. K., Kumar, D., and Sharma, C. 2018. Assessment of Metal Accumulation in the Vegetables and Associated Health Risk in the Upper-Most Ganga-Yamuna Doab Region, India. *Am. J. Plant Sci.*, 09(12): 2347–2358.

Gondwal, M., and Verma, R. 2023. A Review on the Status of Ganga River with Reference to its Pollution Sources and Mitigation Policies. Pp. 56–72. *In:* Bhakuni, K. K., Karki, N. S., and Chaturvedi, A. K. (eds).Ecology and Biodiversity of Himalayas, Agra, India.

Goswami, K., and Mazumdar, I. 2016. How idol immersion is polluting the Ganga River in Kolkata, West Bengal: an overview. *Indian J. Appl. Res*. 6(10): 260–463.

Kedzior, S. B. 2015. Pollution and the Renegotiation of River Goddess Worship and Water Use Practices Among the Hindu Devotees of India's Ganga River/Ganga River. *In*: *The Changing World Religion Map*. Dordrecht: Springer Netherlands. 557–576.

Khatoon, N., Naseema, K., Masihur, R., and Husain, K. A. 2013. Study of seasonal variation in the water quality among different ghats of Ganga River Kanpur, India. *J. Environ. Res. Develop.* 1(1): 1–10.

Kumar, D. 2017. River Ganga River–Historical, cultural and socioeconomic attributes. *Aquatic Ecosystem Health and Management*. 20(2): 8–20.

Kumar, P. 2016. Crashing of Abundance and Scarcity: The Tragedy of Water Resources of Kanpur, UP. *Aquat. Ecosyst. Health Manag.* 10(6): 479–492.

Mould, S. A., Fryirs, K. A., and Howitt, R. 2020. The importance of relational values in river management: Understanding enablers and barriers for effective participation. *Ecol. Soc.*25(2): 1–16.

Mukherjee, A. 2022. *Riverine Systems: Understanding the Hydrological, Hydro social and Hydro-heritage Dynamics*. Springer Nature.

Nelms, S. E., Duncan, E. M., Patel, S., Badola, R., Bhola, S., Chakma, S. et al 2021. Riverine plastic pollution from fisheries: Insights from the Ganga River River system. *Sci. Total Environ.* 756: 143305.

NIUA and NMCG. 2021. *Kanpur Urban River Management Plan*, Kanpur Nagar Nigam, Uttar Pradesh, India.

Paarcha, A., Maithani, S., Husain, M., Suhanee, N., and Azad, R. K., 2023. Monitoring Urban Sprawl Using Geo-Spatial Technology: A Case Study of Kanpur City, India, 3(3): 87–99. *In:* Rahman, A., Sen Roy, S., Talukdar, S., Shahfahad (eds) Advancements in Urban Environmental Studies. GIScience and Geo-environmental Modelling. Springer, Cham.

Prakash Kala, C. 2018. Cultural Significance and Current Conservation Practices of the Ganga's Ecosystem and Environment. *Appl. Ecol. Environ. Sci.* 6(4): 128–136.

Shakti, M., and of J., 2021. National Mission for Clean Ganga. *Department of Water Resources, River Development & Ganga Rejuvenation - Government of India*, 21 (December 2017), 1–28.

Shukla, A. K., Ojha, C. S. P., Garg, R. D., Shukla, S., and Pal, L. 2020. Influence of Spatial Urbanisation on Hydrological Components of the Upper Ganga River Basin, India. *J. Hazard. Toxic Radioact. Waste*, 24 (4).

Shukla, R. 2020. Drinking Water, Sanitation and Water-borne Diseases. *Econ. Pol. Weekly*. 50(23):136-140.

Simon, M., and Joshi, H. 2022. Story of the Ganga River: Its Pollution and Rejuvenation. *In*: Mukherjee, A. (eds) *Riverine Systems*. Cham: Springer International Publishing, 21–55.

Singh, R. P. 2001. Effect of wastewater disposal and extent of industrial pollution in and around Kanpur, Uttar Pradesh, India. *Bull Eng. Geol. Env. Springer-Verlag.* 60: 31–35.

Sinha, R. K. 2014. Ecology of the Ganga River - Issues and challenges. *National Seminar on Impact of Technology on Society*: 292–313.

Swaroop, A., Sharma, S., and Kumar, K. 2023. Reviving the Soul of Ganga: A Comprehensive Review of the Development of River Front Ghats. *Int. J.Multi. Res.* 5(2): 1–6.

Troch, M. 2018. Impact of urban activity on Ganga River water quality and ecology: case study Kanpur. Masters Thesis, Universiteit Gent and Universiteit Antwerpen, Belgium, Germany. 1–90.

UN Water, 2023. The United Nations World Water Development Report 2023: partnerships and cooperation for water. *Handbook of Water Purity and Quality*.

Zingraff-Hamed, A., Serra-Llobet, A., and Kondolf, G. M. 2022. The Social, Economic, and Ecological Drivers of Planning and Management of Urban River Parks. *Front. Sust. Cities*, 4: 907044.

8

Walking Through Memories of Waterways

Marluci Menezes,[1,*] *Juliana Valentim Harayashiki*[2] and *Silvia Mikami Pina*[2]

1. Introduction

Freshwater is one of the most altered and threatened ecosystems in the planet, despite their fundamental role in all aspects of life. Healthy water ecosystems can contribute to climate and flood regulation, nutrient cycling, and groundwater recharge, which are key to ecological resilience. Water ecosystems also provide habitats for many species, drinking water, irrigation, transportation, food, and recreation. However, the strong increase in population, agribusiness, urbanization, industry, pollution, and increasing pressure of climate change are degrading the ecosystems' ability to recover and provide those functions and ecosystem services. If this rate persists, by 2050, more than half of the world population will be at risk, with poor and marginalized communities disproportionately affected (United Nations 2018). Nevertheless, socioeconomic inequality is not the only one to emerge amidst the consequences of climate change, given that women and children are 14 times more likely to die in natural disasters than men (WEF 2020), making it also a matter of gender and age.

The "Age of Water Insecurity" (Tvedt 2021) and increasing instabilities in water cycles and precipitation patterns may lead either to prolonged droughts or floods. In any case, the results tend to be the same in terms of the general availability of fresh water when facing extreme weather events: scarcity. Once more than half of the

[1] National Laboratory for Civil Engineering (LNEC), Lisbon, Portugal.
[2] University of Campinas (UNICAMP), Program of Architecture, Technology and City, Campinas, Brazil.
Emails: j234648@dac.unicamp.br, 0000-0002-6123-8923;
smikami@unicamp.br, 0000-0002-4727-5197
* Corresponding author: marluci@lnec.pt, 0000-0001-7031-0053

DOI: 10.1201/9781032711737-8

world's population lives in river basins shared by two or more countries, the risk of water crisis escalating into actual conflicts becomes increasingly plausible (Tvedt 2021). This is particularly true given the lack of robust regulatory institutions and binding international agreements.

In ancient history, the fall of several great civilizations that have been credited to cultural decadence or foreign invasions, is now also believed to have been heavily influenced by changes in the hydrological cycle (Tvedt and Oestigaard 2016, Tvedt 2021). The Indus region faced several difficult years, marked by severe flooding. Additionally, the migration of the Beas River away from Harappa urban center intensified water supply challenges. The rain-based Mayan civilization suffered successive droughts, which weakened the political and economic power of their leaders. Repeated hardships involving the Nile River over the years contributed to the downfall of the Egyptian dynasties. When analyzing these contexts, the political dimension of water becomes evident. It highlights the critical role of protecting waterway ecosystems in promoting peace, prosperity, and to fostering equality – both within societies and between nations.

Although waterways have a long-standing historical and political connection with cities, their importance is not always reflected in modern urban landscapes. River spaces are often neglected and forgotten by collective memories. As they lose visibility in everyday urban life, people's awareness of their relevance diminishes, along with their connection to waterbodies. Memory, composed of multiple references and meanings, depends on social interests and requires constant reactivation. This happens through sharing among people and through stimuli that preserves its connection to the physical space (Menezes and Silva 2023). Walking invites reconnecting with waterways landscapes and their memories. Walking enhances the perception of places through a sensory spatial experience. This perception is shaped by the way people move around and by the memories, which influence sensations (Degen and Rose 2012).

Walking in a city is an aspect of everyday life, it facilitates essential travel and contributes positively to health and well-being. Walking provides a worthwhile opportunity to explore, experience, and comprehend urban landscapes, encouraging contemplation and the development of innovative ideas. While walking, spatial encounters with locations, their names, and associated memories might generate impromptu and creative stories (Menezes 2019). Walking allows individuals and groups to explore urban spaces' diverse characteristics through activities linked to various intentions, meanings, and times. Walking generates diverse spatial narratives that challenge conventional space organization. This reflects a phenomenological view of landscape, grounded in lived experiences and affective intensities (Edensor 2022).

Walking enables exploration of unfamiliar surroundings, such as hidden urban waterways. The slow speed of the walker can contribute to building connections with waterways. Remembering is an instrument for these connections and results in a deeper grasp of the importance of urban waterways. Walking in and through the city can be seen as immersing oneself in the senses and strategically engaging with waterways.

The chapter reflects on the importance of strolling through the memories of waterways hidden by urban development. Three urban waterways cases from different cities – Lisbon (Portugal), Yokohama (Japan), and Santos (Brazil) – where walking is used to recover memories of these bodies of water. The chapter starts by discussing the historical process that led to the massive invisibility of waterways in the cities. It then explores how walking is a sensory experience that activates memory and reinforces emotional connections with waterway landscapes. Finally, this chapter is completed by cases from different cities (Figure 1).

Figure 1: The urban waterways cases in [a] Lisbon (Portugal), [b] Yokohama (Japan), and [c] Santos (Brazil). Source: Authors, 2023, 2016, 2024.

2. Watercourses Throughout History

Perceptions and treatment of the water courses have undergone intense transformations over time. Although many strategic aspects to be considered in the choosing of a good and prosperous permanent settlement have always stacked up in places by rivers, making it almost an obvious choice, for a long time, only temporary settlements appeared on riverbanks (UNESCO 2023). This is because the waterways, source of vitality and force of destruction, carves its character into land through shifting courses, flooding, drying up – its volatile nature imprints the essence of ephemerality (Pinto and Kondolf 2015).

The first civilizations to thrive were capable of understanding and harnessing the potential of rivers, coexisting in harmony with their natural cycles (Tvedt and Oestigaard 2016, Tvedt 2021). In Mesopotamia and the Nile River Valley, several civilizations flourished by setting prosperous agriculture on floodplains during the dry season, knowing that after the flood season, those lands would become fertile again. The floods and retreat of the Nile River were sacralized and explained in the myth of the death and rebirth of gods; the river was revered as a miracle in the middle of the desert – an eternal source of water that intertwined nature and culture.

With the consolidation of trade and transportation routes, the protection and control of strategic places along the rivers became even more important, as they rose to centers of power. As cities developed in some of these centers of power, the level of control over nature, the advancements of water-related technologies also increased, with more sophisticated bridges, aqueducts, diversions of rivers, collection of waters, rectification of water bodies, etc. (Tvedt and Oestigaard 2016, Tvedt 2021). In Rome, during the Pax Romana, the city flourished, showcasing infrastructure and control over water even in artistic ways, through the fountains. Even nowadays, the numerous public fountains in the city enchant visitors from all over the world.

After the Industrial Revolution, however, the dynamics changed completely. Rapid urbanization and population growth aggravated existing issues like the discharge of sewage into water bodies (Tvedt and Oestigaard 2016, Tvedt 2021). The practice has always been around, but the unprecedented scale led to an extreme level of degradation, from which the rivers were not able to recover on their own, and neither were the thriving industrial cities assessing effective alternatives (Tvedt 2021). Instead, these open sewers started to be culverted to be put out of sight, and rectified, diverted – so as "not to disturb" the cities' development. The advent of railways and highways diluted even more the decaying importance the rivers once had, by taking away the last fundamental socioeconomic role they used to have, which was transportation. Floodplains and riverbanks that had not already been changed were ultimately converted into traffic lanes and roads.

2.1 Relationships Between Walking, Landscape, and Memory

Over the last centuries, urban mobility has been supported by vehicle optimization and travel efficiency. In our contemporary world, GPS systems exemplify how urban mobility is driven by efficiency and the logic of individual vehicles, prioritizing the reduction of time between points A and B. This type of approach frames the unfolding landscape as nothing more than a time interval between destinations. The speed of several transportation modes establishes a continuum of transitional space. Nevertheless, when walking, people have time to rest, gaze and question oneself about what is in front, because the contemplated landscape is full of meanings. Even though walking is still assumed as the most primary means of getting from one place to another (Middleton 2011) – something people just do –, it also serves other purposes. Walking can be a means to recover the rhythm of ancestors, to break the frenetic schedule of the system, to get away from the usual routine, and to modify our perception of the territory (Medina and González 2023).

Prevalent types of research branching from transportation, such as forecasting future walking patterns and trends, implementing walking policies, and conducting cost-benefit analyses – tend to strengthen the functionalist approach to walking (Tolley et al. 2001, Sælensminde 2004). This perspective often overlooks the heterogeneity and plurality of different pedestrian experiences (Cresswell 2006). Sennett (1990) frames walking as a way for strangers to negotiate public urban space, unpredictable encounters that are also democratic and civilizing. Like Certeau (1980), for Sennett, walking is a form of urban emancipation that opens democratic possibilities, resistant to top-down imposed order on urban spaces. Dardel (2015) understands that the landscape is not something just to be seen. The landscape is, in its essence, where people realize their own existence, where the dimensions of being are discovered. Moreover, this is done by experiencing, by walking with affection, with fondness. Following theories of embodied cognition, perceptual consciousness is grounded in the human body, enabling the sensory detection of shapes, smells, tastes, and textures that contribute to a meaningful experience of the world (Suguimoto 2014, Dardel 2015). The affection is due to a fundamental relationship with the Earth, to which Dardel calls geographicity, which impels people to intentionally interact with it. Yi-Fu Tuan (1990) calls this feeling topophilia.

Sandeville (2010) observes that landscape is a shared experience – shaped by natural processes and human activities and enriched through memory and narrative. This experience carries plurality, contradictions, and tensions brought by walkers who traverse it. The act of walking facilitates capturing the plurality of urban space, reflecting diverse, juxtaposed, and differentiated uses, practices, meanings, and temporalities. It plays an active role rather than merely observing events (Edensor 1998). The act of walking can be compared to language: both are creative processes that allow improvisations, connections, shortcuts, and thousands of decisions to be made at the moment (Certeau 1980, Butler 2008). Walking is an act of resistance against hegemonic impositions, unfolding as a sensitive and creative engagement with places, enriching the spatial experience (Certeau 1980, Menezes and Mateus 2017, Menezes 2019).

Urbanization has deeply transformed the relationship between people and the waterways' landscapes, weakening bonds between citizens, waterways and urban life (Harayashiki et al. 2025). Memory is a potential resource for connecting people with urban waterways (Menezes and Silva, 2023). The idea of connection emerges from the social dimension of watercourses, a topic discussed across both hydrology and ecology (Kondolf and Pinto 2017). The memory underscores the importance of urban waterways as both a physical and metaphorical mediator. Social memory refers to the way lived experiences are transmitted and articulated across generations, shaping group identity and sustaining cultural reproduction. Without regular renewal, memory tends to fade, and the less satisfactory aspects are dismissed (Halbwachs 1950, Connerton 1989, Fentress and Wickham 1992, Yavuz 2017, Song et al. 2021, Menezes and Silva 2023). Butler (2008) studied how memory and oral history shape perceptions of the Thames riverscape by creating sound walks that were freely shared online and in public venues. Over 18 months, around fourteen thousand people experienced these walks. Surveys by questionnaires and interviews revealed that participants felt empathy, nostalgia, and a stronger awareness of the landscape. Walking fostered a personal connection with the scenery, enriched by imagination, sensory and engagement with local landmarks. This intimate experience of place leads to the creation of memoryscape*s*, where the articulation of memories, emotions and personal narratives, result in a deep sense of place attachment. Basu (2018) introduces the notion of "palimpsest memoryscape" to describe how memories tied to physical and symbolic landscapes coexist and overlap. While his analysis centers on post-conflict territories, he emphasizes natural elements, objects, and rituals as material mediators of historical memory. This perspective invites a reading of hidden waterways—concealed by urbanization or erased from contemporary planning—as layered memoryscapes, where traces of past uses, affects, and social meanings endure and evolve, revealing a fluid, subterranean dimension of collective memory, continually reshaped in the present.

The recovery and restoration of waterway landscapes is a complex matter because new contexts, territories, and places have been woven into the same landscapes (Cianci and Mondelli 2018). This multiplicity of times and dynamics cannot be ignored. There is no ideal moment in the past to which the landscape should return, erasing everything that followed. It will not always be possible to dig up, uncover the watercourses that have been culverted and hidden from our view. In these cases, recovery occurs through everything else – they become history, symbols, traditions, traces, cultures, experiences, meanings, sensibilities, narratives, and memories.

In the following section, three cases will be discussed, in which different dimensions of waterways landscapes are perceived through walking and memory, activating and being activated by a multiplicity of elements, agents, and times.

3. Three Cities, Several Rivers, and Many Memories

3.1 Lisbon, Portugal

The abundance of water shaped the city's evolution amidst hills and valleys; Lisbon has been always influenced by the delta of the Tagus River. The waterbodies network originated a complex structure of river basins of small dimensions. As the urbanization intensified, many rivers and creeks were culverted. This process led do their invisibility within the urban landscape, as these water bodies became hidden beneath from the city's surface. Despite being hidden, traces of their presence can still be found within the built environment, subtly tracing the footprint of waterways inscribed in the city's structure. Menezes and Costa (2023) conducted a survey on water-related toponyms, identifying places' names that directly or indirectly refer to the historical water bodies. Their research focused on several Lisbon's parish councils, highlighting how these traces serve as cultural reminders of the city's watercourses.

The toponym references related to water are diverse and rich, including direct mention of specific water bodies by name, as well as references to water-related elements embedded in the urban landscape. These include water features, hydraulic infrastructure, modes of fluvial transport, landscape features. Terms such as bridge, garden, well, fountain, and pier reflect the widespread presence of waterways in the city and highlight their integration into urban life. For example, the terms "Fonte", "Fontainha", "Chafariz", "Bica", are all commonly translated as in "fountain" in English. However, in Portuguese, each term has a distinct nuance: Fonte refers to a natural water source; Fontainha denotes a small or modest fountain; Chafariz normally describes a decorative structure, where aesthetic function dialogues with the functional aspect; and Bica generally refers to a spout or water outlet with a dynamic flow.

After mapping these signs across Lisbon, walking tours were used to identify and record water-related elements embedded in the city's urban landscape. This chapter focuses on the Alcântara area, whose privileged proximity to the Tagus River positioned it once as key region for maritime trade and industry throughout the city's history, but since 1980s the industrial activities have been declining. Since then, the process of social and economic devitalization has given way to a dynamic social, economic, and even constructive renewal. This transformation has been accompanied by a rise in gentrification, caused by targeted building rehabilitation strategies (Fonseca 2016). Alcântara is also the name of a river that originates in the Benfica Valley, in the north of Lisbon, and flows into the Tagus River. The name Alcântara originates from the Arabic word "al-qantara", meaning bridge. An existing bridge was demolished in the 19th century. The Alcântara River was culverted in 1940s to provide space for roads, an elevated expressway, and railway lines. Existing paintings and tile panels depict key elements, such as the bridge of Alcântara and the aqueducts (Figure 2).

With the aim to supplying drinking water to the city of Lisbon, the Águas Livre Aqueduct was built in the 18th century over the Alcântara Valley (Figures 3 and 4). Feat of engineering, the aqueduct has endured to this day and stands out in the landscape for its impressive arcades. It is a landmark of Portugal's cultural heritage. Walking along the section of the aqueduct that crosses the valley provides an experience to capture the complex landscape of Alcântara: an interplay of textures and tones, the sound of wind, the pulse of the frenetic car traffic crossing the valley,

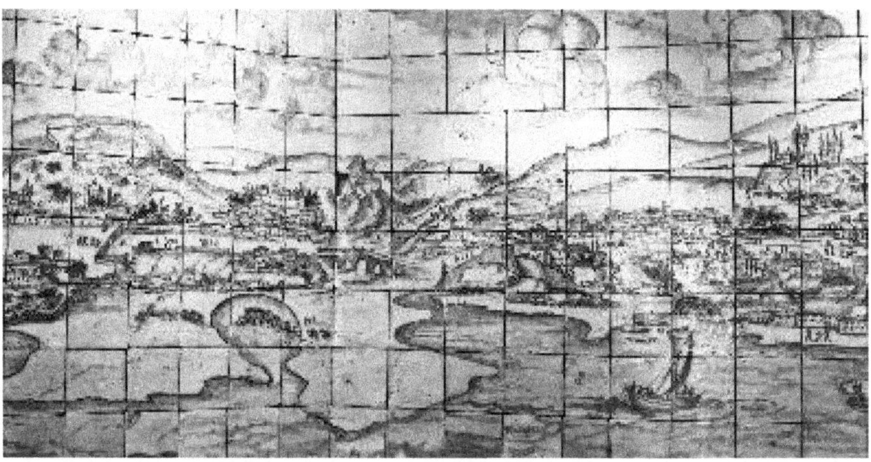

Figure 2: A tile panel dated back to 18th century depicts a historic Lisbon scene with the demolished Alcântara Bridge. The tile panel is currently in the National Tile Museum. Source: Wikimedia Commons, https://commons.wikimedia.org/wiki/File:Ponte_de_Alc%C3%A2ntara_(Grande_Panorama_de_Lisboa,_MNAz).png, under license CC-BY-SA-2.0.

Figure 3: View of the Alcântara Valley, Lisbon, drawing from 1830 by André Monteiro da Cruz, the collection of the National Museum of Ancient Art. Source: Wikimedia Commons, https://upload.wikimedia.org/wikipedia/commons/1/1b/Vista_do_vale_de_Alc%C3%A2ntara%2C_Lisboa%2C_c._1830%2C_MNAA_01.jpg, under license CC-BY-SA-4.0.

the clear sensation of being between the progressing city and the urban park Monsanto (created in the 1930s). The Alcântara River is hidden in this view, although from certain points along the aqueduct, the Tagus River frames the landscape.

The presence of water – explicit or suggestive – is inscribed in local toponymy and engraved on stone plaques embedded in buildings. The name Alcântara is also used for the neighborhood, a viaduct, the main road, and the local public transport station. This trend is also currently evident with new residential condominiums targeting high-income households, reflecting ongoing urban transformations and socio-spatial dynamics.

One of the city's wastewater treatment plant and disinfection facilities, which discharge into the Tagus River, was built in the Alcântara Valley in the late 20th century. In the 21st century, the management of urban water cycle became aligned with the circular economy principles. This change, combined with the modernization of infrastructure, led to a new name: "Fábrica de Água" – Alcântara Water Factory (Águas do Tejo Atlântico 2025). In 2014, the Lisbon municipality approved the Alcântara Valley Urbanization Plan, emphasizing its ecological and scenic importance (CML 2014). Since then, the region has been promoting the reuse of non-potable water for irrigation and the strengthening of soft mobility infrastructure. The aim is also to benefit from the landscape and environmental features provided by proximity to the Monsanto Urban Park. Furthermore, the aim is to strengthen

Figure 4: View of Alcântara Valley, Lisbon. Source: Authors, 2023.

the valley's connection to the Tagus River, integrating these natural features into a cohesive ecological framework.

Walking, now enabled by pedestrian-friendly trails that connect the Monsanto Forest Park to the Tagus River, and framed by green spaces, evokes the hidden water system beneath people's feet. At certain points along the trails, the silhouette of the wastewater treatment plant—now covered by a green roof—emerges in the landscape. Its presence evokes memories of the sharp odor that once hung in the air when someone came near. The landscape feels renewed, with a natural atmosphere that nearly hides the surrounding heavy transport infrastructure. Improvements in the valley's natural drainage system are evidence of an emerging urban sensitivity toward waterways. Yet, walking on the trail unfolds a conflicting reflection with landscapes and green surfaces coexisting with transport infrastructure, heavy traffic, viaducts, train stations, highways – making the valley a dynamic corridor rather than a quiet haven.

The requalification of the valley counts not only for environmental aims, but also for territorial branding and social imaginaries and for the potential displacement pressures. Once viewed as a marginal or merely functional area, requalification efforts now shape the physical and symbolic remapping of the area. Is the newfound celebration of water a strategy for resilience or/and a subtle tool for reconfiguring? Yet, while walking through the valley, the presence of water resonates beyond the visual: it triggers a perceptual memory, evoking sounds, smells, textures, and emotional imprints left by those who lived in its proximity. The stream once crossed barefoot, by informal paths and daily routines, still carries remnants of local memories and lived experiences, shaped by scarcity and improvisation.

3.2 Yokohama, Japan

Ooka River Basin runs in the heart of Yokohama, a city located South of Tokyo, belonging to the Tokyo Metropolitan Region. This basin is made of four rivers: Ooka, Hino, Horiwari, and Nakamura, and one bypass that connects the Hino River to the flood retention reservoir of Konan District and into the ocean. The bypass has open and culverted sections. In 2018, the urbanization of the river basin was already extremely high, reaching 95%, accentuating a historical trend of a high degree of morphological modification of urban waterbodies.

Japan has a long history of riverscape interventions, prior to the industrial revolution, using natural waterbodies, irrigation canals, and rice paddies as infrastructure in balance with riparian ecosystems. However, once the Meiji Era (1868–1912) started, the traditional riverscapes values would not resist the intense urbanization processes under western influences. After two hundred years of isolation, Japan reopened the international port in Yokohama to the world. The city became a gateway to new and foreign influences, ranging from gastronomy, attire and accessories to advanced science, engineering, and technology. Japan assimilated western technological paradigms, replicating the same problems, and later, adopting similar solutions: mono-functional grey infrastructure. During the Showa Era (1926–1989), the effort to deal with pollution and flooding resulted in canalization and rectification of the majority of urban waterbodies. In 1990, the urban waters were pretty much under control, but lifeless, until new movements and legislative changes happened, in pursuit of restoration of urban rivers.

The 1999 "Nature-Oriented River Work" law, enacted by the River Bureau and the Ministry of Construction, marked the start of a major shift. This process culminated in 2003 with the enactment of the "Nature Restoration Law", led by the Ministry of Environment. In Yokohama, as usual, changes happened earlier than in the rest of the country, even before the enactment of the first law in 1990. In 1987, the city of Yokohama pioneered the restoration of the Itachi River. The urban plan that included the Ooka River Promenade is also from 1980.

Walking along Ooka River during cherry blossom season triggers personal and collective memories, evoking the past events that happened in the community (Figure 5.a). It is also a longstanding Japanese cultural tradition, with poems, paintings, woodblocks, collective imagination, social codes, and special nuances of sociability (Harayashiki 2018, Harayashiki et al. 2025). The season revolves around the beauty of ephemeral things, the appreciation of a singular moment unfolding, fleeting like so many things in life. During the evenings, there is a local festival at Gumyoji Shoutengai, centered at Sakura Bridge, that expands over the riverbank sidewalks, sometimes taking over the entire streets with food vendors, and large mats with families and groups of friends placed over the spots they think will give them the best view of cherry flowers. Indeed, it is quite an endeavor and responsibility to find the perfect sitting/viewing spot, with some people arriving hours in advance to secure a good spot for the rest of the group. The usually sober, organized, and even austere streets of Japan, assume another dimension during these laid-back and seemingly chaotic evenings, which fill the same streets along the river with so much life, laughter, and playfulness. Furthermore, numerous memories are evoked.

[a]

[b]

[c]

Figure 5: Yokohama: [a] People at crossing stomps on the water level of Ooka River Promenade during Cherry Blossom season, [b] Nakamura River, and [c] Bus stop on Horiwari River. Source: Authors, 2016.

Many sections of the promenade evoke a peaceful sense of safety, and calmness, almost like it can slow down a little the dynamics of usual life that go at full speed, maybe just a block or two away from there. Nevertheless, some places go one step further than just slowing down, and for that, someone must go beyond the riverbanks: the hidden walks at the water level. Occasionally, there are stairs leading to these hidden walks. No one staircase is exactly the same as the other, while some seem to exude an almost forbidden atmosphere, there are maybe one or two that resemble bleachers, open and inviting those who dared to go down to sit and wait a while longer before they go all the way. The hidden walks are made of short paths, sometimes it is possible to cross to the other side of the river to access another walk - through a small stomp, but the paths are mostly disconnected from one another. They are digressions from the main paths on street level. The narrow paths, the uneven and rough surfaces, the grass and weeds growing out of the cracks of the pavements and of the inclined stone walls of the stabilized riverbanks. Urban sounds are fading away, and the sounds of nature are becoming more and more perceptible in the atmosphere. The riverbanks have the appearance of stonewalls, but they are ultimately a finishing element made out of texturized concrete, a reference to the well-known materiality of historical Japanese riverscapes. The actual structural elements are made of earthquake-proof steel-concrete components. These modest hidden walks are, perhaps, the closest things to reconstructing the memory of an ideal, primordial, ancient, or "natural" sense of rivers, because they offer elemental sensory input: the sound of water, the freshness in the air, the roughness of the stones and the dirt on the ground, the movement of the grass in the wind.

The only other river in the system with stairs leading to the water level is the Horiwari River. Instead of a path, it leads to an anchored boat, at least it used to, nowadays the entrance to these stairs is closed and prohibited. The Horiwari River, almost a straight line flowing in the direction of Nakamura River, used to be connected to a network of small canals in the delta of Ooka River which no longer exists, and accommodate commercial transportation. The canal was closed on the ocean end because of a private marina, and going towards Nakamura meant being buried in the open air beneath the expressway that runs above (Figure 5.b). This left the Nakamura River in a strange situation, present and absent at the same time, always in the shadows. The Horiwari River exists between these dead ends, one physical, and one metaphorical, out of its function, disconnected from its initial canals network.

Walking along the Horiwari River is uncomfortable and loud: along the banks there are busy roads with several lanes. People pass through several bus stops with nothing to signal their location: no seating, no shelter against the sun or the rain, no shade from trees. The only amenity found in one of the bus stops was an old, solitary chair on the right bank, put there by someone who cared enough (Figure 5.c). The solitary chair was a protest in the middle of that arid traffic zone, so is the Horiwari River, confined to the same conditions, yet drawing people to its waters. Even though forbidden, it is common to find on the closed stairs old and middle-aged men fishing. Despite discomfort and transgressions, small traditions endure, echoing the waterway's resilience.

3.3 Santos, Brazil

Located on an island in Brazil, the city of Santos was home to the country's main coffee export port. However, the last decades of the 19th century were marked by epidemics that decimated part of the population and prevented the port from functioning properly. Following a series of plans and projects, the sanitary engineer Saturnino de Brito and his team presented a sanitation plan that included a system for separating sewage and rainwater. The plan was implemented in two phases. The first phase involved constructing a series of open surface macro-drainage channels, which were considered ideal for a flat city like Santos. The second phase consisted of building and expanding sewage collection networks and underwater outfalls. Six macro-drainage channels were implemented; the first was inaugurated in 1907, with the last completed in 1927 (Andrade 1992). This project is regarded as one of the most significant achievements in Brazilian sanitary engineering. Three other channels (called channels 7, 8, and 9) capture and conduct water flowing down from the hills, thereby preventing flooding.

The entire set of equipment, layouts, and urban forms was part of a comprehensive plan for the city's expansion towards the beachfront and the south-eastern part of the island. This plan set a new standard for urban and sanitary conditions in Brazil, particularly concerning technical soundness, aesthetic appeal, and environmental quality. The road system incorporated the layout of existing streets and created squares or roundabouts at intersections and the ends of avenues, with water channels running alongside the avenues. This prevents monotony and gives certain places a unique identity. The urban ensemble of channels, pavements, and densely wooded avenues build important corridors through which the sea breeze can reach areas further inland, providing relief from the heat. The squares and gardens, particularly the 7 km long beach gardens, have become the city's main landmark, contributing to the extensive sand strip that is responsible for its popularity with tourists.

This system is so important that it has become a spatial reference point strongly linked to the identity of the residents. The channels, which are distributed at regular intervals along the beachfront, have become a guide for locations and movements (Figure 6a, b and c). The channel numbers, from 1 to 6, are still used today to indicate the position for people travelling around Santos Bay. The channel system, with its abundant trees, pavements, bridges, and walkways, is like Haussmann's boulevards and opens possibilities for new social practices and environmental awareness. These optical devices intentionally transformed the city into a landscape for the first time, and water in an urban environment became an aesthetic value for the first time in Brazilian cities. That system brought a sense of proximity and an unexpected feeling of security and protection around the water. The tree-lined channels are places for walking, meeting up, relaxing, and enjoying leisure activities. The channels are part of the daily lives of residents and a reference point for tourists. Recently, cycle paths were built along some channels, such as Channel 6, adding another dimension to the channels, as cycling is becoming increasingly important for urban mobility.

Saturnino de Brito's network of channels transformed Santos into more than just a healthy city. It changed the relationship between citizens and urban water, bringing new uses, appropriations, and memories of experiences and perceptions.

[a]

[b]

[c]

Figure 6: Santos: [a] Channel 4 flowing to the sea, [b] Channel 2, and [c] Channel 6. Source Authors, 2024, 2025, 2024.

Through his work on the Santos' channels, Saturnino introduced the concept of urban riverscapes, doing more than just improving sanitation. He also opened up possibilities for appropriation and fostered a positive and profound connection to water by connecting the banks to their uses, the urban fabric, and the pedestrian scale. This established opportunities for aesthetic education through experience and affection for water (Harayashiki and Pina 2024).

It is possible to cross the entire city along a waterbody, stimulating your perception, senses, and imagination. Although they were all built at the same time and in the same way, unlike natural bodies of water, the channels have become their micro geographical structures, with unique atmospheres and surrounding landscapes that take on different characteristics and ambiences. Some channels maintain a more homogeneous riverscape, while others show different phases of the city and the socioeconomic differences. Channel 3, for instance, is characterized by expensive residential buildings near the waterfront in an area known as the Rich Neighborhood. As it approaches the historic center, it is flanked by lower buildings with no front setbacks and a variety of uses, including residential, workspace and retail. The lack of some of the original trees, which have not been replaced, alters the original conception of the channels as public promenades.

The channels were only declared historical heritage sites in 2006 by the State and Municipal Heritage Councils as a strategic act to forbid initiatives and plans to cover the waterways. This highlights the necessity for continuous awareness-raising and educational initiatives concerning various issues of urban waterways. One such possibility is to restore these landscapes through memory, personal narratives, experiences, expanding the knowledge about the territory as a collective and continuous act through an inclusive participatory process. In addition to raising awareness, this could stimulate new ideas for urban plans that are more sensitive to water and integrate and aggregate the potential of such infrastructure to play in landscapes and territories.

4. Final Notes

The chapter addressed walking as a sensitive spatial practice, examining its role in activating memory and reconnecting with waterscapes. Based on three urban cases, Lisbon, Yokohama, and Santos, the study reflected on how walking can be an important ally in reclaiming the memory of waterways. Walking becomes a narrative gesture yet appropriated by memory. It creates spatial statements and reveals subtle traces. It also reinforces the connections between people and waterscapes. The chapter began with a brief discussion of the historical factors that led to the invisibility of urban waterways. Then addressed the theoretical frameworks for the interplay between memory, walking, and landscape. The idea that walking activates sensory memories and creates bonds with the present through objects, natural elements, and sociocultural references was highlighted, giving rise to the concept of memoryscape.

The analyzed cases identify contexts and processes that contributed to river's invisibility. Despite their differences, they all demonstrate the neglect of waterways throughout urban growth. Walking recalls memories that emerge. They are awakened by readings or narratives and help also to perceive the still present waterscape. This landscape interacts with urban fabric, transportation infrastructure, buildings, people, greenery, and biodiversity. Walking forms memoryscapes, with materials, natural, social, and symbolic elements connecting past and present, allowing reflection on what was and what is seen, anticipating new futures, potentially more sensitive to water bodies. Activated memories go hand in hand with social engagement for more sustainable cities.

Walking through Alcântara shows how technical, political, and social engagement for more sustainable initiatives, desirable and adapted to contemporary environmental challenges, effectively becomes an integral part of the memoryscape. At the same time, as a narrative gesture that articulates with the perceptual dimensions of memory, walking provokes reflections about a possible future heralded by green and blue infrastructure. These often appear to be more greenwashing, exposing the risks of erasing the complexity of landscape. In Yokohama, urban growth and technical-scientific westernization have weakened the traditional relationship between culture and nature. Waterways have become invisible, often leaving only traces. Walking helps to recover these losses. It creates ephemeral experiences that strengthen resilience and integrate the memoryscape, reconnecting people with local waterscapes. In Santos, the water infrastructure for sanitary purposes did not ignore the scenic value of water. The quality of urban and landscape planning was decisive. Water has become a central element in the development of public spaces. Memoryscapes were increased through valuing walking. At the same time, the case of Santos is also a warning, waterways can disappear under urban pressure. Walking restores memories linked to nature, ritual, and symbolism. The channels became an urban, cultural, and environmental heritage. Their resilience is demonstrated through being heritage-listed and new ways of appropriation through walking, in a kind of continuous reinvention of the riverscapes memory.

Walking activates memory and stimulates the generation of urban narratives. These contribute to creating of water memoryscapes and offering a powerful tool for reconnecting people with natural, cultural, social, and symbolic landscapes. Moreover, it fosters sensitivity and critical reflection on the processes of erasing landscapes, as well as the importance of their conservation as heritage.

Acknowledgements

This study is part of the RUN Project – Citizen Science and Co-Creation for Regeneration or Urban Rivers and Mitigation of Risks, funded by the Ibero-American Cyted Program (agreement no. 420RT0008), and part of the Doctoral Thesis in Architecture, Technology and Cities at the University of Campinas (UNICAMP) by Juliana Valentim Harayashiki.

References

Águas do Tejo Atlântico. 2025. Fábrica de Água. https://www.aguasdoTagusatlantico.adp.pt/content/fabrica-de-agua.

Andrade, Carlos Roberto. 1992. A peste e o plano: o urbanismo sanitarista do Engenheiro Saturnino de Brito. Master's diss., Universidade de São Paulo. https://repositorio.usp.br/item/000736963.

Basu, Paul. 2018. Palimpsest Memoryscapes: Materializing and Mediating War and Peace in Sierra Leone. In A Museum Studies Approach to Heritage, edited by Sheila Watson, Amy Jane Barnes, and Katy Bunning, 59:19. London: Routledge. https://www.taylorfrancis.com/chapters/edit/10.4324/9781315668505-59/palimpsest-memoryscapes-paul-basu.

Butler, Toby. 2008. Memoryscape: Integrating Oral History, Memory and Landscape on the River Thames. In People and Their Pasts: Public History Today, edited by Paul Ashton and Hilda Kean, 223–239. London: Palgrave Macmillan. https://repository.uel.ac.uk/item/8649z.

CML - Câmara Municipal de Lisboa. 2014. Plano de Urbanização de Alcântara. Aprovado pela Deliberação n.º 318/AML/2014, de 2 de dezembro. Lisboa: CML. Documentos técnicos elaborados entre 2011 e 2012. https://www.am-lisboa.pt/301000/1/000494,000002/index.htm.

Certeau, Michel de. 1980. L'invention du quotidien. Paris: Gallimard.

Cianci, Maria Grazia, and Francesca Paola Mondelli. 2018. Hidden River: Recovering the Memory of a City. In Proceedings of World Heritage and Knowledge XVI International Forum, Roma, Italy, June 2018, edited by Carmine Gambardella. https://iris.uniroma3.it/handle/11590/341046.

Connerton, Paul. 1989. How Societies Remember. Cambridge: Cambridge University Press.

Cresswell, Tim. 2006. On the Move: Mobility in the Modern Western World. London: Routledge.

Dardel, Éric. 2011. O homem e a terra: natureza da realidade geográfica. Translated by Werther Holzer. São Paulo: Perspectiva.

Degen, Monica, and Gillian Rose. 2012. The Sensory Experience of Urban Design: The Role of Walking and Perceptual Memory. *Urban Studies* 49 (15): 3271–3287. https://doi.org/10.1177/0042098012440.

Edensor, Tim. 1998. The Culture of the Indian Street. In Images of the Street: Planning, Identity and Control in Public Space, edited by Nicholas R. Fyfe, 205–221. London: Routledge.

Edensor, Tim. 2022. Landscape, Materiality and Heritage: An Object Biography. Singapore: Palgrave Macmillan. https://link.springer.com/book/10.1007/978-981-19-7030-6.

Fentress, James, and Chris Wickham. 1992. Social Memory. Oxford: Blackwell.

Fonseca, Henrique Manuel. 2016. Representações sociais sobre mudança urbana no bairro de Alcântara. Master's diss., ISCTE – Instituto Universitário de Lisboa. https://repositorio.iscte-iul.pt/bitstream/10071/12527/1/Tese_Doc_Final_Completo.pdf.

Halbwachs, Maurice. 1950. La mémoire collective. Paris: Press Universitaires de France.

Harayashiki, Juliana Valentim, Marluci Menezes, and Silvia Mikami Pina. 2025. The (In)Visibility of the River in the Perception and Representation of Landscape in Santa Bárbara d'Oeste (Brazil). Cidades, Comunidades e Territórios 50: 19–39. https://doi.org/10.15847/cct.40931.

Harayashiki, Juliana Valentim, Silvia Mikami Pina, and Evandro Monteiro. 2024. Dimensões públicas das paisagens fluviais urbanas hegemônicas. Anais do VIII Encontro da Associação Nacional de Pesquisa e Pós-graduação em Arquitetura e Urbanismo (ENANPARQ), Rio de Janeiro. https://www.even3.com.br/anais/enanparq8/851908-DIMENSOES-PUBLICAS-DAS-PAISAGENS-FLUVIAIS-URBANAS-HEGEMONICAS

Harayashiki, Juliana Valentim, Silvia Mikami Pina, and Evandro Ziggiatti. 2025. Os Rios na paisagem e na vida: por uma paisagem urbana infraestrutural, afetiva e simbólica. In Atas da 11ª Conferência Internacional da Rede Lusófona de Morfologia Urbana – PNUM, Sintra, Portugal, September 2023, 189–195. Lisboa: ISCTE. https://www.dinamiacet.iscte-iul.pt/_files/ugd/e54c2c_9e7afa0f2c964376b39a9132d82020d2.pdf.

Harayashiki, Juliana Valentim. 2018. The Study of the Influence of Ooka River Basin Water Bodies in the Urban Morphology of Yokohama. Master's diss., Yokohama National University. https://www.urban.ynu.ac.jp/english/theses/2018-03.html.

Kondolf, G. Mathias, and Pedro J. Pinto. 2017. The Social Connectivity of Urban Rivers. Geomorphology 277 (15): 182–196. http://dx.doi.org/10.1016/j.geomorph.2016.09.028.

Medina, Somoza, and Lois González. 2023. Walking as a Cultural Act and a Profit for the Landscape: A Case Study in the Iberian Peninsula. GeoJournal 88: 2171–2186. https://doi.org/10.1007/s10708-022-10745-x.

Menezes, Marluci, and Dória Costa. 2023. Memórias dos Rios e Ribeiras na Toponímia da Cidade de Lisboa. In *Rios Urbanos na Ibero-América: Casos, Contextos e Experiências*, edited by Carlos Smaniotto Costa, Marluci Menezes, Montserrat Pallares-Barbera, Gabriela Pastor, Eliana do Pilar Rocha, and Klinge O. Villalba Condori, 325–339. Lisbon: Lusófona University Press. https://doi.org/10.60543/ecati/b6af-7210.

Menezes, Marluci, and José Diogo Mateus. 2017. Walking as Tactile Method in Urban Planning and Design. In The Making of the Mediated Public Space, edited by Carlos Smaniotto Costa and Konstantinos Ioannidis, 65–74. Lisbon: Lusófona University Press. https://doi.org/10.24140/2020-sct-vol.4.

Menezes, Marluci, and Sara Silva. 2023. O Papel da Memória na Recuperação (Imaginária) da Paisagem Fluvial Urbana. In *Rios Urbanos na Ibero-América: Casos, Contextos e Experiências*, edited by Carlos Smaniotto Costa, Marluci Menezes, Montserrat Pallares-Barbera, Gabriela Pastor, Eliana do Pilar Rocha, and Klinge O. Villalba Condori, 341–356. Lisbon: Lusofona University Press. https://doi.org/10.60543/ecati/em3y-zc66.

Menezes, Marluci. 2019. Caminhar como contributo metodológico do cocriar cidade. CES Contexto – Debates 24: 12–19. Universidade de Coimbra. https://www.ces.uc.pt/publicacoes/cescontexto/ficheiros/cescontexto_debates_xxiv.pdf.

Middleton, Jennie. 2011. Walking in the City: The Geographies of Everyday Pedestrian Practices. Geography Compass 5 (2): 90–105. Blackwell Publishing Ltd. https://doi.org/10.1111/j.1749-8198.2010.00409.x

Ministry of Environment. 2004. Law for the Promotion of Nature Restoration, last modified December 9th. https://www.env.go.jp/content/900452866.pdf

Ministry of Land, Infrastructure, Transport and Tourism. 2012. Transition of policy for river environment in Japan. https://www.mlit.go.jp/river/basic_info/english/pdf/conf_06-1.pdf

Pinto, Pedro J., and Mathias Kondolf, G. 2015. Rivers in Urban History: Tales of Fear, Harmony, Destruction, and Opportunity. In Proceedings of 2nd International Conference on Integrative and Sustainable Rivers, Lyon, France, June 2015.

Sælensminde, Kjartan. 2004. Cost–Benefit Analyses of Walking and Cycling Track Networks Taking into Account Insecurity, Health Effects and External Costs of Motorized Traffic. Transportation Research Part A: Policy and Practice.

Sandeville, Euler. 2010. A paisagem do município como território educativo. In Município que Educa: Fundamentos e Propostas, edited by Paulo R. Padilha, Sheila Ceccon, and Priscila Ramalho. São Paulo: Instituto Paulo Freire.

Sennett, Richard. 1990. The Conscience of the Eye: The Design and Social Life of Cities. London: Faber and Faber.

Song, Shuang, Shuai Wang, Bojie Fu, Yuanyuan Li, Yanhua Zeng, and Zhenyao Shen. 2021. Improving Representation of Collective Memory in Socio-Hydrological Models and New Insights into Flood Risk Management. Journal of Flood Risk Management 14 (1): e12679. https://doi.org/10.1111/jfr3.12679.

Suguimoto, Flávia Tiemi. 2014. Paisagens lúdicas. PhD Thesis, Universidade de São Paulo. https://www.teses.usp.br/teses/disponiveis/16/16135/tde-19122014-171507/pt-br.php.

Tolley, Rodney, Les Lumsdon, and Karen Bickerstaff. 2001. The Future of Walking in Europe: A Delphi Project to Identify Expert Opinion on Future Walking Scenarios. Transport Policy 8 (4): 307–315. https://doi.org/10.1016/S0967-070X(01)00026-9.

Tuan, Yi-Fu. 1990. Topophilia: A Study of Environmental Perception, Attitudes, and Values. New York: Columbia University Press.

Tvedt, Terje, and Terje Oestigaard, eds. 2016. A History of Water. Series III, Volume 3: Water and Food from Hunter-Gatherers to Global Production in Africa. London: I.B. Tauris.

Tvedt, Terje. 2021. Water and Society: Changing Perceptions of Societal and Historical Development. London: Bloomsbury Publishing.

UNESCO. 2023. River Culture – Life as a Dance to the Rhythm of the Waters. Paris: UNESCO Publishing. https://unesdoc.unesco.org/ark:/48223/pf0000382775.

United Nations. 2018. Sustainable Development Goal 6: Synthesis Report on Water and Sanitation. New York: United Nations Publications. https://sustainabledevelopment.un.org/content/documents/19901SDG6_SR2018_web_3.pdf.

WEF – World Economic Forum. 2020. Global Risks Report. Geneva: World Economic Forum. https://www.weforum.org/publications/the-global-risks-report-2020/.

Yavuz, Iimak. 2017. An Urban Memory Lost in Amnesia: Riverscape of Ankara. In Proceedings of ICONARCH International Congress of Architecture and Planning (I), 303–315. Konya, Turkey, May 2017. https://iconarch.ktun.edu.tr/index.php/iconarch/article/view/184.

9

Collaborative Efforts in Urban River Monitoring

The Role of Digital Citizen Science

Joaquín Cochero,[1,*] *María Agostina Pecile,*[1]
Alejo Fabián Bonifacio,[2,*] *Milagros Bengolea,*[3]
Micaela Zambrano,[2] *Juan Manuel Brito,*[2]
María Mercedes Nicolosi Gelis,[1] *María Belén Sathicq,*[1]
Micaela Ailén Mujica,[1] *María Laura Ballesteros,*[2]
Andrea Hued,[2] *Silvana Halac,*[7] *Rocío Soledad Pazos,*[1]
Maricel Centi Ferrei,[8] *Marina Tagliaferro,*[2,4]
María Dalponte Ayastuy[5,6] and *Diego Torres*[5]

1. Introduction

As urbanization increases, rivers within city landscapes are subject to significant environmental pressures. These include pollution, habitat destruction, and the alteration of natural water flows, all of which impact the ecological health of urban waterways (DeNicola et al. 2015, Walsh et al. 2005). Traditional monitoring and conservation efforts, often led by government agencies or scientific institutions, are constrained by limited resources and capacity, leaving many urban rivers under-monitored and vulnerable to degradation. In these environments, "citizen science" has grown substantially in the past decades. This approach, which is referred to with many other terms, including "participatory science" or "community science" to define a broader profile of participants (Haklay et al. 2021, Cooper et al. 2021), leverages public participation in scientific research, and thanks to digital technologies is now more accessible than ever. The proliferation of smartphones, social media platforms, and online databases plays a role in the democratization of science, allowing everyday

Authors' affiliation given in the end of the chapter.

DOI: 10.1201/9781032711737-9

citizens to contribute to meaningful research (Conrad and Hilchey 2011, Bonney et al. 2014).

Citizen science has proven especially relevant in urban river conservation, where collective action is essential for sustainable outcomes. By engaging local communities, digital citizen science projects can help monitor water quality, document biodiversity, identify pollution sources, and track environmental changes over time (Silvertown 2009, Hecker et al. 2018a, Allen et al. 2019). The role of digital citizen science in urban river conservation is twofold. First, it enhances the volume and quality of data available to researchers, helping to fill gaps in traditional monitoring programs (Buytaert et al. 2014). Second, it deepens community engagement, fostering a greater understanding of socio-environmental challenges while empowering citizens to become advocates for their local ecosystems. Incorporating the local knowledge of people living near rivers adds diverse perspectives to the knowledge-building process, allowing for a more holistic approach to managing and conserving these ecosystems.

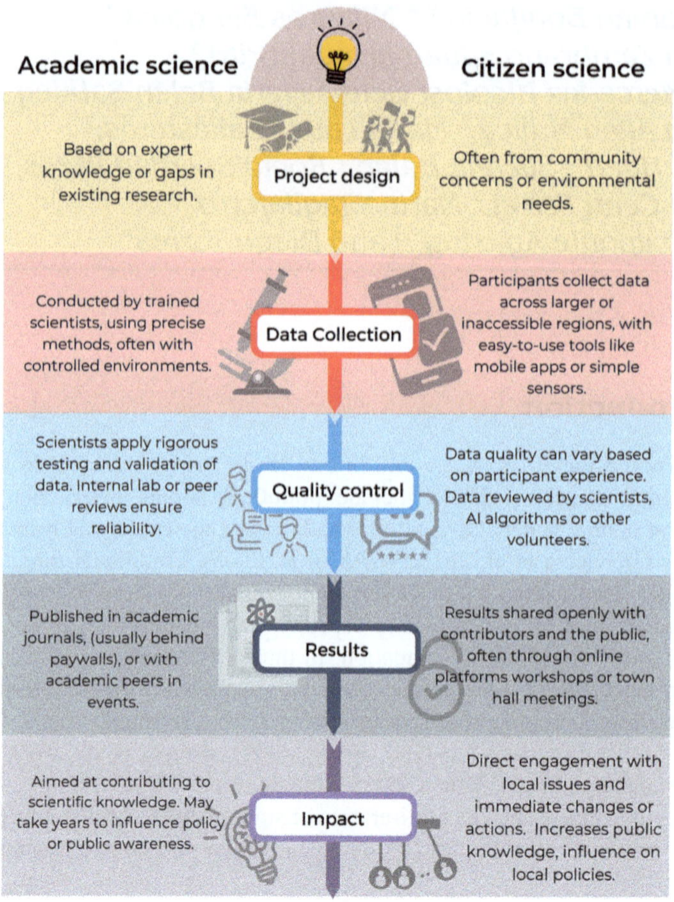

Figure 1: Summary flow chart comparing the research conducted by traditional scientific academies (universities, research centers, etc.) and research initiated by community or citizen science.

In these collaborative efforts, participants can actively take part in any or all of the stages of organized scientific projects (Bonney *et al.* 2009), which distinguishes them from traditional scientific projects led exclusively by academia (Figure 1). Starting with data collection, citizen scientists use tools such as smartphones and monitoring kits to gather information on water quality, pollution sources, and biodiversity. This data is then commonly uploaded to an online platform, where it is aggregated and organized in a centralized database. Researchers, policymakers, and community members can access this data in real-time, facilitating collaborative analysis and discussion. These insights inform decision-making processes, guiding urban planners, environmental agencies, and advocacy groups in formulating conservation strategies and policies (Buytaert et al. 2014).

Citizen science promotes collaborative monitoring where local governments, NGOs, and academic institutions partner with citizens. By integrating diverse stakeholders, projects can harness broader expertise, while fostering greater transparency and public trust in the data (Conrad and Hilchey 2011).

2. The Digital Revolution in Citizen Science

With the rise of digital mobile technologies, science has become increasingly accessible to the general public. This shift, often referred to as the "democratization of science" allows citizens to contribute to meaningful research in real-time, even without formal scientific training (Haklay 2013). Smartphones equipped with GPS, cameras, and various sensors, combined with user-friendly apps, help people collect and submit environmental data daily. Social media platforms and online forums have further amplified the reach of these efforts by providing spaces for participants to share findings, engage in discussions, and collaborate on conservation initiatives across geographical boundaries. Moreover, cloud-based databases and interactive platforms make it easy for citizen-collected data to be uploaded, stored, and analyzed by scientists and policymakers (Wiggins and Crowston 2011). These tools let people share information instantly, making it easier to spot environmental problems as they happen - like pollution discharges, algae outbreaks, or dead fish in the river. Urban environments present unique challenges for ecological monitoring due to their complex landscapes and rapid changes driven by human activities. But they also provide opportunities for citizen science initiatives to thrive; since urban rivers are more accessible to the public compared to remote landscapes, internet access is more readily available, and the larger populations in cities offer a greater pool of potential participants (Newman et al. 2012). With more people concentrated in urban areas, it also becomes easier to mobilize a diverse group of volunteers for environmental monitoring, and the urban setting provides better infrastructure for outreach activities and publicity efforts, allowing citizen science projects to engage with local communities more effectively through social media, public events, and collaborations with local organizations, schools, and media outlets (Conrad and Hilchey 2011). One major advantage of citizen science is the ability to monitor localized events that may be missed by periodic, centralized efforts. For example, citizen scientists can document acute pollution events, such as sewage overflows after heavy rains, or spot sudden changes in biodiversity due to invasive species, all in real-time (Buytaert et al. 2014).

Innovative monitoring techniques have also emerged from citizen science. Projects like FreshWater Watch and Drinkable Rivers (Table 1) utilize low-cost testing kits distributed to participants to measure parameters such as water turbidity, pH levels, and nutrient content. Open hardware tools are also being used increasingly, allowing participants to build and customize their own sensors, and entire communities have formed around the development of open tools (e.g., Gathering for Open Science Hardware, or GOSH). These data can provide a more comprehensive picture of water quality across an urban watershed, revealing patterns that might go unnoticed by a single research team with limited sampling points. Another critical contribution is the ability of citizen science to capture fine-scale, place-based data over time. Volunteers regularly record data at specific points within their communities, building detailed records of changes in water quality, habitat conditions, and species composition. These longitudinal datasets provide critical insights into the seasonal and annual variability of urban river systems, supporting both short-term management efforts and long-term conservation planning. A common concern about citizen science, particularly in environmental monitoring, centers around the quality and reliability of data collected by non-experts. In urban river conservation, ensuring data accuracy is critical for informed decision-making and long-term management. However, recent advancements in digital tools, coupled with strategic project design, have significantly improved the rigor of citizen-collected data, making it increasingly comparable to data gathered by professional scientists (Kosmala et al. 2016).

Initiatives such as FreshWater Watch and PreserVamos (Table 1) have active training programs that provide participants with the skills and knowledge needed to perform consistent and accurate data collection, while Thames21 has a series of "*waterblitz*" events, where participants use water samplers and kits for ecology surveys to monitor specific habitats in a few hours (Whitelaw et al. 2003). Drinkable Rivers has an onboarding period where participants receive online training and weekly assignments. In general, workshops, online tutorials, and field guides ensure that volunteers understand the sampling protocols, how to use monitoring equipment, and the importance of adhering to standardized procedures. Another key strategy to improve data quality is validation through expert review. In many projects, citizen science data is regularly cross-checked by professional scientists or automated systems. This process involves flagging outliers, verifying unusual data points, and using algorithmic filters to catch inconsistencies (Aceves-Bueno et al. 2017). In some cases, experts even participate directly in the data collection, providing ongoing feedback and support to volunteers, which further enhances the quality of the data.

Gamification and peer review mechanisms have also been implemented in some citizen science projects to boost engagement and data reliability. Creating a reward system (badges, points, leaderboards, etc) encourages participants to follow best practices and improve their data contributions. Peer review allows volunteers to cross-check each other's observations, fostering a sense of accountability and improving overall data integrity (Bonney et al. 2014).

Finally, data aggregation and statistical analysis can help account for individual inaccuracies. By pooling data from large numbers of participants and applying statistical methods to assess trends and averages, projects can smooth out occasional errors and reveal the larger patterns needed for urban river management (Dickinson et al. 2012).

3. Some Case Studies of Urban River Citizen Science Initiatives

Numerous citizen science initiatives focus on urban rivers ranging in scale from small, community-driven efforts to large, international collaborations (a non-exhaustive list in Table 1). Although each project is tailored to offer unique objectives that respond to local or regional interests, there are some common strategies for initiative designing, tool development and training, incorporating participants and results sharing. Some center on hands-on activities like water quality testing and biodiversity assessments, while others leverage digital platforms to gather and share data in real-time. Common objectives include monitoring water quality, tracking the impacts of restoration efforts, supporting educational initiatives, and advancing advocacy work. Most of these initiatives are motivated by an academic team, usually linked to a university or scientific institution. However, the community participating in the project can engage from the beginning or in specific activities throughout its development.

Participation in citizen science projects can vary significantly. In contributory projects, participants primarily focus on data collection, contributing observations

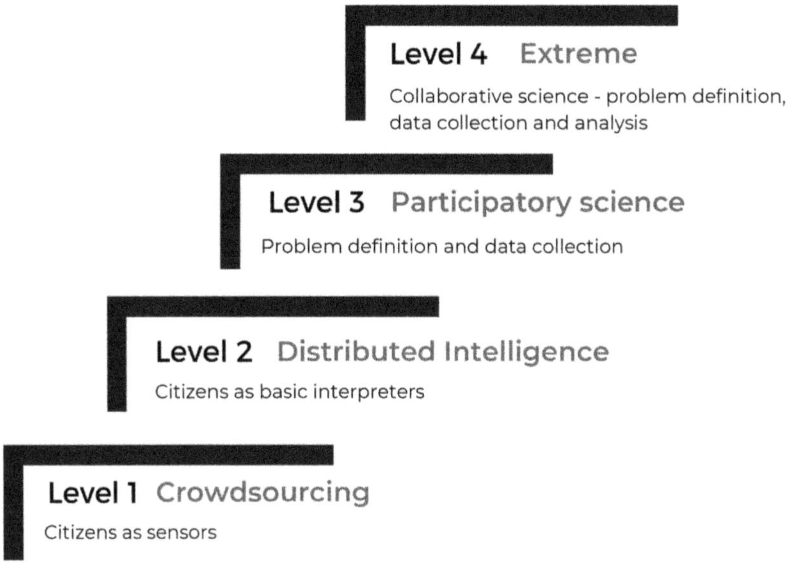

Figure 2: Classification of citizen science initiatives based on the degree of participation by community members. Source: modified from Hacklay, 2013.

or measurements without further involvement. Collaborative projects expand the role of the public, involving them not only in data collection but also in aspects like data analysis, developing methodologies, and shaping objectives. Co-created projects take this a step further, encouraging participants to engage in all stages of the process, from project design to decision-making and implementation. A different categorization is based on the levels of participation (Haklay 2013) and is shown in Figure 2.

Ensuring data accuracy is a recurring concern across these initiatives, especially when working with non-expert participants in dynamic urban environments. Many of the projects listed in Table 1 have addressed this through a combination of standardized protocols, participant training, and expert validation workflows. For example, FreshWater Watch and Drinkable Rivers provide clear step-by-step instructions and calibration procedures alongside their kits, while PreserVamos includes photo-based verification, enabling experts to review reported observations. Some initiatives, like Riverfly Monitoring, rely on biological indicators such as macroinvertebrates, which are cross-validated by trained volunteers or reviewed during joint sampling campaigns. Others, like Thames21, conduct periodic group "blitz" events were multiple teams sample simultaneously under supervision, reducing variability. These strategies not only improve the reliability and reproducibility of the data, but also strengthen community confidence in their own contributions, reinforcing the value of collective monitoring.

3.1 Freshwater Watch

FreshWater Watch is a global citizen science initiative focused on monitoring freshwater ecosystems, including urban rivers. Launched by the Earthwatch Institute, this project empowers volunteers worldwide to collect data on water quality, contributing to a comprehensive global database. The project exemplifies how citizen science can be effectively scaled to address environmental challenges at both local and global levels. In the context of urban river conservation, FreshWater Watch offers participants simple yet scientifically robust tools to assess critical indicators of water health, such as nutrient levels (nitrogen and phosphorus), turbidity, and the presence of algal blooms. Participants are provided with training workshops or online tutorials that teach them how to use the testing kits accurately and record their observations using a mobile app.

One of the standout features of FreshWater Watch is its subscription-based model, which allows volunteers to receive testing kits regularly. This structure ensures that participants remain engaged over the long term, allowing for the collection of temporal data that tracks changes in water quality over months and years. The continuous nature of data collection is particularly valuable in urban river contexts, where water quality can fluctuate dramatically due to stormwater runoff, industrial discharges, and other anthropogenic factors.

Data collected by FreshWater Watch participants is uploaded to a centralized platform, making it accessible to researchers, policymakers, and other citizen scientists. This data-sharing model encourages collaborative analysis, where volunteers can compare their findings with those of other regions, helping to identify

Table 1: Urban river citizen science projects in different countries, their main objectives, the tasks and participant interaction, products and results, the approximate number of participants and their website.

Project Name	Country	Main Objective	Year Started	Participant Interaction & Tasks	Products/Results	Approx. Number of Participants	Website
FreshWater Watch	Global	Monitor freshwater health and engage communities in conservation.	2012	Participants use **water testing kits** to monitor river and stream health by measuring nutrient levels and water quality.	Global datasets on water quality; educational materials	~30,000 globally	FreshWater Watch
Drinkable Rivers	Netherlands & Belgium	Assess water quality for drinkability and promote public awareness.	2017	Volunteers test water quality for drinkability using simple physical and chemical tests, and submit results via a mobile app.	Interactive maps of water quality; public reports	~1,000	Drinkable Rivers
PreserVamos	Argentina	Establish community-driven monitoring of urban river habitats.	2018	Local communities are trained to monitor urban rivers, reporting river habitat characteristics via a mobile app. Water quality kits are also provided to specific stakeholders.	Monitoring reports; open access maps; community engagement initiatives	~1,000	PreserVamos
Riverwatch	USA	Monitor and improve river health through citizen engagement.	2005	Citizen scientists monitor water quality, track invasive species, and document river health using field kits and data platforms.	Data-driven policy recommendations; annual reports	~5,000	Riverwatch
Thames21	UK	Improve waterway quality and community involvement in river health.	2000	Volunteers engage in litter collection, habitat restoration, and water quality testing, contributing to the restoration of urban waterways.	Habitat restoration projects; annual clean-up events	~7,000	Thames21
Streamkeepers	Canada	Enhance stream health through community monitoring and education.	1999	Citizens perform stream surveys, water quality assessments, and help with habitat restoration in urban watersheds.	Training materials; water quality databases	~2,000	Streamkeepers

Table 1 cont. ...

...Table 1 cont.

Project Name	Country	Main Objective	Year Started	Participant Interaction & Tasks	Products/Results	Approx. Number of Participants	Website
Bristol Avon Rivers Trust (BART)	UK	Restore and enhance the river ecosystem through community engagement.	2010	Volunteers monitor the **water quality** of urban rivers and help in **invasive species removal** and **habitat conservation** efforts.	Habitat improvement reports; community education	~1,200	BART
Waterkeepers	Global	Protect water bodies through community monitoring and advocacy.	1999	Local communities monitor rivers for pollution, engage in advocacy efforts, and participate in clean-up activities.	Legal actions against polluters; educational resources	~45,000 globally	Waterkeeper Alliance
Stream Health monitoring	New Zealand	Enhance urban stream health through citizen monitoring.	2011	Volunteers help with **water quality monitoring**, **stream surveys**, and participate in **stream restoration** activities.	Published water quality data; public restoration events	~3,000	Stream Health Monitoring
Riverfly Monitoring Initiative	UK	Monitor river invertebrates as an indicator of water quality.	2007	Volunteers monitor the health of rivers through **invertebrate sampling** and **water quality tests** to detect pollution.	Annual reports; pollution alerts	~1,000	Riverfly Monitoring
CoCoRaHS - Urban Rivers	USA	Monitor rainfall and its effects on urban river flooding.	1998	Participants collect **rainfall data** and report on **flooding events** that impact urban rivers, contributing to hydrological models.	Flood prediction models; historical rainfall data	~25,000 globally	CoCoRaHS
CoAct / ¿Qué pasa Riachuelo?	Argentina	Environmental justice and governance of a highly urbanized watershed	2020	Co-design of a platform which includes environmental and social issues through workshopes.	Online platform with community-built maps	~500	Que Pasa Riachuelo

patterns or anomalies that might indicate broader environmental issues. The project's reach across urban rivers in different cities provides a powerful tool for comparing how urbanization impacts freshwater ecosystems globally.

3.2 Drinkable Rivers

Drinkable Rivers is a global citizen science project involving 65 organizations across 22 countries. It engages people in monitoring water quality using hands-on methodologies and accessible tools. Participants use simple kits to test things like nutrients in the water, how clear or cloudy it is, its acidity, and bacteria like *Escherichia coli*, which provide insights into the overall health of local rivers. The data collected by citizens is then validated by scientists, creating a collective open-access database to track progress toward restoring drinkable rivers.

Community participation is central to these projects, facilitated through local hubs including schools, visitor centers, environmental groups, and water agencies. These hubs not only distribute monitoring kits and provide instructional materials but also organize events and regular measurement sessions. By targeting diverse groups, including children and local leaders, the program fosters long-term community stewardship. Additionally, the Drinkable Rivers movement builds awareness through initiatives like river walks, which highlight the cultural and ecological significance of rivers, and networks such as "Mayors for Drinkable Rivers," which encourage municipalities to take coordinated action. These global and local collaborations help create a shared responsibility for water quality, transcending geographical and political boundaries. This holistic approach - combining scientific monitoring, community involvement, and advocacy - enables widespread participation in the preservation and restoration of river systems, driving a global movement toward healthier watersheds.

3.3 PreserVamos

The PreserVamos project, initiated by the United Nations Development Programme (UNDP) in Argentina along with a pre-existing citizen science initiative called AppEAR (Cochero 2018), is an example of how citizen science can be integrated into urban river conservation efforts. In this initiative, participants can monitor freshwater habitats, particularly in urban environments, by answering questionnaires in a mobile application for Android® devices. All data collected is macroscopic and can be recognized without any previous training, and each survey is accompanied by photos that help the specialists with the data validation process. It specifically focused on monitoring the ecological health of urban rivers in four municipalities in Argentina (San Antonio de Areco, Balcarce, and Mercedes in the Buenos Aires province, and La Granja in the Córdoba province). The project aimed to empower local communities to participate in environmental monitoring while exploring different strategies for engaging the public and local governments in long-term river management. Throughout the PreserVamos project, two different engagement strategies were tested: top-down and bottom-up approaches. Each approach aimed to increase local participation, but the methods of implementation and their effectiveness varied significantly. In San Antonio de Areco, Balcarce and Mercedes, a top-down strategy was tested for a year, while in La Granja a bottom-up strategy was used.

Top-Down Approach: The top-down strategy relies on leadership and coordination from local governmental and institutional actors. In this model, local environmental agencies and municipal authorities, along with the academic team of the project belonging to two universities, served as the primary drivers of the initiative. The academic team proposes the monitoring framework, defines the project's main objectives, and designs the tools that are used by participants, including promotional and communication materials. The local environmental agencies are in charge of the recruitment, by reaching out to the public through official channels, such as environmental education programs and local councils (Figure 3., red triangle). In this approach: a) The academic team designs the main aspects of the project, including tool design and organizational structure, and is the primary decision maker; b) The local government also acts as the facilitator and coordinator of the project; c) Participants are primarily involved in data collection and d) Monitoring efforts are aligned with municipal environmental goals and national conservation policies (Figure 3, top right triangle).

In PreserVamos, the top-down approach showed several advantages, particularly in terms of resource allocation and organizational structure. Government leadership provided a clear framework for participants to follow, ensuring that data collection was standardized and aligned with environmental policy goals. Additionally, the involvement of local authorities helped increase the visibility of the project, securing the tools needed for conducting monitoring efforts and facilitating the communication channels with key stakeholders in the territory. However, this approach also had limitations. Community engagement was often limited to predefined tasks, with less opportunity for active participation in decision-making. This occasionally led to a sense of detachment among participants, as their contributions felt constrained by the structure imposed by the authorities. Moreover, public interest waned after the initial efforts, as citizens felt like passive contributors rather than active agents in the project.

Bottom-Up Approach: Conversely, the bottom-up strategy focuses on community engagement and public campaigns of involvement. In this approach, the project is designed by community members and neighbors aiming to address a specific territorial need, and emphasizes direct outreach to environmental groups, schools, and neighborhood organizations. Community involvement is encouraged through public workshops, events, and media campaigns to raise awareness about the importance of river health. The role of academic sectors and environmental agencies in these initiatives is usually as technical resources, but they are approached by community-driven projects as support and do not usually take part in the decision process (Figure 3, bottom left triangle).

In this bottom-up approach: a) Local communities initiate the project based in a territorial environmental issue and participate in every stage, from defining monitoring objectives to interpreting data and proposing actions; b) The project heavily relies on informal networks, community leaders, and volunteers to organize activities; c)

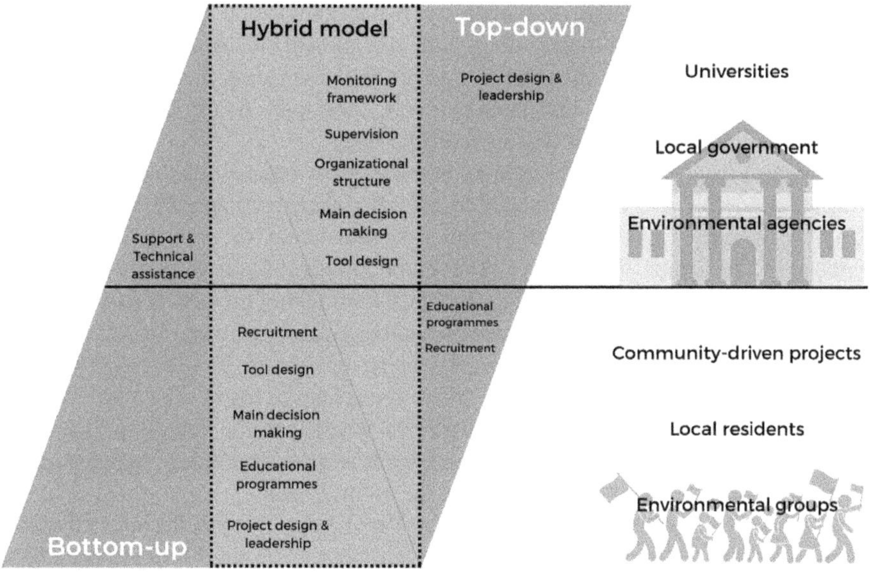

Figure 3: Bottom-up (bottom left triangle) and Top-down (top right triangle) models considered in strategy design for citizen science projects. The Hybrid model (center box) considers elements of both strategies, where the initiatives are citizen-initiated and academic institutions provide the organizational structure and mentoring.

Educational workshops are used to train community members in scientific methods and water quality assessment, increasing local ownership of the project; d) Media campaigns and social media play a crucial role in mobilizing citizens and spreading awareness about urban river issues.

The bottom-up approach succeeds in fostering a stronger sense of ownership and engagement among participants. By involving local communities from the outset, the project cultivates deep-rooted interest in river conservation and encourages citizens to take on leadership roles within their neighborhoods. Participants quickly took ownership of the project, expanding its reach to other areas and creating more dynamic, adaptive solutions as they became personally invested in the outcomes. However, the bottom-up model faced challenges in coordination and sustainability. Without strong institutional support, maintaining long-term commitment and ensuring consistent data collection was often difficult. This model thrives particularly when there is strong community interest with local citizens fully embracing and taking ownership of the project. In La Granja (Córdoba), for instance, the initiative faced setbacks in the year following local elections, as changes in government disrupted continuity, making it challenging to sustain the effort over a longer period.

A Hybrid Model - The Reciprocal Model: The “reciprocal citizen science model”, introduced by the Institute of Transdisciplinary Discoveries (ITD, University of Pécs, Hungary) at the “International Transdisciplinarity Conference” (Sík et al. 2021), combines the advantages of top-down and bottom-up approaches. It is based on citizen-initiated research ideas and is citizen-led; universities or scientific institutions provide support, especially in methodology and equipment. Briefly, the projects are citizen-initiated along with recruitment, the academic institutions provide the organizational structure and design the monitoring framework, but both at the community and the institutional levels participate in the decision-making process and tool development (Figure 3, center box).

It differs from traditional bottom-up approaches by co-creating the project along with the academic and environmental institutions; their participation and scientific coaching increase the organization and systematization of data, which can be further used for scientific purposes. The methodological accuracy in this approach is enhanced by scientific networking, the decision-making process is shared and if the citizens need it, the institution can provide equipment as well. To achieve this, a comprehensive *mentoring program* has to be developed for the initiative; but differently from a top-down approach where the scientific institutions train citizens in data collection and other activities, participants get specific mentoring according to their needs. After an initial assessment, similar to a placement test, the mentoring program's organizers decide what training the mentee needs.

One of the key takeaways from the PreserVamos project is that the most effective approach is this "hybrid model" - where government institutions provide resources and structure while local communities lead locally organized activities. It can leverage the organizational strength and resources of the top-down approach, ensuring that participants are well-equipped, and that data collection meets scientific standards, while harnessing enthusiasm and localized knowledge of the bottom-up approach that allows for more community-driven solutions and long-term engagement. It was also noteworthy that an enhanced collaboration between municipal governments, academic institutions, and local groups, can create a shared sense of responsibility for urban river health.

The effectiveness of the hybrid model varies depending on the local governance structure, community capacity, and institutional culture. In municipalities with well-established environmental departments and prior experience in participatory governance, such as San Antonio de Areco or La Granja, the model has shown to thrive scientific institutions provide structure and tools, while empowered local communities drive monitoring and advocacy. In contrast, in territories where institutional fragmentation or political turnover are common, the hybrid approach faces greater challenges. For example, in Mercedes, although strong initial community engagement catalyzed participation, changes in local leadership disrupted coordination, underscoring the importance of long-term institutional anchoring. Similarly, in contexts with limited digital access or low scientific literacy, additional investment in training and facilitation is required to bridge the gap between academic expectations and community needs.

4. The Impact of Citizen Science on Policy and Conservation

The collaborative nature of citizen science fosters a sense of ownership and accountability among participants, which can lead to increased advocacy for environmental protection and sustainable practices. Furthermore, the integration of citizen-generated data into policy discussions not only enhances the scientific basis for decisions but also ensures that local knowledge and community concerns are recognized and addressed (Conrad and Hilchey 2011, Fischer 2000).

Using citizen science data in policy development can enhance transparency and public trust. By involving community members in data collection and analysis, policymakers can demonstrate a commitment to participatory governance. This collaborative approach encourages stakeholders to engage in the policymaking process, resulting in more informed and inclusive environmental policies (Schade et al. 2021, Hecker et al. 2018b).

4.1 Data-Driven Decision Making

One of the key strengths of citizen science lies in its ability to generate large volumes of data, often over extended periods, that can inform policy and guide conservation efforts (Fraisl et al. 2022). By involving communities in collecting environmental information, these projects offer timely and local data that helps manage urban rivers more effectively and flexibly. Decision-makers can use these data to track pollution trends, assess water quality, and evaluate the effectiveness of restoration efforts. Beyond informing environmental regulations, citizen science data can directly contribute to urban planning processes, particularly in contexts where rivers intersect with land use, infrastructure, or public health. For instance, mapping of flood-prone areas or informal waste dumping sites by citizens can guide the placement of green infrastructure, buffer zones, or waste management improvements. In Latin America, participatory data has been used to justify environmental education programs in schools near degraded waterways, or to condition urban development permits on ecological restoration actions. When integrated into municipal Geographic Information Systems (GIS) or urban observatories, these datasets help identify environmental inequities and prioritize interventions in underserved neighborhoods.

In many cases, data collected by citizen scientists has prompted direct policy interventions. Citizen-collected data on urban river pollution in the UK has been used by the Environment Agency to enforce water quality standards, leading to stricter regulations and pollution control measures. Similarly, in the U.S., the FreshWater Watch project has contributed to local watershed management strategies by identifying areas with nutrient runoff issues that required targeted mitigation efforts. The PreserVamos initiative has helped define new channels of communication between citizens and the local environmental authorities, along with the definition of a "green alert" system to integrate public observation in alerting the municipalities of sudden environmental issues.

4.2 Stakeholder Coordination

Effective coordination among stakeholders can also be facilitated by digital frameworks that centralize data sharing and communication. Collaborative dashboards allow scientists, policymakers, and citizens to jointly access and interpret monitoring data. These platforms often include role-based access, communication channels for feedback, and visualization tools that support decision-making. Frameworks like the ones proposed by CoAct or EU-Citizen. Science also emphasize ethical governance structures, ensuring that local voices are represented in environmental planning discussions.

While digital tools can significantly enhance coordination, many successful citizen science initiatives rely on face-to-face communication and locally embedded structures to ensure effective stakeholder engagement. In projects without a strong digital component, coordination often occurs through regular community meetings, participatory workshops, and multi-stakeholder roundtables, where citizens, scientists, and policymakers co-develop goals, monitoring protocols, and action plans. These in-person interactions foster trust, mutual understanding, and social cohesion, which are critical for sustained collaboration. For instance, river monitoring groups like CoAct or PreserVamos may hold seasonal assemblies to share findings, discuss pressing environmental issues, and align their actions with municipal planning processes. In contexts where digital access is limited, such as rural or underserved urban areas, local schools, community centers, or NGOs often serve as communication hubs, disseminating materials, organizing training, and acting as intermediaries between citizens and institutions. The presence of community leaders or "environmental ambassadors", as included in many initiatives, plays a crucial role in mobilizing participation and maintaining continuity, especially when institutional support fluctuates.

4.3 The Role of Education and Community Outreach

In projects like PreserVamos and FreshWater Watch, training programs are designed to teach participants how to use scientific tools, such as water testing kits, mobile apps, and online platforms for data submission. Workshops, webinars, and community events also offer a space for citizens to learn about water quality, biodiversity, and the broader ecological impacts of urban river degradation. In these initiatives, education and community outreach proved essential not only for engaging participants in the short term but also for ensuring the long-term sustainability of citizen science projects. Also, the training provided to participants extends beyond the technical aspects of data collection. It includes broader lessons on the ecological importance of urban rivers, the socio-environmental challenges they face, and the significance of continuous monitoring. This comprehensive educational approach helps participants understand the broader context of their work, motivating them to remain involved over time. For example in PreserVamos, participants who received in-depth training and participated in the field activities were more likely to take ownership of the project, organize local outreach activities, and recruit new participants.

5. Key Recommendations for Future Citizen Science Initiatives in Urban Rivers

Urban rivers present a dynamic and accessible environment for citizen science initiatives, but the success and sustainability of these projects depend on strategic planning, clear objectives, and effective stakeholder engagement. Based on the sections above, we summarize several key recommendations for designing future citizen science initiatives.

5.1 Decide Upon Clear Objectives

One of the most important steps in ensuring the success of a citizen science initiative is setting clear and achievable objectives. Defining what the project aims to accomplish from the outset not only improves the organization but also allows for the effective measurement of its impact. These goals should be specific, realistic, and in line with both the conservation needs of urban rivers and the priorities of the watershed's inhabitants.

In addition to setting objectives, it is crucial to engage key stakeholders from the beginning. These stakeholders may include local governments, NGOs, universities, research institutions, local communities, and civilian organizations. Their integration provides diverse perspectives and resources while fostering long-term commitment and collaboration. By securing the support and active involvement of these actors, potential obstacles—such as a lack of technical resources or resistance to environmental policies—can be overcome more easily.

5.2 Identifying Key Actors in Local Communities

To secure long-term commitment from local communities, it is essential to identify and engage key actors within these communities. Certain individuals often play pivotal roles as leaders in their immediate environments, influencing and mobilizing others. These key actors should be aligned with the objectives of the citizen science project to ensure sustained engagement over time. Moreover, it is crucial to consider the "contagion effect" these individuals can have, as their active participation and enthusiasm may inspire and motivate others within the community to join and remain involved in the initiative.

5.3 Select Adequate and Inclusive Tools, While Considering Ethical and Legal Matters

The tools used in citizen collaboration should be accessible to participants, easy to use, and capable of providing accurate, reliable data. Depending on the project's objectives, these tools can range from simple water testing kits to mobile applications for data recording, drones for aerial monitoring, or sensors for continuous data collection. Importantly, the tools selected should align with the skill levels of participants and the technical requirements of the project. When selecting tools, it is also crucial to consider how the data will be collected, stored, and shared. This includes addressing legal and ethical issues that may arise during the process, particularly in urban settings where there may be privacy concerns or restrictions on data sharing. Data privacy and ownership are also important when dealing with environmental monitoring and the participation of citizens. Participants should be informed about how their data will be used, who will have access to it, and what the long-term plans for the data are. Transparency in data management helps build trust and encourages continued engagement from the community.

5.4 Create a Balance Between Top-Down and Bottom-Up Approaches in a Hybrid Model

Both top-down and bottom-up strategies have their merits in citizen science projects, but finding the right balance between these approaches is critical for urban river initiatives. A co-created experience can lead to an effective hybrid model between top-down and bottom-up strategies to maximize the sustainability of the project.

5.5 Evaluate and Adapt

Regularly reviewing the data collection methods, participant engagement, and overall project goals ensures that the initiative remains relevant and effective. In most scenarios, particularly in top-down approaches, there is a dissonance between what the original expectations of the academic team have and what the community needs and there must be room for compromises. Flexibility is key—adapting the project to respond to emerging environmental challenges or incorporating feedback from participants can enhance both scientific outcomes and community involvement.

Developing a system to measure and share the project's impact, both in terms of scientific data and social engagement, can further motivate participants and demonstrate the value of the initiative to policymakers, stakeholders and the academic community. Demonstrating clear, measurable results will help address potential skepticism from researchers by highlighting the scientific rigor and value of citizen science contributions. And incorporating success stories and lessons learned into outreach efforts can help refine methodologies and scale up citizen science projects to address broader environmental issues across urban rivers.

Acknowledgements

We extend our heartfelt gratitude to everyone involved in the various citizen science initiatives that made these projects possible. Particularly linked to the PreserVamos initiative, we are especially grateful to Lorena Moscovich, Maria Verónica Moreno, and Fernanda Aun Castells from Co_Lab PNUD Argentina, and to Guillermina D'Onofrio from the Ministerio de Ciencia, Tecnología e Innovación (MinCyT) in Argentina, who made this project possible.

This publication is based upon work of the Project RUN Citizen science and co-creation for regeneration of urban rivers and mitigation of risks, which has received funding from the Programa Iberoamericano de Ciencia y Tecnología para el Desarrollo CYTED, under grant agreement No 420RT0008, https://www.cyted.org/run-rios-urbanos-naturalizados.

References

Aceves-Bueno, E., Adeyemi, A.S., Bradley, D., Brandt, W.T., Callery, P., Feraud, M. et al. 2015. Citizen Science as an Approach for Overcoming Insufficient Monitoring and Inadequate Stakeholder Buy-in Adaptive Management: Criteria and Evidence. Ecosystems 18: 493–506. https://doi.org/10.1007/s10021-015-9842-4.

Allen, D. C., Kopp, D. A., Costigan, K. H., Datry, T., Hugueny, B., Turner, D. S. et al. 2019. Citizen scientists document long-term streamflow declines in intermittent rivers of the desert southwest, USA. Freshwater Science, 38(2): 244–256. https://doi.org/10.1086/701483.

Andrachuk, M., Marschke, M., Hings, C., and Armitage, D. 2019. Smartphone technologies supporting community-based environmental monitoring and implementation: a systematic scoping review. Biological Conservation 237: 430–442. Elsevier Ltd. https://doi.org/10.1016/j.biocon.2019.07.026.

Bonney, R., Ballard, H., Jordan, R., McCallie, E., Phillips, T., Shirk, J. et al. 2009. Public participation in scientific research: Defining the field and assessing its potential for informal science education. A CAISE Inquiry Group Report. Washington, DC: Center for Advancement of Informal Science Education (CAISE).

Bonney, R., Shirk, J. L., Phillips, T. B., Wiggins, A., Ballard, H. L., Miller-Rushing, A. J. et al. 2014. Next Steps for Citizen Science. Science, 343(6178): 1436–1437. https://doi.org/10.1126/science.1251554.

Bonney, R., Phillips, T. B., Ballard, H. L., and Enck, J. W. (2016). Can citizen science enhance public understanding of science? Public Understanding of Science, 25(1): 2–16. https://doi.org/10.1177/0963662515607406.

Buytaert, W., Zulkafli, Z., Grainger, S., Acosta, L., Alemie, T. C., Bastiaensen, J. et al. 2014. Citizen science in hydrology and water resources: Opportunities for knowledge generation, ecosystem service management, and sustainable development. Frontiers in Earth Science, 2: 26. https://doi.org/10.3389/feart.2014.00026.

Cochero, J. 2018. AppEAR: a citizen science mobile app to map the habitat quality of continental waterbodies. Ecologia Austral, 28(02). https://doi.org/10.25260/EA.18.28.2.0.686.

Conrad, Cathy C., and Krista G. H. 2011. A Review of Citizen Science and Community-Based Environmental Monitoring: Issues and Opportunities. Environmental Monitoring and Assessment, 176(1): 273–291. https://doi.org/10.1007/s10661-010-1582-5.

Cooper, C. B., Hawn, C. L., Larson, L. R., Parrish, J. K., Bowser, G., Cavalier, D. et al. 2021. Inclusion in citizen science: The conundrum of rebranding. Science, 372(6549): 1386-1388. https://doi.org/10.1126/science.abi6487.

DeNicola, E., Aburizaiza, O. S., Siddique, A., Khwaja, H., and Carpenter, D. O. 2015. Climate change and water scarcity: The case of Saudi Arabia. Annals of Global Health, 81(3): 342–353. https://doi.org/10.1016/j.aogh.2015.08.005.

Dickinson, J. L., Zuckerberg, B., and Bonter, D. N. 2010. Citizen Science as an Ecological Research Tool: Challenges and Benefits. Annual Review of Ecology, Evolution, and Systematics, 41: 149–172. https://doi.org/10.1146/annurev-ecolsys-102209-144636.

Fischer, F. 2000. Citizens, experts, and the environment: The politics of local knowledge. Duke University Press. https://doi.org/10.2307/j.ctv11smwd9.

Fraisl, D., Hager, G., Bedessem, B., Gold, M., Hsing, P., Danielsen, F. et al. 2022. Citizen science in environmental and ecological sciences. Nature Reviews Methods Primers, 2: 64. https://doi.org/10.1038/s43586-022-00144-4.

Haklay, M. 2013. Citizen Science and Volunteered Geographic Information: Overview and Typology of Participation. *In:* Sui, D., Elwood, S., and Goodchild, M. [eds.]. Crowdsourcing Geographic Knowledge. Springer, Dordrecht. https://doi.org/10.1007/978-94-007-4587-2_7.

Haklay, M., Dörler, D., Heigl, F., Manzoni, M., Hecker, S., Vohland, K. et al. 2021. What is citizen science? The challenges of definition. Frontiers in Ecology and in the Environment 21(5): 244–250. https://doi.org/10.1002/fee.2635.

Hargitai, E. G., Sik, A., Samoczi, A., Hathazi, M., and Bogdán, C. 2022. Bottom-Up and Reciprocal Citizen Science: Untapped Resources of Novel Ideas. Preliminary Experiences of a Citizen Science as Public Engagement Program. Revista Lusófona de Estudos Culturais (RLEC)/Lusophone Journal of Cultural Studies (LJCS) Revista Lusófona de Estudos Culturais, 9(2): 6.

Hecker, S., Haklay, M., Bowser, A., Makuch, Z., Vogel, J., and Bonn, A. [eds.]. 2018a. Citizen Science: Innovation in Open Science, Society and Policy. UCL Press. http://www.jstor.org/stable/j.ctv550cf2.

Hecker, S., Bonney, R., Haklay, M., Bowser, A., Makuch, Z., Vogel, J. et al. 2018b. Innovation in citizen science – Perspectives on science-policy advances. Citizen Science: Theory and Practice, 3(1). https://doi.org/10.5334/cstp.114.

Kosmala, M., Wiggins, A., Swanson, A., and Simmons, B. 2016. Assessing data quality in citizen science. Frontiers in Ecology and the Environment, 14(10): 551–560. https://doi.org/10.1002/fee.1436.

Newman, G., Wiggins, A., Crall, A., Graham, E., Newman, S., and Crowston, K. 2012. The future of citizen science: Emerging technologies and shifting paradigms. Frontiers in Ecology and the Environment, 10(6): 298–304. https://doi.org/10.1890/110294.

Schade, S., Pelacho, M., van Noordwijk, T., Vohland, K., Hecker, S., and Manzoni, M. 2021. Citizen science and policy. *In:* Vohland, K., Land-Zandstra, A., Ceccaroni, L., Lemmens, R., Perelló, J., Ponti, M. et al. [eds.]. The science of citizen science (p. 529). Springer Nature.

Silvertown, J. 2009. A new dawn for citizen science. Trends in Ecology & Evolution, 24(9): 467–471. https://doi.org/10.1016/j.tree.2009.03.017.

Walsh, C. J., Roy, A. H., Feminella, J. W., Cottingham, P. D., Groffman, P. M., and Morgan, R. P. 2005. The urban stream syndrome: Current knowledge and the search for a cure. Journal of the North American Benthological Society, 24(3): 706–723. https://doi.org/10.1899/04-028.1.

Wiggins, A., and Crowston, K. 2011. From conservation to crowdsourcing: A typology of citizen science. Proceedings of the 44th Hawaii International Conference on System Sciences, 2011. https://doi.org/10.1109/HICSS.2011.207.

[1] Instituto de Limnología "Dr. Raúl A. Ringuelet", Consejo Nacional de Investigaciones Científicas y Técnicas (CONICET), Universidad Nacional de La Plata (UNLP). Boulevard 120 #1437. Argentina.
Emails: apecile@ilpla.edu.ar;mercedesnicolosi@ilpla.edu.ar;mbelen@ilpla.edu.ar;
micaelamujica@ilpla.edu.ar;rpazos@ilpla.edu.ar

[2] Instituto de Diversidad y Ecología Animal, Consejo Nacional de Investigaciones Científicas y Técnicas (CONICET), Facultad de Ciencias Exactas, Físicas y Naturales (UNC), Universidad Nacional de Córdoba (UNC). Av. Vélez Sarsfield 299, X5000 JJC, Córdoba. Argentina.
Emails: micaela.zambrano@unc.edu.ar,juan.brito@unc.edu.ar; mlballesteros@unc.edu.ar;
andrea.hued@unc.edu.ar; azulmarinita@gmail.com

[3] Investigación en Psicología y Ciencias del comportamiento - Instituto de Investigaciones Psicológicas (IIPsi), Consejo Nacional de Investigaciones Científicas y Técnicas (CONICET) Universidad Nacional de Córdoba (UNC). Boulevard de la Reforma y Enf. Gordillo Gómez, Facultad de Psicología . Córdoba, Argentina.
Email: mbengolea@mi.unc.edu.ar

[4] Centro Austral de Investigaciones Científicas, Consejo Nacional de Investigaciones Científicas y Técnicas (CONICET), Bernardo Houssay 200, . Ushuaia, Tierra del Fuego. Argentina.
Email: azulmarinita@gmail.com

[5] LIFIA, CICPBA-Facultad de Informática, Universidad Nacional de La Plata (UNLP). Calle 50 y 120, La Plata, Argentina.
Email: mdalponte@unq.edu.ar

[6] Departamento de Ciencia y Tecnología, Universidad Nacional de Quilmes (UNQ). Roque Sáenz Peña 352, Bernal, Argentina.
Email: dtorres@lifia.info.unlp.edu.ar

[7] Centro de Investigaciones en Ciencias de la Tierra, Consejo Nacional de Investigaciones Científicas y Técnicas (CONICET), Universidad Nacional de Córdoba (UNC). Ingeniero Ismael Bordabehere y Haya de La Torre, Ciudad Universitaria, Córdoba X5016GCA, Argentina.
Email: silvana.halac@unc.edu.ar

[8] Facultad de Ciencias Naturales y Museo (FCNyM, UNLP). Calle 60 y 122, La Plata, Argentina.
Email: mcentiferrei@gmail.com

* Corresponding authors: alejo.bonifacio@unc.edu.ar; jcochero@ilpla.edu.ar

10

Land Tenure Regularisation in Areas of Socio-Environmental Conflict

Struggling for the Arrudas River (Belo Horizonte, Brazil)

Nagayamma Aragão,[1,*] *Luciana Souza Bragança*[2] and *Alana Costa de Oliveira*[2]

1. Socio-Environmental Conflicts vs. Land Tenure Regularisation

Urban expansion in environmentally hazardous areas have become a defining feature of informal urbanisation processes in different Latin American cities. In Brazil, particularly in metropolitan regions such as Belo Horizonte, such informal settlements often emerge in environmentally vulnerable zones, such as hillsides, riverbanks, water protection areas, and other risk situations (Fernandes 2021). In Brazil, settlements resulting from illegal land take are called *ocupação* (lit. occupation) because they are not planned or regulated by authorities and thus exist without legal title or secure tenure (IRDC 2021). The case of the Arrudas River illustrates the spatial dynamics and social drivers behind the *occupation* (land take)

[1] Universidade Lusófona, Department of Architecture and Urban Planning and Interdisciplinary Research Centre for Architecture, Urbanism and Design (University of Lisbon), Lisbon, Portugal.
[2] Universidade Federal de Minas Gerais, Escola de Arquitetura, Belo Horizonte, Brazil.
Emails: lubraganca@gmail.com, 0000-0001-5707-624X; alanacoli@ufmg.br, 0009-0008-3632-8370
* Corresponding author: nagayamma.aragao@ulusofona.pt, 0000-0002-2207-9587

DOI: 10.1201/9781032711737-10

processes, where rapid expansion overlaps with socio-environmental threats. In this context, it is important to note that such settlements do not arise spontaneously or evolve gradually, rather, they are the result of deliberate actions undertaken by organised groups with a clearly defined objective of securing property rights. Vacant properties, whether parcels of land or buildings, are the primary targets of these groups, which mobilise collectively to assert claims to housing, with the ultimate aim of achieving land tenure regularisation and securing long-term property rights. These settlements are frequently justified by *occupiers* (dwellers) as the only viable alternative in the face of structural housing shortages and spatial segregation. However, their location in areas prone to flooding, landslides, or water contamination raises significant challenges for urban governance and public health. The question of land tenure regularisation is key in such conflict situations, as it is not only a legal mechanism for securing inhabitant's right to remain in situ (IRDC 2021), but it is also a political and social process that mediates access to urban space and citizenship rights. The informal settlement Paulo Freire, sprawled along the Arrudas River exemplifies the land occupation movement in Belo Horizonte, especially in recent decades. Land occupation is a phenomenon that arises in a context of long-standing social inequality and disparities and in the access to land and housing. This situation was exacerbated in the 1960s and 1970s, when a growing urbanisation rate intensified the processes of territorial exclusion and informal housing. Named in honour of the educator Paulo Freire, the settlement grew in the context of a housing crisis in Belo Horizonte, when several socially vulnerable and homeless families organised themselves to occupy vacant land along the Arruda and Capão dos Porcos river valleys, an area that is located next to a protected area (Area for Permanent Preservation APP). An APP is a legally protected area surrounding water bodies and other environmentally sensitive areas.

Inequalities in the production and appropriation of the urban spaces, resulting from an uneven access to spatial resources, provoke socio-environmental conflicts, in particular in working-class neighbourhoods. These conflicts reflect strains between the needs of low-income populations, environmental preservation, and interests of public and private agents (Pereira and Ramalhete 2017, Aragão and Smaniotto 2023, D'Orey et al. 2023). Spatial injustice emerges in informal settlements, where communities and individuals face significant challenges due to the lack of housing and livelihood opportunities, proper infrastructure, and suffer from the limited access to essential services and failure of the state to fulfil its basic functions. Conversely, promoting spatial justice involves recognising these disparities and implementing inclusive urban policies that ensure equitable access to land, infrastructure, and services for all residents. In such context, land tenure regularisation emerges not only as a legalistic tool but as a tool of spatial justice and environmental reconciliation.

A land occupation movement like the settlement Paulo Freire, while born of necessity, reflects also a broader claim to urban space and environmental justice. It underscores the dwellers' struggle not just for shelter but for full citizenship. As Lopes (2020, p. 341) states, *"clandestine urbanisation is the physical face of a broader phenomenon of social transformation, which accompanies the trend towards home ownership (...) it has become imperative to establish specific mechanisms to bring these clumsy situations on the front line of urban planning".* Conflicts between housing demands and environmental preservation intensify the strains the community is facing. To solve them, a balanced and rights-based approach is required, one that recognises the legitimacy of informal settlements while promoting environmental protection. The solution lies in adopting inclusive urban policies, investing in upgrading and in infrastructure, effective land tenure regularisation and active community participation in urban planning. The marginalisation of vulnerable populations exacerbates conflicts, especially when their attempts to occupy land for housing are treated as an illegal act, resulting in criminalisation and forced evictions. In Brazil, the antagonisms between the struggle for housing and environmental protection, leads to a problematic dichotomy. These two basic rights - the right to adequate housing and the right to a healthy environment, should not be viewed as opposing forces but integrated into urban policies that reconcile sustainability and human needs (Bragança et at. 2023).

This chapter explores the tensions between the right to adequate housing and the right to a healthy environment, focusing on the often-overlooked dimension of land tenure regularisation in environmentally fragile urban areas. It addresses a "blind field" in urban planning, which emerges due to neglect of relevant conflicting areas and aspects in the planning of urban space. This is because of a lack of holistic and integrated approaches, resulting in a disconnected view of urban life ignoring local social, political, and economic dynamics. On the other hand, urban planning s often seen as serving the interests of powerful economic groups, particularly in ways that undermine environmental goals and deepen inequalities.

Grounded in the conceptual framework of the social production of space, this chapter discusses the impact, limits and conflicts that emerge in the practice of land tenure regularisation in informal settlements. It analyses the social production of space and the interrelations between society and nature, focussing on three planning dimensions: (1) the context of urbanisation and rivers, (2) new frameworks for land tenure regularisation in informal settlements, and (3) the right to the city and the relationship between humans and nature. The chapter addresses emerging conflicts that need to be tackled towards strengthening urban resilience. Taking case of the settlement Paulo Freire (Figure 1), where the *Parque das Ocupações do Barreiro*, an urban park is being co-created with the inhabitants. The park project also integrates environmental policy and land tenure regularisation in a multifaceted strategy, which ensures not only to mitigate risks, but also to support the political, spatial, and ecological integration of marginalised communities.

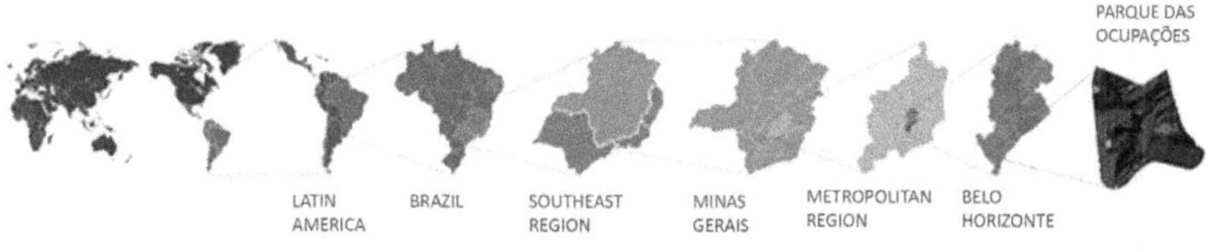

Figure 1: Location of the Park Ocupações do Barreiro. Source: Natureza Política, 2023.

2. Methodology

The research adopts a qualitative, interdisciplinary, and participatory methodology grounded in critical urban theory and socio-environmental justice. Data collection was conducted through a combination of analysing legal documents, field observation, semi-structured interviews, participatory mapping, and community workshops. These activities were developed in collaboration with the Research Group *Natureza Política* of the Federal University of Minas Gerais and in close partnership with local residents and social movements - following principles of action research and co-production of knowledge. Geospatial and legal data were analysed to assess the overlapping dimensions of land tenure insecurity, ecological degradation, and urban planning policies. The participatory activities aimed to integrate residents' experiences and perceptions into the construction of collective strategies for land tenure regularisation, while fostering environmentally just and socially inclusive urban transformations. The methodological framework emphasises the dialogue between academic research and grassroots knowledge, aiming at providing evidence for policy recommendations rooted in context of urban populations struggling for the right to the city. The analysis has thus sought to advocate for a more nuanced understanding of informality, territorial justice, and ecological urbanism within the broader framework of the debate on land tenure regularisation and spatial justice.

2.1 The Arrudas River and Parque das Ocupações as a Case Study

The case study focuses on the informal settlement *Paulo Freire*, located in the Arrudas River valley, where the urban park *Parque das Ocupações do Barreiro* is being developed. *Paulo Freire* settlement is situated in the western municipal administrative region of *Barreiro* and is characterised by environmental vulnerability and ongoing disputes over land tenure. In 2008, social movements occupied parts of the valley, giving rise to what is now referred to as the *Vale das Ocupações* (Valley of Occupations), with several informal settlements, such as Camilo Torres, Irmã Dorothy, Horta, and Nelson Mandela, as well as some industrial facilities (Figure 2). In this area the social movement *Movimento de Luta nos Bairros, Vilas e Favelas (Movement for struggling in Neighbourhoods, Villages, and Slums* (*www.mlbbrasil.org*) is very active. These settlements and facilities being located at the border of an environmental protection area along the Arrudas River, generate overlapping demands and spatial conflicts highlighting the pressure between urban growth and ecological preservation. Since 2016, the Research Group *Natureza Política*, through the Project *Parque da Ocupação,* has been working with the residents of the settlements Paulo Freire and Eliana Silva, with the focus on promoting the rights to the city and land tenure regularisation.

Even with these informal settlements, a substantial portion of the basin area remains unoccupied and thus permeable. This includes two springs located near the main headwaters of the *Arrudas River*. Protecting this area is essential for maintaining the region's hydrological balance and thus contributing to reduced flood risks and enhancing urban environmental quality across the *Arrudas River* basin. The biome of the region of *Barreiro* is *Cerrado*, it is a savannah with the richest biodiversity in the world (wwf n.d.). The Arruda River is largely unaffected by anthropogenic modification, as it has not been channelled or strengthened. The river landscape, the riparian vegetation and local fauna remain relatively well preserved. According to the city's Master Plan, the area is designated as a Municipal Environmental Preservation Zone (Zona de Preservação Ambiental Municipal).

3. Collaborative Land Tenure Regularisation as a Tool for Socio-Environmental Justice in Informal Settlements

Land tenure regularisation is essential for ensuring legal security and access to rights. It refers to the process of formalising and documenting land rights, based on the right to remain in situ where possible. The land tenure transition shifts the settlement from informal, often illegal occupation systems to formal, registered land ownership. However, the registered land tenure process must consider socio-environmental complexities and requires dialogue between public authorities, local communities, and environmental experts. This is crucial to resolve conflicts fairly and to identify solutions that respect both needs of social groups and the ecological sustainability.

In all, the current situation in *Vale das Ocupações,* which includes the settlement Paulo Freire, socio-environmental conflicts and land tenure regularisation efforts

Figure 2: The Arrudas River basin and the informal settlements. On the right the *Paulo Freire* neighbourhood. Source: Natureza Política archives, 2017.

Figure 3: Spatial agency and landscape transformation in Paulo Freire: grassroots responses to environmental challenges. Source: Natureza Política & Jardins Possíveis, 2023.

demonstrate the need for a collaborative approach. Through the engagement of the community, experts and public authorities, it was possible to mitigate environmental impacts in areas that are already occupied and thus guarantee the right to remain in situ and promote social inclusion. To respond to social exclusion, the Project *Parque da Ocupação*, which provides technical support for land tenure regularisation in the *Arrudas River* basin, aims to prevent soil sealing, enhance landscape and improve the natural water flow (Bragança 2024). It also identifies sustainable solutions developed by local residents and connects these practices with the broader goal of ensuring the right to adequate housing for all (Figure 3). Managing social disparities, preserving local culture and identities, as well as tackling informality in settlements, are factors that can weaken the project's goals (Wojnowska-Heciak 2019). In this context, territorial action recognises both latent conflicts and sustainable alternatives developed by vulnerable populations, rejecting unilateral 'awareness-raising' or 'empowerment' approaches. Instead, the focus is on mutual and collaborative participation, creating connections and networks based on shared knowledge and interests (Bragança et al. 2023). The Project *Parque da Ocupação* plays a key role in the land tenure regularisation process offering technical guidance, legal support, and

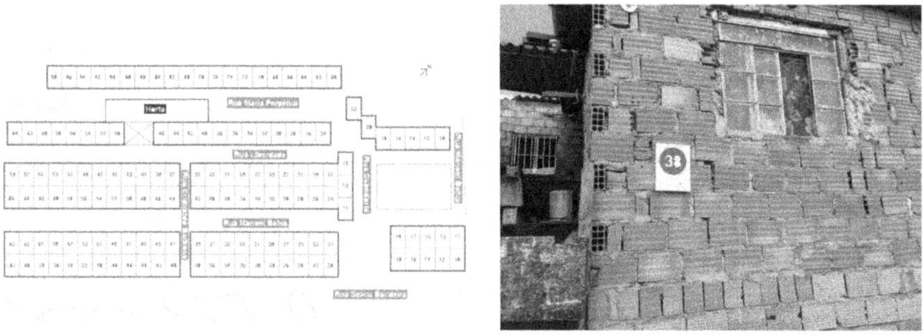

Figure 4: Schematic layout of a cadastral survey for the plots' numbering. An address is a first step in making the plots legitimate and building a positive identity to the residents. Source: Natureza Política & Jardins Possíveis, 2023.

sustainable planning approaches. It also encourages the development of inclusive public policies to improve living conditions of residents. The goal is to support fair and responsible urban planning through collaboration between public authorities and the local community, helping to secure the right to the city.

The Project *Parque da Ocupação* addresses the full scope of urban development across the settlements in the watershed of the Arruda River. It creates a tool for environmentally conscious urbanisation and fosters alignment with public institutions (Figure 4). This process is complex and marked by both progress and setbacks, where land take and built houses are often negotiated, established, or dismantled over time. A clear and effective communication between public authorities, local communities, and environmental stakeholders is crucial for resolving urban socio-environmental conflicts. This process supports fair and sustainable urban development that upholds the right to the city and access to adequate housing for all.

4. Legislation on Land Tenure Regularisation in Brazil and Challenges from an Environmental Perspective

Inequalities in land ownership has significantly deep historical roots in Brazil. The land ownership is concentrated in the hands of an agrarian elite since the colonial period, when the *sesmarias* policy (Menendes Motta 2005) excluded indigenous and traditional communities from land ownership. The situation has worsened with the spread of export agriculture and rural-urban migration, which resulted in property speculation and increasing inequality in land ownership (Cruz 2023). The Land Law *(Lei de Terras)* of 1850 was an important milestone in regulating property once it introduced the notion of private ownership through purchase instead of simple occupancy. However, its implementation ended up with the expansion of latifundia and deepening social exclusion (Martin 2022). An effective land reform is still a serious issue in the country, with a large part of the population struggling for land for housing and/or cultivation (IPEA 2005). Prior to the 1960s, Brazil's development was assessed based on urbanisation and industrial growth by focusing exclusively on economic growth (Veiga 2005).

In 2009, Brazil advanced in the regularisation of land tenure through the Law 11.977/09, which was part of the social housing programme "*Minha Casa, Minha Vida*" (My house, my life) (Brazilian Government 2023). Nevertheless, in 2016, the Provisional Measure 759/2016 enacted following the impeachment of the former president Rousseff, reinstated the individual land titling and land privatisation. Later the measure was codified as Law 13.465/2017, which attracted criticism for promoting environmental injustice, especially in ecologically sensitive areas, such as the case of the Arrudas River in Belo Horizonte. Achieving sustainability requires balancing the demands of the real estate market with social interests, especially concerning social housing and informal employment sectors. Public authorities play a fundamental role in advancing sustainable urbanisation, and addressing the environmental, social and cultural needs of cities, as defined by the *Lei do Estatuto da Cidade* (City Statute Law 2001), and emphasised by Farias and Montes (2024). The concept of sustainable cities encompasses ensuring access to adequate housing, basic sanitation, infrastructure and public services, while aligning them with the social responsibility inherent in land use (Law 10.257/2001). Furthermore, Article 1.228 of the Brazilian Civil Code, when defining the property rights, underscores that land ownership must fulfil both social and environmental functions. It also sanctions abuse that compromise ecological balance and social welfare.

The social function of property is a fundamental legal principle, asserting that ownership rights must be exercised not solely for the individual proprietor's benefit, but also in alignment with the broader interests of society and environment. This principle emphasises the obligation of landowners to contribute to social equity, environmental sustainability, and collective well-being. Consequently, it imposes limitations on the private property rights, which cannot supersede collective and environmental considerations. This concept is intrinsically linked to sustainable development and the advancement of socio-spatial justice (Machado 2008, Pereira and Ramalhete 2017). Land-based conflicts, particularly between the right to housing and the right to a healthy environment are exacerbated by irregular occupancy of protected areas, such as the Permanent Preservation Area, which are often perceived as readily accessible sites, including for luxury real estate projects. This situation highlights the frequent marginalisation of environmental issues in urban planning processes. Since 2012, the Law 12.651/2012 (Forest Code) has established rules to safeguard native vegetation and designating Permanent Preservation Areas along the watercourses. This legislation prohibits the construction of buildings within a 30 m buffer zone surrounding any waterbody, as would be the case of the Arrudas River. In 2021, through the Law 14.285/2021, the responsibility for establishing and managing such protection areas in urban settings was transferred to municipal authorities, thereby weakening the protection of watercourses furthermore. Due to urban growth and the resulting environmental pressures, along with an increasing global awareness of the impacts of human activities on climate and water resources, Brazil developed several environmental policies and legislation aimed at regulating and mitigating such effects, particularly since the 1930s and more intensively from the 1980s onwards.

The Demographic Census 2010 defines "subnormal settlements" as informal neighbourhoods typically located in environmentally vulnerable zones, such as steep slopes and river floodplains (IBGE 2010). Although these settlements are often situated in proximity to central urban areas, facilitating access to employment opportunities and infrastructure, they are distinguished by inadequate living conditions and the presence of socially marginalised groups. The growth of informal settlements is closely linked to internal migration towards major metropolitan areas, motivated by the pursuit of a better quality of life. Furthermore, the Brazilian Federal Constitution of 1988 enshrines the right of all citizens to a healthy environment and stipulates that urban policies comprehensively endorse the social functions of property and the city while ensuring the well-being of its inhabitants. It also recognises housing as a fundamental social right and states that environmental effectiveness must guide the country's economic development. The Law 13.465/2017 (Urban land regularisation) sets legal, urban planning, environmental and social steps towards incorporating informal neighbourhoods into urban land use planning and formally recognising and documenting the legal ownership of land, particularly for those occupying informal settlements. The land tenure regularisation enforcement on the banks of watercourses, was strengthened by the Resolutions 302 and 303 of CONAMA (National Environmental Council 2002a,b), which laid down rules, parameters and limits of permanent preservation areas towards mitigating the illegal occupancy in protected areas. In 2012, the Law 12.651/2012 expanded land tenure regularisation in permanent preservation areas (Locatelli 2020). The Law 13.465/2017 was enacted with the objective of protecting areas of significant environmental interest while facilitating the land tenure regularisation of informal settlements. This legislation introduced changes to the legal framework for land tenure regularisation in consolidated urban areas, redefining the process under two categories: REURB-S focussed on areas of social interest and REURB-E focussed to areas where the residents have better financial capacity to bear the costs of urbanisation (Locatelli 2020).

Land tenure regularisation extends beyond issuing titles (Figure 4), as it includes the provision of public services, sanitation, and transportation. The processes face specific challenges in the areas of environmental preservation. In these areas technical studies are imperative to inform decision-making, in particular regarding protection measures and the implementation of compensatory measures, given that the conservation of natural resources is prioritised, as stipulated by Law 12.651/2012 (Forest Code). In 2021, the enactment of Law 14.285/2021 amended the Forest Code to address for the regularisation of constructions situated along the riverbanks and lakeshores within urban areas. This legislation delegates to municipal authorities the responsibility for defining the buffer zones along the banks of rivers, streams, lakes, and lagoons within their jurisdiction. However, this delegation may undermine the safeguarding of waterways in urban areas due to the lack of a comprehensive set of local regulatory instruments enabling municipalities to effectively enforce the law. Where applicable, studies, analyses, and regularisation procedures concerning buildings located in preservation zones must adhere to the provisions established by both the Forest Code and Law No. 13,465/2017, as further detailed in CONAMA Resolution No. 369/2006 governing land regularisation.

Alongside constraints pertaining to land tenure regularisation and environment protection issues, the rapid advancement in recent years of a liberal policy agenda prioritising 'economic freedom' over urban reform and socio-spatial inclusion has introduced more challenges. This shift in priorities tends to undermine efforts aimed at equitable urban development and environmental justice. Land tenure regularisation being approached in a more legalistic and traditional way, with a greater focus on juridical and notarial aspects instead of a broader and better integrated paradigm, reflects a more limited view of the subject (Fernandes 2023). This process has gradually broken down the legal-urban-environmental order. Thus, reconciling housing and environmental preservation as a way of promoting urban reform and socio-spatial inclusion becomes an even greater challenge.

4.1 The Arrudas River

Multiple structural factors contribute for the settlement of low-income groups on underutilised and environmentally unsuitable urban land. Fernandes (2021) identifies several contributing factors to this phenomenon, such as restricted access to land linked to poverty, fragmented and poorly implemented public policies, weak enforcement of urban and environmental legislation, an exclusionary formal property market, and deficient urban governance. The combination of these factors renders urban development inadequate in responding to the demands of this socio-economic group. A recurring criticism of urban planning is its failure to effectively integrate key policy domains—including land use, housing, environmental protection, transport, taxation, and financing. This lack of integration contributes to the vulnerable status of urban rivers, intensifying the socio-environmental conflict between the water bodies and informal settlements.

Even with the Urban Land Regularisation (REURB) serving as a strategic approach to incorporate informal settlements into the formal urban framework, the territorial inclusion continues to be a considerable challenge in Brazil. The REURB process facilitates infrastructure enhancement and addresses issues of social exclusion, but the lack of an effective housing policy has led to the occupation of areas of risk and for environmental preservation, thereby perpetuating cycles of precarious living conditions and socio-economic vulnerability, as exemplified by the settlements along the Arrudas River (see Figure 5). Within this context, the implementation of land tenure regularisation initiatives is crucial to ensure access to adequate housing and infrastructure, fostering social justice and sustainable development at the local scale. Thus, it is imperative to incorporate the dynamics of river basins into urban land management and urban planning processes to effectively guide urban growth. The maintenance and preservation of riparian forests, alongside their integration within the urban fabric, play a crucial role in the environmental management of the Arrudas River, providing sustainable and multifunctional solutions to the urban and social challenges faced by the region.

The Project *Parque da Ocupação* proposes to incorporate natural elements such as green areas, urban gardens and rainwater drainage systems in land regularisation efforts, thereby fostering environmental sustainability. These interventions are crucial for effective rainwater management, erosion control and mitigation of soil sealing, as well as creating recreational spaces that enhance the quality of life for the riverine community. Measures such as the (re)naturalization of the banks of the

Figure 5: The settlement Paulo Freire. Source: Natureza Política & Jardins Possíveis, 2023.

Arrudas River and the establishment of urban gardens serve as a mitigation strategy to promote biodiversity and encourage sustainable utilisation of urban spaces. The challenge remains in reconciling land tenure regularisation with environmental conservation along the *Arrudas* River, while ensuring simultaneously the right to adequate housing without compromising water resource integrity. Despite advances in legislation, adjustments are needed to deal with the complexity of land use and environmental issues, considering the importance of urban rivers for quality of life. The Project *Parque da Ocupação* is forcing a discussion on environmental safeguarding and urban sustainability through a comprehensive urban development plan for the entire catchment area of the *Arrudas* River.

5. Land Tenure Regularisation, Social Participation and the Development of the *Parque das Ocupações* along the Arrudas River

The main pillar for the Research Group *Natureza Política* is the engagement of social actors in urban planning and design processes. The Project *Parque da Ocupação* began by mapping the watershed along the Arrudas River, understanding it as a dynamic and open laboratory. It aims to identify the existing uses and elements in the territory by using the map as a political tool. In conventional planning, the relationship among design, implementation and management is typically perceived as distinct and sequential phases. However, in self-built informal neighbourhoods, it is essential to align the project's future closer to the immediate timeframe of implementation, which is characterised by urgency and necessity. Such temporal disjunction reflects a capitalist perception that follows a linear logic geared towards control and efficiency, often disconnected from the collective realities of informal neighbourhoods (Lopes et al. 2021, Bragança et al. 2023). The project considers the entire watershed of the Arrudas River, which serves as a framework for reflecting on land tenure regularisation and the consequences for nature preservation. The research programme has been developed collaboratively with the local community, although this has not been without conflict or contestation. 'Doing together' is the basis of the project, but adopting collective practices requires more flexibility, adaptability and a non-linear process, which must adapt to the unpredictable and urgent nature of the process.

Indeed, reclaiming (occupying) land is not always a collective act, and the profile of settlers are plural and constantly changing. The sense of community usually arises when there is a collective effort to build urban facilities, bringing people together towards a common goal. Uncertainty about land ownership and the possibility of being removed from occupied space is often a challenge for such collectiveness (Lopes et al. 2021, Bragança et al. 2023). Land appropriation in environmentally sensitive areas, such as along watercourses, is a source of conflict and poses a significant societal challenge. Participatory processes can provide a pathway to address these challenges. Parque das Ocupações emerged from such a conflict: in 2013, a resident illegally built a house within the protected area, prompting legal action. The dispute became a catalyst for rethinking local land-use patterns, as it triggered a collective negotiation process involving residents, social movements, the university, and public authorities. As a result, the settlement was ultimately regularised and accepted. This incident highlights the need to better protect natural space, but on the flip side, it fostered collaboration between the university and a local social movement, the *Movimento de Luta nos Bairros, Vilas e Favela*s (MLB). The immediate result of this collaboration was the implementation of a sewage system to prevent further water contamination. In 2015, the MLB established another settlement where the fighting for land integrated environmental concerns. This helps turn Parque das Ocupações into a symbolic and tangible space of socio-environmental resistance. The goals for the Research Group *Natureza Política* include securing land tenure rights, incorporating an environmental agenda, engaging residents with researchers in promoting the development of community spaces, such as an orchard, playground and vegetable garden. Improvements in sanitation facilities and drinking water supply were also in the agenda. In 2017, the project became part of the Arrudas River Basin Subcommittee and participated in public tenders focused on springs preservation and afforestation efforts. The Project *Parque da Ocupação* was showcased at the VI Sustainability Biennial competition and was awarded for its methodology, centred on integrating vegetation with the natural drainage patterns.

In the second half of 2017, the plots in the settlement Paulo Freire became a house number and signs were installed indicating to the new Occupation Park. Thereby, a narrative based on water, vegetation, fauna and the struggle for housing could be reinforced. In 2018, the COPASA (Water and Sewage Sanitation Company) and the NGO *Boi Rosado* donated eighty tree seedlings, including fruit and native species, to spruce up the streets. With the residents an action plan was organised to prepare the planting and to care for the saplings. In 2019, a meeting with the project's stakeholders led to the formation of four working groups, focused on mobilisation, urban infrastructure (drainage and paving), afforestation, and waste recycling. Each one operating with a certain degree of autonomy, with the university providing support for all. Within these groups, discussions around land tenure regularisation gradually emerged, particularly as it is related to the broader goals of urban improvements and environmental sustainability. Among the groups, the afforestation group has been the most active, playing a central role in fostering community engagement and dialogue about securing land tenure rights.

The COVID-19 pandemic has caused a significant slowdown in the implementation of the *Project Parque da Ocupação* and the advancement of environmental, water, and urban planning agendas. The public health crisis, coupled with inadequate access to public services, has intensified the vulnerability of local populations, many of whom were already experiencing severe deprivation. As a result, the outbreak intensified the socio-economic difficulties encountered by residents, leading them to prioritise immediate needs, such as health and basic living conditions, often at the expense of long-term concerns such as environmental preservation and urban infrastructure development. The pandemic-induced crisis has not only hindered the progress of structural projects but has also encouraged further the illegal occupation of inadequate areas, such as riverbeds and zones designated for environmental conservation. The increase of social pressure is jeopardising the progress previously made by in terms of sustainable land management and ecological preservation. The expansion of such informal settlements not only increases environmental risks but also highlights the urgent need to revisit debates on sustainable urbanisation, calling for the integration of housing and environmental policies in order to mitigate negative impacts and enhance the quality of life of concerned communities.

Currently, the *Project Parque da Ocupação* is undergoing reactivation and is developing tangible action for advancing land tenure regularisation, a crucial measure towards formalising and enhancing the living conditions of the population. In line with this initiative, the City Council of Belo Horizonte is developing the *Jatobá* Urban Restructuring Plan, which aims to integrate sustainable urbanisation and land tenure regularisation efforts. The plan represents an opportunity to secure land rights, while at the same time, it facilitates the structural improvements necessary for environmental protection and enhancing urban quality. In this context, the *Research Group Natureza Política*, which has been working in the Eliana Silva and Paulo Freire neighbourhoods since 2016, was invited to collaborate directly with the City Council and the local community to present their joint endeavours. The academic and community partnership strengthens the link between research, university extension, and teaching activities, which is fundamental to legitimising the residents' demands in the process of designing sustainable and inclusive solutions. Recognising the important work done over the years gives the community an active voice in designing public policies to meet their needs. Among the issues discussed are the clearance of riverbeds, a key measure for mitigating environmental and flood-related risks, and urban improvements aimed at basic infrastructure and sanitation.

Designating protected area and adjusting land use patterns are fundamental measures for balancing housing demands with environmental preservation. Although the process is in Project *Parque da Ocupação* at an early stage, it has the potential to build a growing commitment between public authorities and civil society in the search for integrated solutions that combine land tenure rights with the conservation of natural resources and improving the quality of life for residents. This initial phase marks the beginning of a long process of negotiations and implementation involving multiple actors and interests. The continuity of this effort depends on effective coordination between the different government agencies, academia and the community. This transdisciplinarity ensures that proposed solutions are sustainable and capable of tackling the complex challenges involved urbanisation of vulnerable areas.

6. Final Reflections

The preliminary findings from the *Project Parque da Ocupação* along the Arrudas riverbanks emphasise the importance of land tenure regularisation and participatory urbanisation in mitigating socio-environmental risks in informal settlements, as exemplified by the settlement Paulo Freire. The local context shows the tension between the challenge of providing affordable housing for low-income groups and the need of environmental conservation. Furthermore, the results highlight the importance of integrating sustainability principles into urbanisation processes to efficiently address vulnerabilities experienced by the riverine community. The interaction between community-based housing movements and environmental protection frameworks reflects a broader dilemma in urban governance, how to reconcile the right to adequate housing with the imperatives of ecological sustainability. As demonstrated, the land take in environmentally sensitive areas challenge conventional paradigms of urban planning, exposing institutional gaps in addressing the dual urgency of housing provision and environmental preservation. Although land tenure regularisation is often viewed as a pathway to legal and infrastructural inclusion, existing legal instruments have proven inadequate in responding to the socio-ecological features of these territories. Overcoming this mismatch requires regulatory innovation and cross-sectoral planning approaches capable of recognising the legitimacy of popular claims without undermining the preservation and restoration of ecological functions. In this context, urban resilience must be reconceptualised not merely as an environmental or infrastructural objective, but as a socially embedded process.

The analysis of the Arrudas River basin highlights the complex interplay of social, economic and environmental factors in the context of informal urbanisation and securing land rights. The *Project Parque da Ocupação* shows how social participation can be a driving force for more inclusive and sustainable initiatives, and thus for risks mitigating and promoting social justice. The socio-environmental conflicts in the neighbourhood reflect a historical inequality, where communities seek not only for adequate housing, but also to live in a safe environment. Poor housing condition and inadequate public policies, force vulnerable populations to settle sensitive areas such as the Arrudas riverbanks. To this respect, securing land rights is understood as an integral process that guarantees social and environmental rights, and not only a scheme to formalise property rights. The *Project Parque da Ocupação* is an innovative response to local challenges. In a joint effort of communities, experts and public authorities, it aims to incorporate responses to housing problems into environmental protection. This reiterates the need for taking forward the right to the city in places facing major contradictions. The implemented measures, like urban gardens and the recovery of the Arrudas River besides contributing to rainwater management and preservation of biodiversity, improving thus the local quality of life, act also as barriers against the further expansion of informal settlements in a disorderly manner. An active participation of the community in planning efforts is key for the success of such initiatives. Respectful and collaborative dialogue is essential, as it values residents' local knowledge in the co-construction of solutions that address both their social rights and environmental concerns.

Urban policy frameworks should protect civil rights, ensuring that efforts to secure adequate housing are harmonised with environmental conservation, ultimately benefiting vulnerable populations and ecological systems alike. Promoting of land tenure regularisation and participatory planning processes help on the one side, to reduce the risks of unregulated urban growth. On the other side, it fosters the development of more equitable and sustainable cities. The Arrudas River basin has the potential to become a paradigmatic showcase of this transformation, with a concerted commitment among public authorities, civil society, and local residents. An inclusive dialogue emerges as a critical mechanism for advancing fairness and sustainability. The reactivation of the *Project Parque da Ocupação* in conjunction with the City Council's *Jatobá* Urban Restructuring Plan bring fresh hope. The efforts to secure and regularise land ownership and improve infrastructure highlights the importance of collaborative processes in order to meet the needs of residents and promote sustainable development. Community engagement and institutional support remains crucial to meet the challenges of sustainable development.

The case of the Arrudas River shows how important is recognising and valuing community's capacities for self-organisation. The community and the environment potentialyl serve as factors in co-creating inclusive and sustainable solutions. The future of *Vale das Ocupações* and its communities depends on the continuity of the dialogue and an effective implementation by collaborative processes. Nature-based solutions could be a promising concept to articulate both vertical and horizontal actions. This study is a preliminary contribution, posing critical questions and underscoring the need for more in-depth analysis to advance understanding of how social inclusion can reconcile with the protection and restoration of urban rivers in the context of land tenure regularisation.

Acknowledgements

The research is based on the scientific mission within the framework of research projects *RUN | Naturalized Urban Rivers* funded by Cyted Ibero-American Program for Science and Technology for Development (grant no. 420RT0008, https://cyted.org/es/run) and *ECS | European Citizen Science* funded by the Horizon Europe Framework Programme for Research and Innovation (grant no. 101058509).

References

Aragão, N., and Smaniotto Costa, C. 2023. A regeneração de rios urbanos, paralelos entre Portugal e África – Desafios e práticas. In C. Smaniotto Costa, M. Menezes, M. Pallares-Barbera, G. Pastor, E. P. Rocha and K. O. Villalba Condori [eds.]. Rios Urbanos na Ibero-América: Casos, Contextos e Experiências, Cultura & Território 6: 27–45. Lisbon: Lusófona University Press. https://doi.org/10.60543/ecati/8tgb-ym88.

Bragança, L. S. 2024. Possible Gardens and the waters in the city: research and teaching. Cadernos de Pós-Graduação em Arquitetura e Urbanismo 24 (1): 169-184. https://doi.org/10.5935/cadernospos.v24n1p185-199.

Bragança, L. S., Lopes, M. S., Grossi, G. D. B., and Pereira, A. D. F. 2023. A água e o Parque das Ocupações: Belo Horizonte (Brasil). pp. 141–159. *In:* Smaniotto Costa, C., Menezes, M., Pallares-Barbera, M., Pastor, G., Rocha, E. P., and Villalba Condori, K. O. [eds.]. Rios Urbanos na Ibero-América: Casos, Contextos e Experiências. Cultura & Território, 6: Lusófona University Press, Lisbon. https://doi.org/10.60543/ecati/caj8-6d21.

Brazilian Government 2023. Brazil's Minha Casa, Minha Vida Program. https://www.gov.br/secom/en/latest-news/2023/brazil2019s-minha-casa-minha-vida-program-is-back-delivering-over-2-700-homes-across-six-brazilian-states.

Cruz, A., and Ghidorsi, J. 2023. A influência das sesmarias na estrutura fundiária do Brasil República. Rev. Int. Consinter Direito. https://revistaconsinter.com.

D'Orey, A., Aragão, N., Smaniotto Costa, C., and Duarte, T. 2023. A review of urban river regeneration projects: practices and challenges. *In:* Smaniotto Costa, C., and Aragão, N. [eds.]. Understanding and transforming the territory: new approaches and perspectives. Cultura & Território, 5: 113–153. Lusófona University Press, Lisbon. https://doi.org/10.60543/ecati/zc65-5z76.

Farias, J. A., and Monte, M. 2024. Dimensões da sustentabilidade em plano diretor urbano. Revista Meio Ambiente e Sustentabilidade 13(26): 34–52. https://doi.org/10.22292/mas.v13i26.1144.

Fernandes, E. 2006. Principles, bases and challenges of the National Programme to Support Sustainable Urban Land Regularization in Brazil. *In:* Huchzermeyer, M., and Karam, A. [eds.]. Informal settlements: a perpetual challenge?: 62–83. UCT Press/Cordaid, Cape Town.

Fernandes, E. 2011. Regularization of informal settlements in Latin America (Policy Focus Reports). Lincoln Institute of Land Policy, Cambridge, USA.

Fernandes, E. 2023. Desafios da regularização fundiária urbana no contexto da Lei Federal no 13.465/2017. O Social em Questão 1(55): 241–260. https://doi.org/10.17771/PUCRio.OSQ.61455.

Finisterra. 2017. Nota Editorial. Revista Finisterra 52(104). https://doi.org/10.18055/Finis12619

Furlan, D. L. S., and Fróes, A. C. S. 2020. Ocupações de áreas ambientalmente frágeis em Almirante Tamandaré/PR: o direito à moradia em face da preservação do meio ambiente. Cad. Metrópole 22: 579–600. https://doi.org/10.1590/2236-9996.2020-4811.

Gonçalves, R. S. 2009. Repensar a regularização fundiária como política de integração sócioespacial. Estudos Avançados. 23(66): 237–250. https://revistas.usp.br/eav/article/view/10421.

IBGE - Instituto Brasileiro de Geografia e Estatística (2012). Censo Brasileiro de 2010. Rio de Janeiro: IBGE.

IPEA - Instituto de Pesquisa Econômica Aplicada. 2005. Radar social: Moradia, Brasília: Ipea.

IDRC - International Development Research Centre. 2021. The Right to Housing for Residents of Informal Settlements. https://www.make-the-shift.org/wp-content/uploads/2021/07/THESHIFT-plainreport-v2_compressed.pdf.

Locatelli, P. A. 2020. A atuação homeostática e interagente do Ministério Público polímato na regularização fundiária urbana. *In:* REURB: Regularização Fundiária Urbana: aspectos teóricos e práticos: 57–76.

Lopes, D. 2020. Áreas Urbanas de Gênese Ilegal: Problemas Suscitados por um Regime Legal Excepcional que Teima em Perpetuar-se no Tempo. De Legibus: Revista de Direito.

Lopes, M. S. B., Bragança, L., Deusdedit, M., and Rena N. 2021. Biopotent social technology: occupations park and university extensions. *In:* Kanoun, O., and Viehweger, C. [eds.]. Frontiers of Science and Technology. https://doi.org/10.1515/9783110584455-001.

Martin, J. S. 2022. Lei de Terras: a reafirmação da estrutura latifundiária no Brasil. https://www.politize.com.br/lei-de-terras/.

Menendes, M. M. M. 2005. The sesmarias in Brazil: colonial land policies in the late eighteenth century. e-Journal of Portuguese History 3, 2: 1–12. https://doi.org/10.26300/wh44-9x32.

Pereira, M., and Ramalhete, F. 2017. Planeamento E Conflitos Territoriais: Uma Leitura na ótica da (in) justiça Espacial. Finisterra 52 (104). https://doi.org/10.18055/Finis6972.

Veiga, J. E. 2005. Desenvolvimento Sustentável: O Desafio do Século XXI. Editora Garamond, Rio de Janeiro.

Wojnowska-Heciak, M. 2019. The Naturalness of the Vistula Riverbank's Landscape: Warsaw Inhabitants' Perceptions, Sustainability 11(21). https://doi.org/10.3390/su11215957.

WWF World Wilde Life (n.d.) Cerrado - facts. https://www.worldwildlife.org/places/cerrado

Additional sources - legislation

CONAMA - Conselho Nacional do Meio Ambiente. 2002a. Resolução 302 - 20/03/2002. Dispõe sobre os parâmetros, definições e limites de Áreas de Preservação Permanente de reservatórios artificiais e o regime de uso do entorno. https://www.legisweb.com.br/legislacao/?id=98315.

CONAMA 2002b. Resolução 303 - 20/03/2002. Dispõe sobre parâmetros, definições e limites de Áreas de Preservação Permanente. https://www.legisweb.com.br/legislacao/?id=98313.

CONAMA 2006. Resolução 369 - 28/03/2006. Dispõe sobre os casos excepcionais que possibilitam a intervenção ou supressão de vegetação em Área de Preservação Permanente - APP. https://www.legisweb.com.br/legislacao/?id=104080.

Federal Constitution of Brazil 1988. Brasília: Planalto. https://www.planalto.gov.br/ccivil_03/constituicao/constituicao.htm.

Law no 601 - 18/08/1950. Dispõe sobre as terras devolutas do Império. Brasília: Planalto. http://www.planalto.gov.br/ccivil_03/LEIS/L0601-1850.htm.

Law no 11.977- 07/07/2009. Programa Minha Casa, Minha Vida. Brasília: Planalto. https://www.planalto.gov.br/ccivil_03/_ato2007-2010/2009/lei/l11977.htm.

Law no 10.257 - 10/07/2001. Estatuto da Cidade. Brasília: Planalto. https://www.planalto.gov.br/ccivil_03/_ato2015-2018/2017/lei/l13465.htm

Law no 12.651 - 25/05/2012. Código Florestal. Dispõe sobre a proteção da vegetação nativa. Brasília: Planalto. https://www.planalto.gov.br/ccivil_03/_ato2011-2014/2012/lei/l12651.htm.

Law no 13.465 - 11/07/2017. Dispõe sobre a regularização fundiária rural e urbana, sobre a liquidação de créditos concedidos aos assentados da reforma agrária e sobre a regularização fundiária no âmbito da Amazônia Legal. Brasília: Planalto. Https://Www.Planalto.Gov.Br/Ccivil_03/_Ato2015-2018/2017/Lei/L13465.Htm.

Law no 14.285 - 29/12/2021. Altera as Leis nos 12.651, de 25 de maio de 2012, que dispõe sobre a proteção da vegetação nativa. Brasília: Planalto. https://www.planalto.gov.br/ccivil_03/_Ato2019-2022/2021/Lei/L14285.htm.

MPV no 759 - 22/12/2016. Dispõe sobre a regularização fundiária rural e urbana, sobre a liquidação de créditos concedidos aos assentados da reforma agrária e sobre a regularização fundiária no âmbito da Amazônia Legal. Brasília: Planalto. https://www.planalto.gov.br/ccivil_03/_Ato2015-2018/2016/Mpv/mpv759.htm.

11

Socio-Spatial Representations of the Everyday Uses of the River Cambamba, Luanda

Edson Kiluanje Paulino Lourenço[1] and *Marluci Menezes*[2,*]

1. Introduction

The intense urbanisation of territories has a strong impact on the natural environment, especially in the face of climate change and social crisis. The importance of rivers for the balance of human settlements is undeniable. However, large populations concentrated in limited spaces demonstrate a weak mobilisation towards protecting natural resources. In the case of many African cities, as Luanda, rapid and disorderly urban growth has significantly intensified the pressure on natural resources, leading to a limited supply of water (Odume et al. 2022, Anghileri et al. 2024).

The expansion of informal settlements, the lack of adequate urban planning, socio-economic precariousness, water scarcity, imminent risks and the occurrence of catastrophes, contribute to the environmental degradation. Poor governance and lack of public participation in the managing of natural resources further exacerbate the situation. UNESCO (2023) describes the seriousness of this situation, highlighting the significant decline in renewable water resources in Sub-Saharan Africa. Water bodies have often been neglected in the public policies of many African countries; and Angola is not an exception (Pereira 2011). In Angola, however, there has been

[1] Universidade Agostinho Neto, Luanda, Angola and Universidade Lusófona, Department of Architecture and Urban Planning, Lisbon, Portugal.
Email: edikipaulo80@gmail.com, 0000-0001-5769-3707
[2] Nacional Laboratory for Civil Engineering (LNEC), Lisbon, Portugal.
* Corresponding author: marluci@lnec.pt, 0000-0001-7031-0053
DOI: 10.1201/9781032711737-11

a growing political and technical commitment to addressing issues regarding the water management and quality. This interest has been driven by concerns over water scarcity, challenges arising from inadequate sanitation infrastructure, and the risk of flooding. Nevertheless, when it comes to implementing effective public policies for the sustainable management of water resources, much remains to be done (Santos 2014, Cain 2018, João and Campos 2024). The lack of Angolan studies that consider communities alongside their needs, practices and socio-spatial representations contributes to an institutional gap in citizen engagement. Research into the socio-spatial representations of blue infrastructure elements, such as rivers, remains highly limited.

This chapter seeks to reduce this knowledge gap. This calls for using a comprehensive strategy that considers social and cultural dimensions while emphasising the importance of natural resource sustainability (ICLEI 2018). This exploratory and qualitative study examines the socio-spatial representations of the River Cambamba. It considers it's symbolic, historical, cultural, perceptual, emotional, political, social, and environmental dimensions, and their implications for advancing sustainability. It is assumed that adopting an approach that promotes community engagement fosters social involvement and shared responsibility.

2. Theoretical Framework

The concept of social representation, introduced by Moscovici (1961) in the field of social psychology, seeks to explain how we ascribe meaning to the world around us. This process of constructing meanings shapes the way we engage with objects and carry out everyday practices, ultimately influencing our behaviour (Sancovschi 2007, Jodelet 2017, Sánchez et al. 2021).

Space, understood as a system of meanings, becomes perceptible to its users through a process of selection and valuation of its components (Renier 1982, Menezes 1998, 2002). The ways in which space is perceived, used, and appropriated respect a special collective logic, shaped and perpetuated through the social actors' practices (Pellegrino 1982). The existence of a simulacrum of intellectual operations enables us to revisit the relationship between visible and objective space with its representation. Shared spatial perceptions embody a dynamic process of sensory and cognitive interpretation. This process is articulated though the continual interplay between reality and imagination, and between the symbolic representations and the verbal expressions in everyday life (Jacob 1982).

Representations are perceived as an intrinsic duality of space itself – a subjective interpretation materialised and symbolically expressed within spatial reality (Lévy and Segaud 1983, Silvano 1997, Menezes 1998, 2002). This viewpoint is consistent with the understanding that the city exists simultaneously as a representation and an objective reality (King 1996, Duncan and Ley 1997). Thus, understanding the spatial representation process becomes essential. This knowledge enables to understand the material operations involved in space appropriation, as well as the processes of creating meaning and valuing space.

Piaget and Inhelder (1981) suggest that mental structures govern spatial organisation and are divided into two types of thinking: representative, which is stable and more conservative; and operative, which is more transformative. According to the authors, collective representations of space reflect an emotional bond with it, leading to a symbolic and emotional configuration within the social construction of space. Gomes et al. (2022) argue that this experience directly influences how urban communities assign meanings and values to rivers. Piaget and Inhelder (1981) also suggest that the persistent and conservative character of collective spatial representations influences socio-spatial reproduction. In contrast, change is linked to operative thinking – an active cognitive process that, as the authors argue, enables individuals to reorganise representations and thereby drive spatial transformation. As noted by Lévy and Segaud (1983), changes within space can be endogenous and exogenous. Endogenous changes arise from volunteer actions or from the progressive evolution of the system, resulting in a new social and symbolic order. Conversely, exogenous changes stem from the intention of certain societies to assert dominance or to replace pre-existing normalised spatial configurations (Lévy and Segaud 1983, Menezes 2000).

The concept of collective representation of space can also be approached in two additional ways: as a mapping process, which identifies key elements of a geographic space – such as the city, natural features (e.g., the river), or other characteristics – or more as an abstract perspective, which examines the human perception of space. In the latter case, space is more than just a physical place; it is also a psychological, mental and symbolic one. This understanding of space involves the concepts of lived space (immediate and profound) and distant space (mediated by culture and history) (Pereira 2008). Distinguishing between the near and the distant space enables a deeper understanding of how we absorb information about space and the environment. These aspects are also a basis for investigating the elements that underpin the social construction of space.

In riverside communities, collective representations of water bodies convey the importance attributed to these resources and highlight their influence on local quality of life. These representations provide valuable insights into the relationships between these communities and their physical surrounding and natural environment. They also shed light on the symbolic, social, and cultural dimensions that shape the space experience and meaning (Song et al. 2020, Galvão and Tedesco 2022, Menezes and Silva, 2023). Such understanding deepens our insight into the social relevance and ecological functions that rivers hold within human-environment relationships (Cao and Wantzen 2023). The importance of collective representations of space lies in understanding the society and environment relationship.

3. Research Context

3.1 Urbanisation and Impact on Rivers in Luanda

Urbanisation in Luanda is characterised by a rapid and often disorderly growth, primarily driven by internal migration and the occurrence of informal settlements, known as *musseques*. Many of these informal settlements are situated near rivers, provoking pollution and the degradation of riverbanks. Urbanisation in these areas reflects an unequal distribution of infrastructure, where the lack of proper planning and effective public policies leads to the deterioration of vital water resources (Gorski 2008, Sánchez et al. 2021). These factors have direct consequences on the sustainability of rivers, turning them into waste repositories compromising their environmental quality and increasing flood risks.

In almost half a century since Angola's independence[1], the city of Luanda has undergone a process of transformation marked by unprecedented growth in its infrastructure and population. This growth results in segregated areas with informal settlements, raising concerns among Angolan planners and administrators (Chissola 2015, Guerra 2018). However, due to socio-historical reasons, informality has persisted in the development of its urban landscape. The 2014 Census revealed that Luanda had a population of 6,945,386, with around 35.3% living in informal settlements (INE 2016). As Amaral (1978, 1983) states, the slums (*musseques*) represent the growth of unstructured urban areas, often emerging around riverbanks in a context marked by exclusion and precarious living conditions.

The socio-spatial configuration of Luanda reveals an urbanisation that diverges from conventional standards (Castro and Reschilian 2020). The vast prevalence of informality in land use in Luanda has influence on biodiversity and results in a conflictual and crisis-ridden relationship between nature, society and the city. According to Toni et al. (2023), land use on riverside areas, especially in contexts of urbanisation and intensive agriculture, have a detrimental impact on river water quality. This is evidenced in the case of the Cubal River, in Angola. Surface water sources become contaminated due to the lack of wastewater treatment facilities, endangering ecosystems and public health (Edokpayi et al. 2017). Many of these developments stem from a shortage of housing options and limited access to formal urban infrastructure. As a result, vulnerable populations are often forced to settle in environmentally sensitive areas, such as riverbanks and floodplains. The lack of infrastructure and social support in these communities worsens environmental impacts. Residents often lack the resources and knowhow necessary to implement adequate land use and waste recycling practices.

Understanding the representations of watercourses held by the riverine populations provides a broader perspective of the urban river phenomenon. These representations directly influence patterns of use, impacts, and socio-ecological recognition of rivers, while also shape risk perceptions and environmental responsibility.

3.2 Exploring the River Cambamba Case

According to the Kimbundu-Portuguese Dictionary (Junior n.d.), the noun "Kámbamba" (currently written in Portuguese as "Cambamba") translates to "wand, whip, stick". The corresponding adjective conveys the idea of something "continuous, complete, entire, from beginning to end, uninterrupted, or lasting all day long". Additionally, the term has a botanical connotation, as it can also refer to an "edible fruit". However, the Kimbundu language does not have a formal normative standard.

[1] Angola officially declared its independence from Portugal on 11 November 1975.

From the perspective of anthroponymy and orally transmitted hydronyms, as narrated by residents of the Kilamba-Kiaxi district (crossed by the River Cambamba), the term "kambamba" is understood as a diminutive of "thing", meaning "little thing". According to oral testimonies from elders, the River Cambamba is so named because, in the past, peasants had small agricultural plots along its banks, especially during the rainy season. The crops for subsistence were referred to in Kimbundu as "mbamba", meaning thing. By adding the prefix "ka", a diminutive marker of thing, the term became "kambamba".

The River Cambamba is a relatively stable watercourse that interacts with the rapid expansion of the city of Luanda (Figure 1). This river, which also defines a hydrographic basin, is the largest rainwater drainage basin in Luanda, encompassing the most densely inhabited urban area (DW 2013). Its broad scope highlights its importance: the river basin covers an area of 52.360 ha, crosses key urban districts of Luanda (namely, Cazenga, Kilamba-Kiaxi, Luanda, Talatona, and Viana), and containing a water volume of approximately 52.360 m^3/s (188.496 m^3/h) (Casimiro 2021).

The River Cambamba was selected for this study due to its ecological relevance and the pressing challenges it faces, including rapid urbanisation, irregular land occupation, and pollution. The river is heavily plagued with the accumulation of garbage along its course, particularly around the bridges. Certain sections of the river have controlled drainage systems and are lined with concrete walls to stabilise its course and flow. However, the majority of the river remains unregulated, causing it to widen and frequently overflow during the rainy season. These overflows pose a significant threat to homes located along its banks (DW 2013). The degradation

Figure 1: One of the few stretches of the River Cambamba without houses on its banks. Source: Authors, 2024.

of the River Cambamba is hastened by the lack of environmental education among locals and the lack of adequate sanitation infrastructure.

3.3 Methodological Framework

This study adopts an exploratory and qualitative approach to examining the collective representations of space along the banks of the River Cambamba. The research used information collection techniques, such as image recording, photographic documentation, and conducting interviews with residents.

The spatial visibility aspects were analysed through photographic images. To ensure high-quality images and accurately capture the nuances of land use in the areas adjacent to the river, photographs were taken in the late afternoons of August 2024. A Canon 5D Mark II digital camera equipped with a 50 mm lens was used. As suggested by Asadpour (2013), photography provides a multifaceted means of capturing the complexity of urban environment. It integrates material elements such as buildings and structures, natural and social features, while also evoking cultural memories and perceptions.

For the photographic documentation, specific areas along the river were selected based on criteria such as population density, proximity to the riverbed, and visibility of environmental impacts. Informal settlements, buildings in high-risk areas, and river sections with evident pollution were photographed to provide a comprehensive overview of the dynamics and challenges present along the Cambamba. In addition to visual analysis, ten semi-structured interviews were conducted with residents.

Table 1: Thematic axes and questions for the interviews.

Thematic axis	Questions
Social profile	How old are you? Where were you born? What is your job/professional activity? How long have you lived here and with whom?
Visual and emotional perception of the river	[Showing photographs taken on the river] What do you see here? What catches your attention most? How do you assess the state of conservation of the river (is it well maintained, or does it appear dirty)? How do you feel about your impression of the river?
Perception of change	[Memories of the river through time] Comparing these photos to what you remember of the river in the past, what has changed? Have these changes brought good or bad things to the community?
Perception of the river's impact on people's lives	[Impact of the river on people's lives] How is this water corridor important to your life or to the lives of the people here? How do people who live here use this river on a daily?

Source: Authors, 2024.

The interviews were designed to capture the interlocutors' perceptions of the current state of the River Cambamba and its hydrographic basin. Table 1 displays the thematic axes and interview questions.

The interviews were audio-recorded and transcribed, with prior consent obtained from the participants regarding this method of recording and its use for scientific research purposes. To enhance understanding of the collective representations of space related to the River Cambamba, the interviews were analysed across several dimensions: perceptual and emotional, economic, political and social, cultural and historical, as well as environmental and sustainable.

4. Research Results and Discussion

4.1 Study Results

The interviews conducted resulted in a gender balance, with five men and five women. Their ages ranged from 19 to 63 years. Most were unemployed, although some work in the informal sector as traders or providers of small-scale services. All live in low-income urban areas, in households comprising couples with underage children, while two respondents were adults living with their parents and siblings. Table 2 provides a detailed profile of the respondents.

Figures 2 and 3 illustrate the current situation of the River Cambamba, clearly showing pollution and degradation, with substantial accumulation of plastic waste

Table 2: Profile of respondents.

Reference	Gender	Age	Local	Time living in the area	People they live with	Employment
R-1	Male	40	Malanjino	37 years	Wife and children	Bricklayer
R-2	Male	53	Malanjino	53 years	Children and brother	Unemployed electrician
R-3	Male	39	Malanjino	39 years	Wife	Unemployed aluminium technician
R-4	Female	21	Malanjino	11 years	Husband	Housewife
R-5	Female	26	Malanjino	26 years	Husband	Hairdresser
R-6	Male	19	Catinton	19 years	Parents and siblings	Unemployed
R-7	Male	63	Catinton	More than 20 years	Children	Retired
R-8	Female	21	Catinton	17 years	Parents and siblings	Student
R-9	Female	48	Catinton	10 years	Husband and children	Housewife
R-10	Female	41	Catinton	24 years	Husband and children	Housewife

Source: Authors, 2024.

Figure 2: Section of the River Cambamba with accumulation of garbage and pollution. Source: Authors, 2024.

Figure 3: Section of the River Cambamba with significant pollution. Source: Authors, 2024.

Figure 4: Houses close to the riverbank are at risk of collapsing, in the Kilamba Kiaxi municipality. Source: Authors, 2024.

and debris along the riverbed and its banks. While residents of neighbouring areas are among those who contribute to the problem, they also express concern about the river's deteriorating state. All the images capture the evident neglect of cleaning and basic sanitation efforts along the river, a situation corroborated by the statements gathered with the interviews.

"*This water here (...) is not preserved at all, no one knows why. But this water mostly comes from Maria Pia, from the city, from the sewers*" (R-6, 19 years old).

All respondents emphasised that the current local environmental conditions severely affect the quality of life, as they must deal with precarious hygiene conditions and the lack of sanitary infrastructure. Nevertheless, when asked to compare the present state of the river to the past, their responses varied. Some recall the past as being cleaner, in contrast to the current situation.

"*The state of conservation of the river is dirty (...). In the past, machines were used to clean it. Currently it is all dirty, children go there to play (...), it does not guarantee safety for us, residents of the Malanjino neighbourhood*" (R-7, 63 years old).

Others consider that the river was dirtier in the past. This difference in perception of cleanliness is likely influenced by the respondents' age, with older individuals tending to idealise the past, viewing it as a time when things were better. While younger participants often accept the current state of the river as unremarkable, viewing it as a typical feature of the urban environment they inhabit.

"*Before, there was a lot more trash compared to now (...), but I see that it was reduced a little, the river is cleaner than before*" (R-8, 28 years old).

Figure 5: Pedestrian bridge in risk of collapsing. Source: Authors, 2024.

Figures 4 and 5 show the erosion alongside the riverbanks and accumulated trash, particularly close to the community buildings in the municipality of Kilamba Kiaxi. The water flow is obstructed, posing in risk to the safety of nearby buildings and the bridge. Despite the fact that to improve the riverbanks stability concrete structures were added along the river.

Riverbank erosion is a serious problem for residents. They express concerns about their safety and the security of their homes. In the collected testimonies, they highlighted the challenges faced during rainy days, as well as the numerous requests for assistance to local government authorities – many of which remain unanswered.

"*When it rains, the river overflows [...] people feel sick [...] and the government does not do anything [...]. The bridge is not safe*" (R-7, 63 years old).

"*The river was a little ditch (...). Now, with the rain (...), it has started to spread (...). There are abandoned houses here (...), crime is also increasing (...). We really live with sadness, nothing is good for us*" (R-1, 40 years old).

"*We ask for help from the government to fix this bridge (...), otherwise we have to pay to cross, or they will push us into the water*" (R-9, 48 years old).

Figure 5 shows the rapid erosion at the base of the bridge. The image also emphasises how the accumulation of garbage disrupts the natural water flow, intensifying flooding and environmental degradation. The respondents, who expressed concerns about their safety and restrictions in mobility, frequently mentioned these issues. Figures 6 and 7 illustrate the uses of the River Cambamba. People discussed and criticised these uses and gave reasons for their opinions.

"*The children also play there, but it is very dirty*" (R-6, 19 years old).

"*If there was no river here, people wouldn't have a place to put it [their garbage], but since the river is close, some people understand (...) that it just needs to rain and that's it. (...) [Whit the rain] and the more it rains, the rest comes, the rest comes; the*

Figures 6 and 7: Aspects of the daily relationships of children and adolescents with the River Cambamba. Source: Authors, 2024.

rubbish goes down (...). Also, we use water for irrigating products (...), [and] *clean waste from the bathrooms (...)*" (R-7, 63 years old).

From the perspective of how the river influences the lives of residents, it is evident that people interact with the River Cambamba in ways that contradict good environmental practices. This is also in contrast with the cultural and symbolic value rivers hold in Angolan traditional culture. This likely arises from a lack of environmental awareness and widespread urban poverty. Even so, the respondents distanced themselves from any sense of responsibility. They attributed such uses to others, despite their critical remarks about how the local communities use the river. Some even acknowledged the potential for using the river water for irrigation or sanitation, though these uses were not openly endorsed.

"[The river] no one uses it, what is it for? The people use it mostly just to throw trash, nothing else, just to throw trash (...)" (R-63 years old).

"*Practically, people come to get water for watering their plants. Others also use it in the bathroom"* (R-8, 21 years old).

4.2 Discussion and Lessons Learned

The study revealed that individuals who have lived near the banks of the River Cambamba for a longer period are more sensitive to the changes made in this body of water. In this sense, the representations shared by the respondents align in highlighting a growing sense of environmental degradation. This is attributed to the poor ecological condition, the increasing discharge of wastewater and garbage, as well as the inconsistent attention given to it by public policies.

Socio-spatial representations reflect how individuals and groups perceive and engage with their environment, shaping the way water resources are used and valued. These representations are constructed through social interactions and communicative processes. In turn, these foster collective perceptions imbued with values and meanings that influence behaviour (Piaget and Inhelder 1981, Pellegrino 1981, Sancovschi 2007, Jodelet 2017, Sánchez et al. 2021). Social interactions and the perception of the river and its surroundings represent a collective vision of the space, continuously created and reproduced by local residents. Understanding the collective representations of space involves both the objectivity and visibility inherent in the environment. It also entails practices of appropriation and perception that manifest through its use (Pellegrino 1982, Renier 1982, Menezes 1998, 2000, 2002).

The study confirmed that visible space (objective) and its representation (subjective) align with the concept of space, emphasising the dual nature inherent in space (Lévy and Segaud 1983, Silvano 1997, Menezes 1998, 2002). The relationship between visible/objective space and its representation indicates a process of sensory apprehension. This process is subsequently reproduced as a mental image and then verbalised (Jacobs 1987). In addition to the photographic images, collective representations of the River Cambamba are predominantly expressed through references to poor maintenance, garbage, sewage, pollution, insecurity and danger. These associations highlight the prevailing negative perceptions of the river within the community.

The collective representations of the River Cambamba are far from associating the water element with cleanliness, sanctity, life, or purification. The idea of river water as something associated cleanliness is largely confined to its role for disposing of waste or unwanted materials. This applies both within the home and directly into watercourse. The perceived value of the river's waters, more closely linked to the ideas of life and environment, is primarily associated with its use for irrigation.

The more stable and conservative nature of collective representations, which facilitates socio-spatial reproduction (Piaget and Inhelder 1981, Lévy and Segaud 1983), is predominantly linked locally with notions of dirt, pollution and insecurity. This association risks undermining any transformative potential for environmental conservation and sustainability. Alongside conveying a general perception of worsening environmental problems, it also reflects the ongoing perpetuation of socio-spatial segregation and exclusion. Any hope for improvement is typically framed as a demand for external intervention, particularly from politics and administrative bodies. Notably, no symbolic mechanisms were identified that could foster an endogenous change or improvement. While this perception of change suggests a certain normalisation of the situation, it also reflects a sense of resignation. The ongoing environmental deterioration becomes trivialised and accepted as an inevitable reality.

The River Cambamba exemplifies how rapid urban growth, informality, poverty, and spatial disorder expose the vulnerability of rivers in such contexts. Patterns of land use in riverside areas, as seen in Luanda, significantly compromises water quality (Toni et al. 2023). This degradation also affects the emotional and symbolic connection people have with the river, influencing the way they assign social and ecological value to it (Cao and Wantzen 2023). This underscores the role of people's emotional experiences with water bodies in promoting their preservation and sustainability. On the one hand, this experience is shaped by people's living conditions, education, and environmental literacy. When conditions are precarious and driven by socio-economic poverty, environmental degradation poses a serious threat to the conservation and sustainability of hydrological resources. This, in turn, becomes a defining factor in the collective representation of space (see Figure 8). On the other hand, such aspects allow us to question the symbolic, emotional, historical, cultural and environmental role of water bodies for those living in precarious conditions near the rivers.

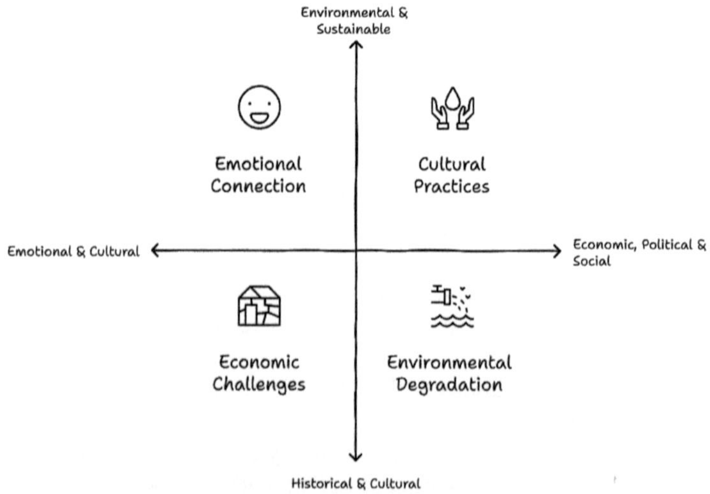

Figure 8: Summary diagram of the role of collective representations of the river in environmental sustainability. Source: Authors, 2024.

In the Cambamba riverside communities, the perceptual-emotional dimension plays a central contribution in understanding people's connection with the environment. This factor may act as either a catalyst or a constraint in the preservation of urban rivers. The hypothesis highlights the role of the perceptual and emotional connection to water bodies. This is seen as essential to valuing and promoting the rational use of blue infrastructure.

Although the emotional aspect influences perceptions of the natural environment, socio-economic and political factors often diminish its importance. The economic, political, and social dimensions tied to collective representations of urban rivers are deeply connected to individuals' positions within society and their access to nearby water resources. The lack of investments in water supply and sanitation infrastructure, coupled with ineffective public policies undermines the urban sustainability. Even so, this issue extends beyond infrastructure alone. Inequality, segregation, and socio-spatial exclusion in access to the city — and, by extension, to education, transport, housing, employment, and other public services — pose a critical challenge to the increasing and safeguarding of sustainability (Sánchez et al. 2021). These factors also jeopardise the capacity of riverside communities to participate in river preservation and management. As a result, some urban interventions prove ineffective, even when designed to reduce flooding and enhance public safety.

In this context, the role of public authorities in safeguarding water resources and promoting environmental education becomes crucial. The inappropriate use of blue infrastructure is often a direct consequence of limited access to drinking water and basic sanitation, alongside the lack of environmental education initiatives. In Angola, especially in densely populated areas like Luanda, the distribution of drinking water and sanitation infrastructure is uneven. This, in conjunction with pollution and poor urban land use, harms sustainability and the peoples' quality of life (Isaac 2013).

Historically, water bodies have played a central role in the daily practices and rituals of traditional communities. Water collection and storage practices, which have been passed down through generations, have preserved cultural identity and local heritage (Song et al. 2020, Galvão and Tedesco 2022). Social representations of urban rivers are also shaped by the history and culture of Angola's riverside communities. The eco criticism explored by Sousa (2023) in analysing the Angolan narrative "Margens e Travessias" (*Margins and Crossings*) by Cardoso (2021) highlights the profound cultural and spiritual significance of rivers for Angolans. Because these natural elements are seen to reflect the soul. Nevertheless, Sousa (2023, 12) also points out that nowadays, "*only a better understanding of the country and the relationship between man and nature and the non-human world can reverse the path of the rivers*". In areas where urbanisation is not yet predominant, the historical and cultural dimension remains a fundamental factor in land use and conservation of natural resources (Waldman 1998, Mbembe 2001, Lourenço and Menezes 2024, Aragão et al., 2025). Traditional practices among Angola's riverside communities exemplify sustainable management of the River Cambamba, demonstrating that modernisation and transformation can coexist with cultural preservation.

However, in cities like Luanda, the rapid urbanisation and poor management of water resources have led to significant degradation of waterbodies. This degradation negatively influences sanitary and environmental conditions, including biodiversity and water quality in urban rivers, while also influencing collective representations of the watercourses (see Figure 9).

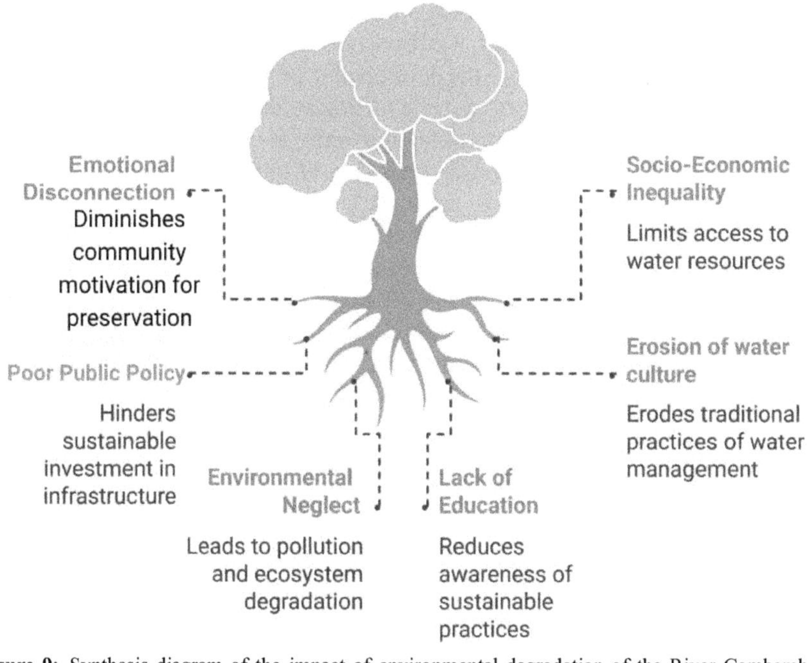

Figure 9: Synthesis diagram of the impact of environmental degradation of the River Cambamba on socio-spatial representations. Source: Authors, 2024.

5. Final Notes

This exploratory study drew on photographic images and associated testimonies. It aimed to highlight the socio-spatial representations of river resources within urban contexts, taking the River Cambamba, in Luanda, as a case study. This approach fosters a more nuanced grasp of how urbanisation and socio-spatial segregation affect watercourses, consequently, people's lives. The study acknowledged the crucial role of collective representations in the processes of meaning-making and socio-ecological valorisation of the river, especially in response to environmental degradation – a phenomenon closely linked to urban poverty.

This study revealed a socio-cultural reality in which the local waterbody is undervalued, clearly reflected in representations of the river, and this shapes meanings, practices and perceptions. Traditionally seen as a vital source of irrigation for crops and vegetable gardens, the River Cambamba has undergone significant changes, such as use for disposal of domestic and sanitary waste, as well the accumulation of garbage along its banks. These actions have led to a decline in water quality, aquatic biodiversity, and the erosion of the riverbanks, compromising its ecological condition. The river also lost its historical, cultural, and socio-economic significance, no longer serving as a space for community gatherings or a connection to Angolan ancestry.

This exploratory and preliminary study, despite its limitations, paves the way for a broader discussion on the conception of the city and its subsystems as living environments. While the study provided valuable insights, its exploratory nature highlights the need for further investigation into the collective representations of the River Cambamba. Additionally, it underscores the importance of exploring how these representations can influence the development of strategies aimed at combating socio-economic poverty and environmental degradation.

Social representations are essential not only in socio-spatial reproduction, but also in fostering transformative aspirations and initiatives aimed at improving environmental conditions. This is particularly true when it comes to people's emotional, symbolic, cultural, social, and ecological engagement, which is vital for driving endogenous changes. Furthermore, understanding the role of these internal changes in triggering exogenous changes, mainly in relation to the city's political and administrative actions, is crucial. This involves proposing new ways to integrate community voices into territorial and environmental planning. In Luanda, as in other similar urban contexts characterised by informal settlements and urbanisation with distinct cultural characteristics, the mobilisation of collective action is limited. As a result, it is challenging to implement effective public measures that promote the sustainability of urban ecosystems. Analytical and intervention tools, such as surveys, assessments, and planning strategies, must be adaptable. They should integrate the diverse knowledge, experiences, and meanings attributed by the various social groups that shape the urban landscape. Better understanding the representations of spatial and social places – especially by actors who are often marginalised in urban development and planning processes – is fundamental. This study reflects a theoretical and methodological approach dedicated to a sensitive understanding of the complexity of urban reality and its nuances.

Acknowledgements

This study is part of the RUN Project – Citizen Science and Co-Creation for Regeneration of Urban Rivers and Mitigation of Risks, funded by the Ibero-American Cyted Programme (agreement no. 420RT0008) and the preparatory work for the PhD theses in Urban Planning at Universidade Lusófona by Edson K. P. Lourenço.

References

Amaral, Ilídio. 1978. Contribuição para o conhecimento do fenómeno de urbanização em Angola. Finisterra 13(25): 43–80. https://doi.org/10.18055/finis2258.

Amaral, Ilídio. 1983. Luanda e os seus "muceques", problemas de Geografia Urbana. Finisterra 18(36): 293–325. https://doi.org/10.18055/Finis2101.

Anghileri, Daniela, Marco Pastori, Patricia Marcos-Garcia, Gunther Umlauf, Ezio Crestaz. et al. 2024. Global Water Challenges in Sub-Saharan Africa and how to strengthen science-policy dialogues on transboundary governance and cooperation. Journal of Environmental Management 365: 121417. https://doi.org/10.1016/j.jenvman.2024.121417.

Aragão, Nagayamma, Carlos Smaniotto Costa, Rubem Domingos, R., and Job Francisco 2025. The Customary Law and the Traditional Leadership Power in Angola—Their Effects on Territorial Planning Issues. Urban Science 9(6): 207. https://doi.org/10.3390/urbansci9060207.

Asadpour, Ali. 2013. Reading urban environment by photo: a critical tool for socio-cultural analysing. Science, Technology and Arts Research Journal 2(3): 153–159. https://www.ajol.info/index.php/star/article/view/98761.

Cain, Allan. 2018. Informal water markets and community management in peri-urban Luanda, Angola. Water International 43 (2): 205–216. https://doi.org/10.1080/02508060.2018.1434958.

Cao, Yixin, and Karl M. Wantzen. 2023. Understanding public perceptions of the urban riverfront as social-ecological systems: the case of Tours (France). Norois 266(1): 47–70. https://doi.org/10.4000/norois.13111.

Cardoso, Boaventura. 2021. Margens e Travessias. Lisboa: Editora Guerra & Paz.

Casimiro, and Anaice Rosa Barros. 2021. Rio Luanda: análise a uma alternativa de tratamento e distribuição de água em Luanda, Angola. Master diss., Universidade do Porto. https://hdl.handle.net/10216/139391.

Castro, José Caléia, and Paulo Romano Reschilian. 2020. O impacto da informalidade na (re) estruturação das metrópoles periféricas contemporâneas: o caso de Luanda, Angola. Scripta Nova 24 (639). https://doi.org/10.1344/sn2020.24.22658.

Chissola, Adilson Aires Albino. 2015. A Influência do processo de planeamento e gestão territorial na produção do espaço urbano. O caso de estudo da cidade de Luanda. Master diss., Instituto Superior Técnico, Universidade de Lisboa. https://scholar.tecnico.ulisboa.pt/records/irhjlyUUanu4uZMvhfJ1JLMDu0M-UTjwsJeV.

Duncan, James, and David Ley, eds. 1997. Place / Culture / Representation. London: Routledge.

DW – Development Workshop. 2013. Diagnóstico do Escoamento Pluvial em Luanda. https://dw.angonet.org/wp-content/uploads/monografia_-_diagnostico_do_escoamento_pluvial_em_luanda_-_maio_2013.pdf.

Edokpayi, Joshua, N., John O. Odiyo, and Olatunde S. Durowoju. 2017. Impact of Wastewater on Surface Water Quality in Developing Countries: A Case Study of South Africa. Water Quality, edited by Hlanganani Tutu: 401–416. InTechOpen. https://doi.org/10.5772/66561.

Galvão, Joana Rita, and Carla Denise Tedesco. 2022. Contributions of environmental perception to sustainability in the buffer zone of a protected area. Ambiente & Sociedade 25. https://doi.org/10.1590/1809-4422asoc20180262r5r1vu2022L4OA.

Gorski, Maria Cecilia Barbieri. 2008. Rios e cidades: ruptura e reconciliação. Master diss., Universidade Presbeteriana Mackenzie. http://dspace.mackenzie.br/handle/10899/26071.

Guerra, Isabel. 2018. O Bairro Prenda em Luanda: Resiliência social ou resiliência urbana? In Optimistic suburbia 3: Researchers' perspective, edited by Ana Vaz Milheiros, 87–108. Coimbra: AMDJAC.

ICLEI – Local Governments for Sustainability's Cities Biodiversity Center. 2018. Sustainable River-Based Urban Planning Guidelines for Sub-Saharan Africa. Cape Town, South Africa. https://africa.iclei.org/wp-content/uploads/2020/01/2018_Publication_UNA_River-guideline-layout.pdf.

INE. 2016. Resultados definitivos, recenseamento geral da população e habitação – 2014. Instituto Nacional de Estatística de Angola.

Isaac, Ana Maria Rebeca Santos. 2013. Caracterização física do sector intermédio da bacia hidrográfica do rio Ngolo, Lubango, Angola. Master diss., Universidade de Coimbra. https://estudogeral.uc.pt/handle/10316/99850.

Jacob, Cristian. 1982. La représentation de l'espace, projet pour une réflexion théorique. In Espace & Représentation, edited by Alain Renier: 197–221. Paris: Les Editions de la Villette.

João, Delfina Dunn, and Júlia de Nazaré Campos. 2024. Usos da água em Angola: um olhar sobre as comunidades rurais no Planalto Central. Revista Internacional em Língua Portuguesa 45: 69–80. https://doi.org/10.31492/2184-2043.RILP2024.45/pp.69-80.

Jodelet, Denise, ed. 1989. Les représéntations sociales. Paris: Presses Universitaires de France.

Junior, A. de Assis. N. D. Dicionário Kimbundo-Português. Luanda: Edição de Argente, Santos & Cia., Ltd. https://archive.org/details/dicionriokimbu00assiuoft/page/8/mode/2up?view=theater.

King, Anthony Douglas, ed. 1996. Re-Presenting the City – Ethnicity, Capital and Culture in the 21st-Century Metropolis. London: Macmillan Press Ltd.

Lévy, Françoise Paul, and Marion Segaud. 1983. Anthropologie de L'Espace. Paris: Centre Georges Pompidou.

Lourenço, Edson Kiluanje, and Marluci Menezes. 2024. Rios e memória coletiva: que lições a apreender acerca das representações sócio-espaciais do rio na obra 'O Outro Pé da Sereia' de Mia Couto? In *Rios Urbanos na Ibero-América: Propostas teórico-conceituais e metodológicas*, edited by Manoel Rodrigues Alves, Camila Ferreira Guimarães, Diego Ríos, María Manuela Silva Zambrano, and Laura María del Rosario Torres, 69–90. São Paulo: Universidade de São Paulo. https://doi.org/10.11606/9788566624410.

Mbembe, Achille. 2001. As formas africanas de auto-inscrição. Estudos Afro-Asiáticos 23(1): 171–209. https://doi.org/10.1590/S0101-546X2001000100007.

Menezes, Marluci, and Sara Silva. 2023. O papel da memória na recuperação (imaginária) da paisagem fluvial urbana. *In:* Rios Urbanos na Ibero-América: Casos, Contextos e Experiências / Ríos Urbanos en Iberoamérica: Casos, Contextos y Experiencias, edited Carlos Smaniotto Costa, Marluci Menezes, Montserrat Pallares-Barbera, Gabriela Pastor, Eliana do Pilar Rocha, and Klinge O. Villalba Condori, 341–356. Cultura & Território, vol. 6. Lisboa: Edições Universitárias Lusófona. https://doi.org/10.60543/ecati/em3y-zc66.

Menezes, Marluci. 1998. Representações colectivas e reformulações socio-espaciais no bairro da Madragoa. Sociedade e Território, 25/26: 89–105. https://shorturl.at/vXRcS.

Menezes, Marluci. 2000. Do espaço ao lugar: do lugar às remodelações sócio-espaciais. Horizontes Antropológicos 6(13): 156–175. https://doi.org/10.1590/S0104-71832000000100008.

Menezes, Marluci. 2002. Espaço: Manutenção, mudança e representação na Madragoa. ITECS 34. Lisboa: LNEC.

Moscovici, Serge. 1961. La psychanalyse, son image et son public. Paris: Presses Universitaires de France.

Odume, Oghenekaro Nelson, Blessing Nonye Onyima, Chika Felicitas Nnadozie, Gift Ochonogor Omovoh, Thandi Mmachaka. et al. 2022. Governance and Institutional Drivers of Ecological Degradation in Urban River Ecosystems: Insights from Case Studies in African Cities. Sustainability 14(21): 14147. https://doi.org/10.3390/su142114147.

Pellegrino, Pierre. 1982. Espace, représentation de l'espace et négociation narrative des représentations (255–258). *In:* Espace & Représentation, edited by Alain Renier. Paris: Les Editions de la Villette.

Pereira, Álvaro. 2011. Água em Angola: a insustentável fraqueza do sistema institucional. Revista Angolana de Sociologia 8: 63–85. https://doi.org/10.4000/ras.519.

Pereira, Michele Batista. 2008. Do próximo ao distante: a construção do conceito de espaço geográfico em crianças do ensino fundamental. Master diss., Universidade Federal do Paraná. https://hdl.handle.net/1884/15659.

Piaget, Jean, and Barbel Inhelder. 1981. La Représentation de L'Espace Chez L'Enfant. Paris: Presses Universitaires de France.

Renier, Alain. 1982. Introduction au colloque: espace, représentation et semiótique de l'architecture. In Espace & Représentation, edited by Alain Renier, 5–33. Paris: Les Editions de la Villette.

Sánchez, Ana Fuentes, Ana Moura Arroz, and Rosalina Gabriel. 2021. Social representations about sustainable development: the perspectives of residents of small islands' cities. Ambiente & Sociedade 24. https://doi.org/10.1590/1809-4422asoc20180212r1vu2021L3AO.

Sancovschi, Beatriz. 2007. Sobre a noção de representação em S. Moscovici e F. Varela. Psicologia & Sociedade 19(2): 7–14. https://doi.org/10.1590/s0102-71822007000200002.

Santos, Carmen. 2014. Water on the African Continent and Its Use and Benefits for Development: A Vision for Angola. New Water Policy & Practice 1(1): 26–30. https://doi.org/10.18278/nwpp.1.1.3.

Silvano, Filomena. 1997. Territórios da Identidade. Oeiras: Celta Editora.

Song, Shuang, Shuai Wang, Bojie Fu, Yuxiang Dong, Yanxu Liu. et al. 2020. Improving representation of collective memory in sociohydrological models and new insights into flood risk management. Journal of Flood Risk Management 14: e12679. https://doi.org/10.1111/jfr3.12679.

Sousa, Sandra. 2023. O olhar ecocrítico de Boaventura Cardoso em Margens e Travessias. Gragoatá 28 (61): e55687. https://periodicos.uff.br/gragoata/article/view/55687.

Toni, Manuel Caca João, Filipe Muabi Vanda Mazebo, and Fernando Vianeque Agostinho. 2023. Impacto dos padrões de uso e ocupação do solo na qualidade da água do Rio Cubal, perímetro sede, município do Cubal, Benguela, Angola. Brazilian Journal of Development 9 (1): 546–559. https://doi.org/10.34117/bjdv9n1-039.

UNESCO. 2023. The United Nations World Water Development Report 2023: partnerships and cooperation for water; facts, figures and action examples. https://unesdoc.unesco.org/ark:/48223/pf0000384659.

Waldman, Maurício. 1998. Africanidade, espaço e tradição: a topologia do imaginário africano tradicional na crônica 'Griot' sobre Sundjata Keita. Revista África 20/21: 219–268. São Paulo: Centro de Estudos Africanos da Universidade de São Paulo. https://doi.org/10.11606/issn.2526-303X.v0i20-21p219-268.

12

From Flooding to (New) Opportunities

The River City Cesena, Italy, is Turning the Tide

Bahanur Nasya,[1,*] *Yilmaz Vurucu,*[2] *Giulia Turci,*[3] *Emma Veronese,*[4] *Silivia Iacuzzi*[3] and *Mary Sarsam*[1]

1. Introduction

1.1 About the Savio River, Cesena (IT)

Cities are facing great challenges and uncertain future scenarios caused by climate change. Local ecosystems can be fatally impacted by as low as a half degree change in the average temperature (EEA 2022). Formulating sustainable strategies to handle an uncertain future requires a strong collaboration between all relevant actors. The concept of sustainability and what it really means is often determined by local planners and decision makers (Brorström, 2015), requiring evaluation based on the case and specific context. The Emilia-Romagna coastal plain in Northern Italy is characterised by wide areas below mean sea level, in part made up of reclaimed wetlands. The shore and back shore are unstable due to rising sealevels. The multi-layered effects

[1] Wonderland, Vienna, Austria.
Email: projectspace@wonderland.cx, 0009-0002-6043-2174
[2] XsentrikArts, Vienna, Austria.
Email: xsentrikarts@hotmail.com, 0009-0000-4584-1872
[3] Municipality of Cesena, Cesena, Italy.
Emails: turci_g@comune.cesena.fc.it, 0000-0003-0930-0838; Iacuzzi_s@comune.cesena.fc.it
[4] University of Bologna, Bologna, Italy.
Email: emma.veronese3@unibo.it, 0009-0004-2658-0274
* Corresponding author: office@wonderland.cx, 0000-0002-5720-910X

DOI: 10.1201/9781032711737-12

of climate change are expected to increase both shoreline and economic instability in the region. The environmental balance will be increasingly impacted in wetlands, nearby the seaside on dunes, and in the countryside via the ecological balance of the rivers, which control the water flows coming from the mountains and hills. The ecological interconnection of all these elements is crucial for the climate resilience of the area. To handle the vulnerabilities of the region, sustainable interventions by the community and decisionmakers are fundamental (Perini et al. 2017). The region is at risk of storms, high surge levels and high-water spring tidal levels. Risk areas include residential areas and important infrastructure (Perini et al. 2016). Natural areas in the region are relatively smaller in scale and size in comparison to urbanised zones. Yet, they emerge as essential parts of the landscape, playing an important role in environment preservation, and consequently for the future resilience of the region (Armaroli et al. 2012). In the specific case of Emilia-Romagna, human interventions on watercourses have been historically significant and continue to be so to the present day, particularly for economic interests linked to seaside tourism (Grottoli et al. 2020). This constellation of risks falls under one of the European working agendas. The EU Directive (2007/60/EC) has forced every EU country to adopt regulations and laws that manage natural risks (e.g., landslides, river floods, coastal erosion, and inundation), given that these risks strongly affect the vulnerability of the highly populated coastal areas. The mentioned EU Directive was adopted on a national level in 2010 with the DLGS42/2010 (Ambiente 2021). Given the growing awareness of an ongoing environmental and climate emergency, the European Commission issued an official communication regarding the necessity to increase the use of Green Infrastructure (GI), later renamed and complemented as Green and Blue

Figure 1: The Savio River and its territorial context. Source: Veronese, 2024.

Infrastructure (GBI). These actions were considered crucial to achieving the goals of the European Climate Change Adaptation Strategy. As a result of this legislation, the care and recovery of river systems were finally recognised as necessary actions for urban planning. Similarly, the EU Adaptation Strategy (European Commission 2021a) continued to define the action field. Specifically, for the Po Basin (to which the Savio River belongs) a 'Plan for the Management of Flooding Risks' was adopted in 2021.

Cesena, a city of around 96.000 inhabitants, is the leader of the Savio Valley Municipalities Union, composed by Montiano and four small villages located along the Savio River - i.e., Mercato Saraceno, Sarsina, Bagno di Romagna and Verghereto. As seen in Figure 1, Cesena is located river-upstream, at a strategic point of access, and acts as a connecting point between the Apennine hilly area and the plains towards the Adriatic Sea coast. In this sense, the municipality's impact on the sustainability of the region is undeniable. Every city claims its own formula for sustainability, yet there is the consensus that decision makers alone cannot make a city sustainable. The quality of the collaboration of residents and public authorities is decisive in local sustainability work. Cities claim to reach sustainability via differing methods and ways (Gibbs and Krueger 2012), and this collaboration has to be enhanced to different layers on which sustainability needs to be improved.

The starting point is in identifying what sustainability means in this local context and determining the existing gaps, followed by the formulation of a vision and the creation of strategies that help put it into practice along the river. This chapter examines the experience inherent in sustainable practices in the city management. It will illustrate the learnings obtained through these practices so far, and outline how the approach and practical implementations continue to change in contemporary projects. To grasp the context, three questions are elaborated in this text:

- What is the vision for the river and within which framework will it be implemented?
- How did and does the city currently do the sustainability work?
- What roles and opportunities are created for the stakeholders?

1.2 Towards Sustainable Approaches for the Savio River Valorisation

1.2.1 Cesena Strategic Axis and Proposed Specific Objectives in the UGP

In 2019, the municipality of Cesena kicked off the participatory process to settle a new Urban General Plan (UGP), responding to Emilia-Romagna Region planning law (LR 24/2017). The process can be considered as being the main city-level tool to guide Cesena towards an urban development that is sustainable and citizens-centred (Demetra 2017). Following three main principles, namely, to avoid urban sprawl, facilitate urban regeneration, and enhance the environmental value of the territory, the plan puts together the vision and goals for the city. As outlined in Figure 2, the strategic vision foresees five main strategic axes and 18 connected objectives

UGP STRATEGIC AXIS (SA) and SPECIFIC OBJECTIVES (SO)	
SA1. Addressing the climate and environmental challenge	SO1.1. Protect the soil as a resource
	SO1.2. Create green and blue urban infrastructures
	SO1.3. Enhance actions towards climate neutrality
	SO1.4. Improve settlements and facilities resilience
SA2. Make the city attractive	SO2.1. Enhance wide-area services and network
	SO2.2. Enhance historical city center and redeveloping public spaces
	SO2.3. Enhance territory fruition
	SO2.4. Strengthen connections and large-scale infrastructures
SA3. Regenerate and valorise the city	SO3.1. Regenerate and qualify the city
	SO3.2. Promote city re-stitching and connections
	SO3.3. Support temporary and innovative uses of spaces
	SO3.4. Qualify the Building Heritage
SA4. Living in the neighborhoods	SO4.1. Improve connections between neighborhoods and fractions
	SO4.2.Enhance urban quality at the neighborhood scale
	SO4.3. Improve accessibility and efficiency
	SO4.4. Housing and new services and collaborative forms for housing
SA5. Valuing the places and landscapes of production	SO5.1. Qualify the sites and landscapes of industrial production
	SO5.2. Qualify the sites and landscapes of agricultural production

Figure 2: The 5 strategic axes and 19 objectives of the UGP. Source: Comune di Cesena 2021.

(Figure 2) to achieve an integrated and systemic planning in the territory (Comune di Cesena 2021).

The Savio River's ecosystem projects tackle the following strategic axis and specific objectives (Comune di Cesena 2021):

- SA1, SO1.2 Create green and blue infrastructure - supporting the grounding of the environmental and ecological value for the Savio River Park in strict collaboration with the smaller Savio Valley municipalities.
- SA2, SO2.3 Enhance territory fruition - facilitating long-distance cycle paths connecting villages along the Savio River (regional and national routes).
- SA3, SO3.1 Regenerate and qualify the city - promoting the city-river relationship, enhancing local community accessibility and fruition.
- SA3, SO3.2 Promote city re-stitching and connections - potentiating the network of green areas also through the Savio River that acts as an ecological and social connector for the entire urban system.

The Savio River is therefore addressed in three of five strategic axes and creates the basis for envisioning the river as a valuable ecosystem boosting the resilience of the settlements and towns. In part three, there will be an exemplary elaboration on how the city executed the sustainability work up until now.

1.2.2 NEB Impact Model and Qualitative Approach to Assess Project Sustainability

Currently, the New European Bauhaus (NEB) stands out in assessing the importance and value of spatial projects for the society. The NEB is an EU policy and funding initiative launched by the European Commission in 2021, aligned with other EU directives and strategies. The policy allows fostering sustainable solutions for transforming the built environment and lifestyles by utilising the green transition framework (European Union 2021). The pillars of the NEB consist of sustainability, inclusivity, and beauty. These aspects are applicable in a diversity of places, including the Savio River projects. Currently, the partners of the CrAFt project investigate such projects across Europe and have developed the evaluation framework NEB Impact Model, which assesses a project's impact on society. The NEB Impact Model is derived from the experience-based knowledge of dozens of reference cities and analyses resulting from interviews with collaborators from the cultural, artistic, and creative sectors, property owners and tenants, and citizens and communities. Consequentially, the model qualified as a suitable evaluation model for the Savio River.

Utilising the NEB Impact Model (Vandevyvere and Wyckmans 2022) allows assessing the work of the municipality concerning the Savio River, as detailed in section three. The model considers five main intervention domains as the pillars of a sustainability (von Wirth et al. 2019), based on the values of New European Bauhaus (NEB). The consortium of partners respect and include the triple values of sustainable (Vandevyvere 2011) development (planet, people, prosperity), while also making sure to enhance the triple values with quality of life and governance. These five pillars are then fortified with 17 impact categories, which are listed in Figure 3. These will be used in section three to identify the actions of the municipality, stakeholders, and citizens.

NEB pillars and impact categories	
Governance	
1	Integrity
2	Participation and co-creation
3	Process quality
Economic performance	
4	Legal certainty and future economic value
5	Total societal cost of ownership
6	Sustainable local embeddedness
Social-Cultural performance	
7	Cultural sustainability
8	Sociability
9	Affordability and inclusivity
10	Effectiveness of services
Healthy living	
11	Indoor environmental quality
12	Outdoor environmental quality
Environmental performance	
13	Sustainable mobility
14	Sustainable land-use
15	Healthy, secured water cycles
16	Circularity - Materials
17	Climate neutrality - Energy

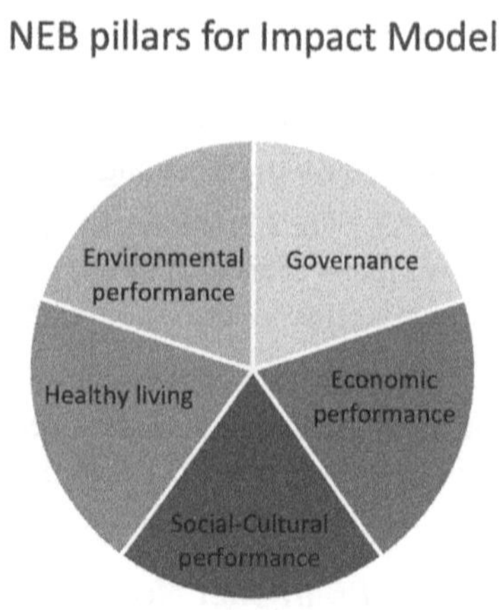

Figure 3: The 5 pillars and 17 impact categories of the NEB Impact Model. Source: Vandevyvere and Wyckmans, 2022.

The 17 categories can help in the assessment of Cesena's experience, while supporting the development of new strategies for and with the Savio River. Based on the NEB Impact Model, it will be possible to discern if the city has a basis to feed from and could do the sustainability work. In line with the UGP strategic axis and goals, four areas of impact are selected for this paper and will be reflected upon in section three. For the meaningful development of the urban strategy, the city must pay particular attention to:

- Climate neutrality and resilience (Section 3.1)
- Sustainable water & land-use (Section3.2)
- Outdoor quality (Section 3.3)
- Participation (Section 3.4)

In the respective sections, the progress achieved will be described and the NEB Impact Model will be utilised to assess the uncovered topics in each field. The basic assessment considers whether the topics have been addressed through existing activities and measures. The purpose is not to assign a rating to the activities, but rather to understand the local framework and ecosystem, identifying the foundational experiences upon which further actions should be developed. In section four, the shortcomings and gaps in local efforts will be identified, and responses to the previously introduced research questions will be provided.

1.2.3 Method

This study employs qualitative analysis as its primary research method, with local activities evaluated using the NEB Impact Model. The four evaluated fields of activity are aligned with the thematic priorities of the UGP developed by the municipality, representing lines of action along the Savio River. These action fields are of considerable relevance, as their effective implementation has the potential to significantly advance sustainability efforts. Climate neutrality and resilience constitute the long-term vision for the municipality. By examining the river-related activities, the study provides a consolidated analysis while establishing a structured approach for future municipal initiatives. The policies stemming from this vision will directly influence water and land resource management, necessitating their integration into a sustainable equilibrium. Accordingly, section 3.2 is dedicated to the spatial dimensions of the vision. The quality of the open spaces play a decisive role in determining feasible actions and uses, particularly connected to specific user groups. Aligning these measures with the broader sustainability objectives will be instrumental in achieving the city's climate goals. Beyond spatial quality and the potential of outdoor areas, the study also evaluates user participation to date, identifying stakeholder expectations and interests that must be harmonised with sustainability targets. In addition to mapping existing capacities, the analysis delves into the weaknesses and barriers in local activities (examined in part four, followed by a series of beneficial conclusions.

2. The Role of the Savio River in the Local Ecosystem

The Savio River, spanning 126 kilometres with a drainage basin of 625 square kilometres, traverses the Emilia-Romagna Region, connecting the provinces of Forlì-Cesena and Ravenna. Within Cesena's territory, the 30-kilometre river axis constitutes a vital ecological resource, providing essential ecosystem services and supporting the local biodiversity significantly. In 2013, Cesena established the "Savio River Park" to enhance its ecological value. However, this initiative was hampered by inadequate governance structures and the absence of a comprehensive management plan. The 2020 political mandate, under the current mayoral administration, repositioned the river at the centre of Cesena's development strategy, particularly regarding the ecological re-connection of the city to its river. This shift was institutionalised through the incorporation of specific river valorisation objectives in the Urban General Plan (UGP) (Figure 2), which proposed the creation of a "protected natural and semi-natural landscape" (Emilia-Romagna Regional laws no. 6/2005, Demetra 2005). This proposal positions the Savio River as a foundation for an ecological corridor and climate-resilient ecosystem, its formal approval remains pending before the regional legislative assembly. Notably, despite the importance of the Savio River as green-blue infrastructure, a coherent strategy and vision for its valorisation has yet to be fully articulated.

The territory of the Savio River has held historical and ecological significance since the Roman antiquity in the 3rd-1st century BC (Brighi 1997). This culturally rich landscape has been extensively documented by scholars like geographer Lucio Gambi (2008) and journalist Antonio Cederna (1998). In the past, the Savio Valley's strategic position as a connector between the Apennines and the Adriatic rendered it vulnerable during the Second World War, when Allied bombing campaigns targeted critical infrastructure, culminating in the construction of a military airfield near Savio (Balsamo 2008). Nevertheless, these disruptions have not diminished the local population's enduring connection to their land and their sustained efforts to improve the environmental quality.

Preliminary empirical analysis of the river's morphological characteristics reveals distinct landscape interactions. The riparian zones are consistently bordered by unpaved roads and multimodal pathways, facilitating access to the primary embankments. The river's course from source at Mount Fumaiolo through the Apennine range to Cesena has profoundly influenced regional settlement patterns and land use practices. Geological studies (Amorosi 2009) confirm that the alluvial plain has experienced intensive anthropogenic modification due to its favourable topography. Within urban areas such as Cesena, engineered embankments interface with residential and industrial developments as seen in Figure 4, while maintaining continuous vegetative cover. Progressing upstream toward the mountainous reaches, the riparian vegetation transition to mature arboreal stands and dense fluvial shrubbery, marking a distinct ecological gradient along the river's course.

The population's longstanding relationship with the river system has been persistently challenged by hydrological volatility, most recently evidenced by a severe flooding event in 2023. The recurrent catastrophes underscore the urgent need

Figure 4: The Savio River in the urban area. Source: Emma Veronese, 2024.

to preserve ecologically intact fluvial systems, while addressing at the same, time the inherent safety concerns. The urban planning priorities necessitate the development of an innovative public-private governance model, which simultaneously safeguards the river's natural integrity and hydraulic security. Such an approach can facilitate community reconnection with this vital resource of the river, while enhancing the technical capacities of volunteer networks, civic associations, and municipal staff in river stewardship. The "Safe and Sustainable City" framework, derived from the Urban General Plan's (UGP) thematic axes, establishes strategic objectives to revitalize the symbiotic relationship between urban development, the Savio River ecosystem, and floodplain restoration. This policy orientation positions the river's ecological and social capital as catalysts for sustainable local development. The proposal foresees a continuous natural park along the river's municipal course. This transformative vision is reconfiguring the waterway as both a functional blue-green infrastructure corridor and an enhanced public space. This integrated planning approach promises to significantly contribute to urban climate resilience, with multiple ecosystem services. At the same time the plan is promoting recreational and cultural engagement with the riparian environment.

3. Local Sustainability Practices

This section examines the practical implementation of sustainable initiatives in Cesena, Northern Italy, and their implications for the future river habitat planning. A systematic review of existing practices serves as a foundation to develop new interventions. The tripartite relationship between the river ecosystem, urban infrastructure, and local community engagement has been analysed on four critical dimensions: (1) climate resilience strategies, (2) sustainable resource management, (3) quality assessment of natural and built environments, and (4) participatory governance models. The subsequent discussion will elucidate the dynamic interrelationship between the Savio River and Cesena's urban development, with particular attention to how this vital watercourse contributes to the region's socio-economic value. Through this examination, the transformative potential was identified.

3.1 Practice 1: Climate Neutrality and Resilience

In the last few years, the Cesena Municipality has been promoting a systemic and integrated planning approach towards a climate neutral and resilient urban environment. To proceed in this direction, a series of emerging needs – such as climate-adaptive and resilient open spaces, a multi-modal and active mobility system, a renovated and energy efficient built environment, as well as a stronger awareness on energy and environmental issues – has to be coordinated and balanced through a thorough inter-departmental collaboration, multidisciplinary competences, as well as structured private-public partnerships with local stakeholders.

The municipality of Cesena has formally ratified several strategic commitments and aligned itself with international networks to address urban transition challenges, as documented in Table 1. This institutional engagement reflects the growing prioritization of environmental considerations within municipal governance, particularly in response to the increasing frequency of extreme weather events. The catastrophic flooding that impacted the Emilia-Romagna region in May 2023 notably affected Cesena, exposing critical vulnerabilities in the current urban framework. Such developments necessitate urgent policy responses, including a fundamental reorientation of urban planning methodologies and environmental management strategies. In this transition framework, two primary intervention zones stick out (Figure 5). The first one is the Vigne-Railway Station district, serving as a testbed for

Table 1: Cesena's connection to national and international sustainability frameworks.

Year	Action	*Resource*	Tasks and Role
2009	Covenant of Mayors		Preliminary target setting
2016	Mayor Adapt	*(European Commission n.d.)*	
2019	SECAP	*Sustainable Energy and Climate Action Plan*	Emission reduction targets, proposal for integrated strategies and actions
2020	Green City Accord	*(European Commisision 2022)*	Italy's first city to target 2030 goals (air quality, water management, urban greening, circular economy, and noise mitigation)
2021	Green City Network	*(Green City Network 2021)*	Italian cities contract towards climate-neutrality
2022	ICLEI network		
2023	Scalable Cities		Observer city
2023	PED-EU-NET	*Cost Action Network*	Case city
2024	Netzero Cities	*Within LET'SGOV Project*	Follower city

climate-neutral urban development; the second is the Savio River corridor, focused on ecosystem resilience and natural capital valorisation.

Climate-neutrality - While Cesena is not among the 100 frontrunner cities committed to carbon neutrality by 2030 (European Commission 2021b), the municipality has recognized the need for a comprehensive and integrated urban vision to advance climate mitigation efforts. At the neighbourhood scale, the Vigne-Railway Station district has been identified as a strategic intervention area for implementing systemic measures aimed at reducing carbon emissions (Figure 5). This transformation will be supported by a combination of European, national, and local funding, facilitating large-scale urban regeneration over the next 5–10 years. The interventions are fourfold. Built environment renewal will address a sustainable retrofitting and redevelopment of existing structures. Energy and mobility systems will deploy flexible and low-carbon infrastructure. With Nature-Based Solutions (NBS) green infrastructure will be integrated in public spaces. Active participation of residents and stakeholders in the transition process is aimed to increase community engagement. These measures collectively aim to establish the district as a testbed for climate-neutral urban development. The re-development of the Vigne-Railway Station district will provide a scalable model for broader municipal and regional sustainability strategies.

Urban resilience - The city of Cesena is leveraging EU-funded initiatives to pilot innovative management approaches and resilience-building interventions along the Savio River corridor (Figure 5). The increasing climate variability manifests through urban heat waves, prolonged droughts, intense precipitation events, and recurrent flooding. In response, the municipality has prioritized climate-adaptive regeneration of urban open spaces. Key mitigation strategies include the implementation of Nature-Based Solutions (NBS), localised greening initiatives, surface de-sealing projects, and Sustainable Urban Drainage Systems (SUDS). These measures aim to reduce climate vulnerability while enhancing local-scale resilience. A significant policy development in this domain is the 2024 *Regulation for Ecosystemic Services* (Comune di Cesena 2024b), enacted under the UGP framework.

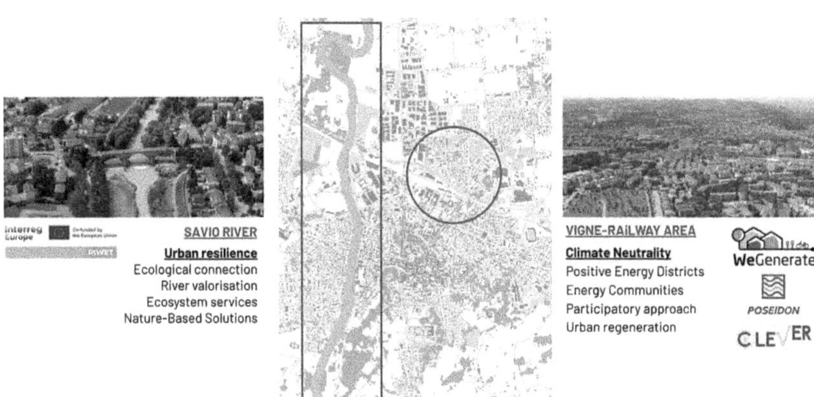

Figure 5: Cesena's pilots towards climate neutral and resilient areas. Source: Giulia Turci, 2024.

This regulatory instrument facilitates climate-responsive urban transformations by institutionalising ecosystem-based approaches to spatial planning. Concurrently, the strategic valorisation of the Savio River as a green-blue infrastructure corridor serves multiple objectives. Strengthening social cohesion is addressed through enhanced public space design, which expands opportunities for outdoor recreation, while preserving and restoring riparian biodiversity. The design has to mitigate urban heat island effects through improved microclimatic conditions. This integrated planning paradigm reflects Cesena's commitment to intertwining ecological preservation with urban development, by systematically incorporating natural systems into the built environment. Synergistic benefits will allow to foster sustainable community engagement, increasing climate adaptation capacity, and safeguarding ecosystem functionality. This way not only long-term urban sustainability, but also the quality of life will be improved.

The existing practices implemented by the municipal authority and its institutional partners have established a robust foundation in governance structures, socio-cultural engagement, and health-related initiatives, as evidenced by the subcategories presented in Figure 6. However, the analysis reveals significant

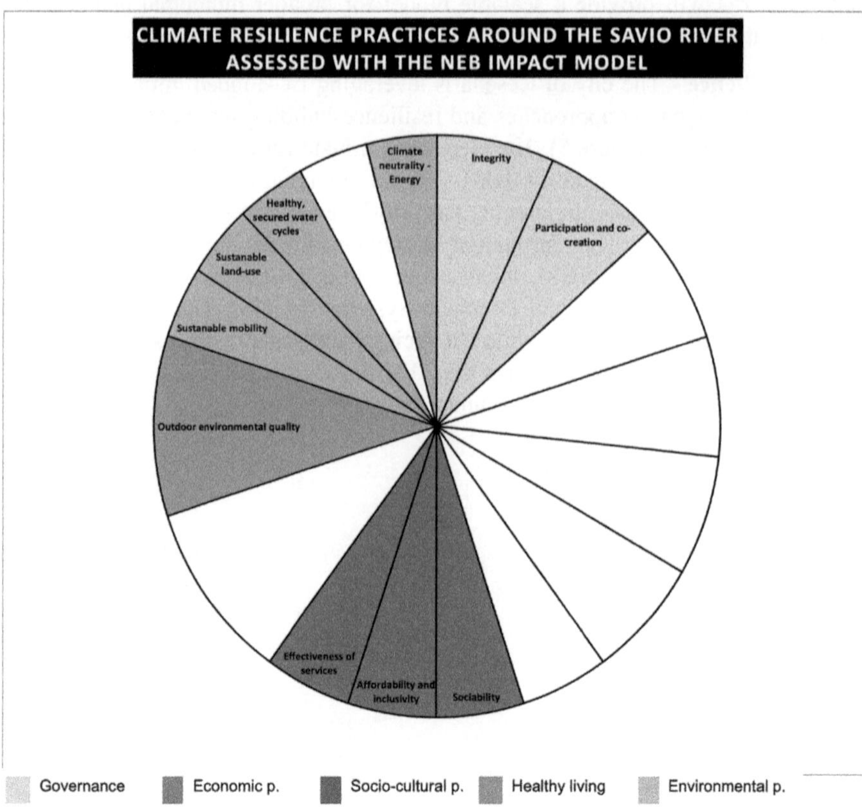

Figure 6: Strengths of the climate actions in Cesena. *The category names are shortened for this graphic. Full names and colours are described in Figure 3.

limitations regarding economic performance and material circularity, as indicated by the empty sectors in the diagrammatic representation.

A critical area requiring improvement concerns process quality within the governance framework. While municipal actors demonstrate commitment to enhancing procedural effectiveness, their capacity to influence external institutions and shape collaborative mechanisms remains constrained. This limitation poses significant challenges for multi-stakeholder and multi-level initiatives (e.g., topics related to regional or national authorities) requiring coordinated action.

Most notably, the current climate action portfolio fails to incorporate economic performance metrics, representing a potentially significant vulnerability to sustain the activities in future. This omission may undermine the medium-term viability and continuity of the Savio River projects, as well as long-term maintenance. This is creating structural weaknesses in project sustainability, suggesting the need for urgent integration of fiscal considerations into climate adaptation planning.

3.2 Practice 2: Sustainable Water & Land-Use

The maintenance of healthy and safe hydrological cycles plays a fundamental role in enhancing both surface and groundwater quality, while simultaneously contributing to climate adaptation, ecological restoration, and flood risk mitigation. Sustainable management of water resources need to be able to encompass potable water, stormwater, and greywater systems. This requires an integrated approach that addresses both environmental and infrastructural dimensions. This necessitates coordinated action among multiple stakeholders, including the Po River Basin District Authority, the Emilia-Romagna Regional Agency for Territorial Security and Civil Protection, the Regional Agency for Environmental Protection (ARPAE), and other relevant institutional actors.

The catastrophic flooding event of May 2023 underscored the urgent need to reform river management strategies to account for evolving climatic conditions. This has prompted a paradigm shift in intervention approaches, with current projects emphasizing climate-adaptive reconstruction rather than simple restoration of pre-existing conditions. The "Special Plan for River Management" (Presidenza del Consiglio dei Ministri 2024), approved in March 2024, establishes strategic guidelines with four objectives. The riverbed needs expansion to accommodate sudden increased flow volumes. The upstream needs flood mitigation measures. Embankment needs redesign which creates a controlled overflow capacity. Additionally, slope stabilisation projects need to be implemented across various topographic zones. Land-use optimisation presents critical considerations regarding functional zoning and landscape-sensitive design.

As noted by geologist Antonio Veggiani (1959, 1971), the Savio Valley has a complex geomorphology, which is characterized by altitudinal variation and diverse soil compositions. The soil types are ranging from coastal sandy-loamy to marly-arenaceous formations. The Cesena plain's clay-dominated soils (Comune di Cesena n.d.) have historically supported agricultural uses. The Savio River corridor can be categorized presently into three primary landscape typologies. The Upstream zone (Borello area) is a transitional landscape between Apennine slopes and initial floodplain. The urban core is adjacent to Cesena's historic centre with significant anthropogenic modification. While the downstream plain is extensively flat agricultural land, with modified hydrology. A notable feature throughout this continuum is the network of pedestrian and cycling pathways ("white roads"), developed through collaborative planning between regional authorities and the Union of Valley Savio Municipalities (I Percorsi del Savio 2024). This infrastructure exemplifies the integration of recreational access with environmental preservation objectives.

The Savio River's historical significance and its designated protection zones demonstrate robust socio-cultural and environmental performance, sustaining high-quality outdoor spaces and well-preserved natural areas. These attributes reflect the enduring value placed on the river's ecological and cultural heritage presented in the Figure 7 with subcategories. While current resource management practices reveal considerable potential, there remains a need for enhanced conceptual frameworks and governance tools to strengthen economic viability and security within the Water Management System, which is indicated by empty sectors of the diagram. A central

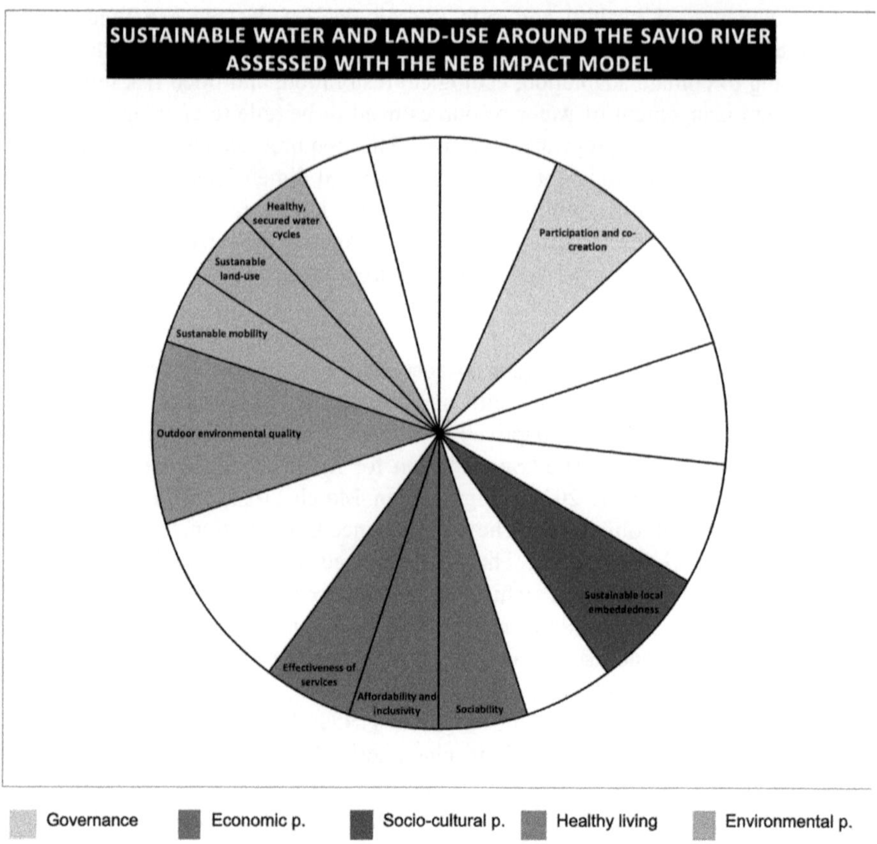

Figure 7: Strengths in the sustainable use of resources in Cesena. *The category names are shortened for this graphic. Full names and colours are described in Figure 3.

tenet of Cesena's approach is the recognition of natural habitats as vital assets, coupled with the active involvement of diverse stakeholders in safeguarding these resources. This principle constitutes a foundational legacy for future governance and stewardship efforts.

3.3 Practice 3: Open Space Quality

The city of Cesena is leveraging EU-funded initiatives to pilot innovative management approaches and resilience-building interventions along the Savio River corridor. The increasing climate variability manifests through urban heat waves, prolonged droughts, intense precipitation events, and recurrent flooding. In response, the municipality has prioritized climate-adaptive regeneration of urban open spaces. Key mitigation strategies include the implementation of Nature-Based Solutions (NBS), localised greening initiatives, surface de-sealing projects, and Sustainable Urban Drainage Systems (SUDS). These measures aim to reduce climate vulnerability while enhancing local-scale resilience. A significant policy development in this domain is the 2024 *Regulation for Ecosystemic Services* (Comune di Cesena 2024b), enacted under the UGP framework. This regulatory instrument facilitates climate-responsive urban transformations by institutionalising ecosystem-based approaches to spatial planning. Concurrently, the strategic valorisation of the Savio River as a green-blue infrastructure corridor serves multiple objectives. Strengthening social cohesion is addressed through enhanced public space design, which expands opportunities for outdoor recreation, while preserving and restoring riparian biodiversity. The design has to mitigate urban heat island effects through improved microclimatic conditions. This integrated planning paradigm reflects Cesena's commitment to intertwining ecological preservation with urban development, by systematically incorporating natural systems into the built environment. Synergistic benefits will allow the fostering of sustainable community engagement, increasing climate adaptation capacity, and safeguarding ecosystem functionality. This way not only long-term urban sustainability, but also the quality of life will be improved.

The environmental quality of open spaces is fundamentally characterized by factors of comfort, health, air quality, and safety - elements that collectively shape their public value. As Jane Jacobs astutely observed, *"...the public place—the sidewalk and street peace—of cities is not kept primarily by the police, ... It is kept by an intricate, almost unconscious, network of voluntary controls and standards among people themselves."* (Jacobs 1961, p. 32) This principle resonates strongly to the community along the Savio River, where outdoor quality has historically emerged through civic engagement rather than municipal oversight. The Savio River waterscape has long served for the citizens as a dynamic centre for recreation and ecological connectivity. Its banks support a range of athletic activities, including cycling, running, and walking. These activities are facilitated by a dedicated riverside path that functions as critical and connecting infrastructure. This pathway not only provides access to sporting facilities (e.g., the Hippodrome area) but also connects key urban nodes, including the city centre and university campus. Beyond its utilitarian functions, the river offers contemplative spaces in contrast to the urban density, increasingly establishing it as an indispensable element of Cesena's cultural and environmental fabric.

Ecologically, the river corridor plays a vital role in regional biodiversity conservation. Its riparian zones host diverse flora and fauna while serving as a hydrological and biological connector between Cesena's green spaces. These attributes amplify its value beyond mere recreation.

The 2023 flood events precipitated profound community impacts, revealing both vulnerability and resilience. In the disaster's immediate aftermath, remarkable civic solidarity emerged, exemplified by the grassroots initiative *"Chi Burdèl de Paciug"* ("The Guys of Mud") as seen in Figure 8 (Casalini and Zanotti, 2023). This volunteer collective embodied adaptive social capital through their debris clearance, resource distribution, and psychosocial support. Such actions not only

Figure 8: The community spirit after the flooding event. Source: Giulia Turci, 2023.

addressed material damage but also reforged community identity. Their community mobilisation demonstrates how environmental shocks can catalyse social innovation and strengthen collective resolve for sustainable recovery.

On the other hand, the flooding precipitated a profound transformation in the community's perception of the Savio River waterscape. The perception of the river was once a locus of leisure and tranquillity, the disaster instilled enduring anxiety, particularly among directly affected populations. The emerged hydrophobia represents a significant rupture in the historical socio-cultural relationship between Cesena's residents and their fluvial environment. Affected communities exhibit detachment behaviours, like reduced waterscape utilisation and altered mobility patterns in riparian zones. These responses mirror post-traumatic place alienation. The new Water Management System needs to address these experiences in the river identity and culture. It needs to combine physical risk mitigation and psychosocial restoration.

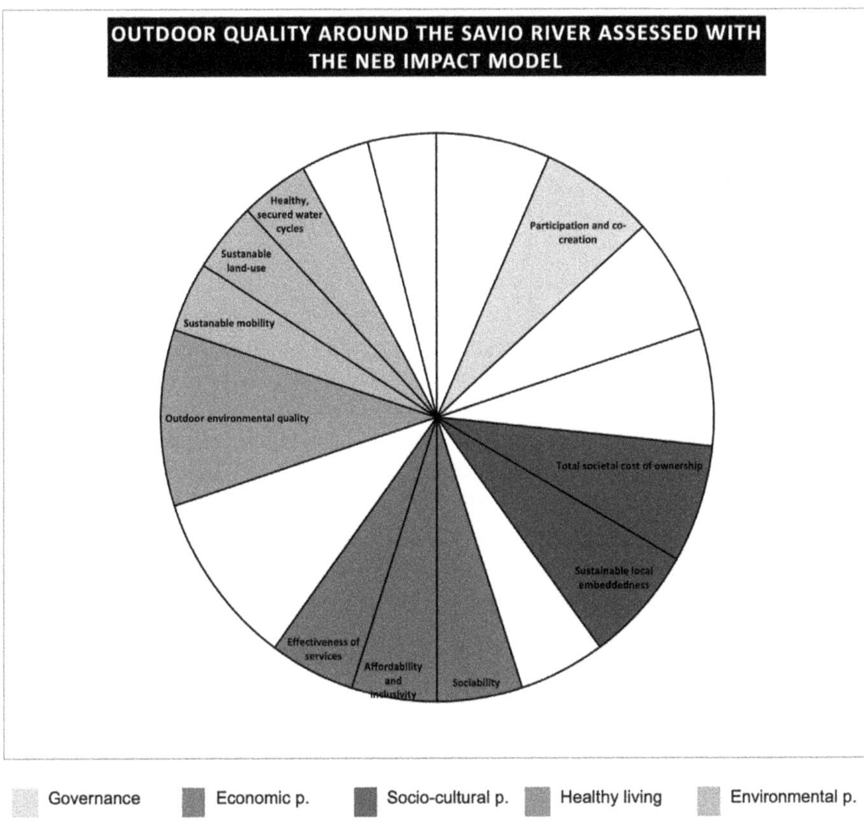

Figure 9: Strengths of the outdoor quality around the Savio River. *The category names are shortened for this graphic. Full names and colours are described in Figure 3.

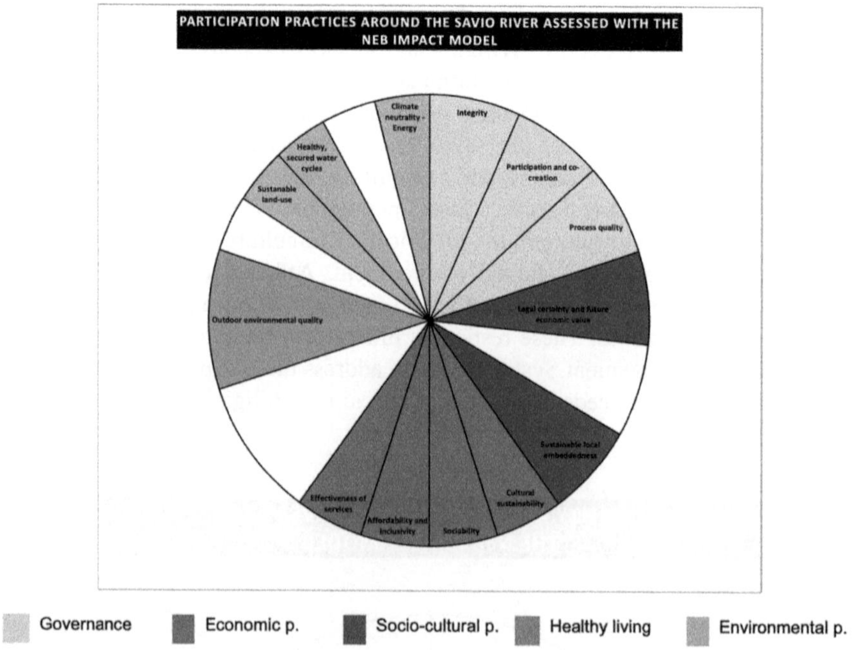

Figure 10: Strengths of participation practices around the Savio River. *The category names are shortened for this graphic. Full names and colours are described in Figure 3.

The outdoor quality of spaces adjacent to the Savio River is strongly shaped by the region's environmental and socio-cultural assets, which collectively foster a high standard of public space liveability. The river corridor benefits from robust ecological functionality and deeply embedded cultural practices, yielding tangible benefits in terms of recreational value, biodiversity, and community well-being. However, governance frameworks remain underdeveloped, constrained by both increasing climate uncertainties and insufficient institutional capacity for landscape stewardship. Climate induced irregularities have exposed vulnerabilities in the current management system. Institutional mechanisms for maintaining and restoring outdoor spaces suffer from fragmented responsibilities, reactive rather than proactive interventions, and limited long-term planning. Despite these structural weaknesses, the region demonstrates exceptional civic engagement in environmental recovery efforts. The profound sense of place attachment among local residents has galvanized collective action, particularly following the 2023 floods, when community-led habitat restoration initiatives emerged as a counterbalance to institutional shortcomings.

This social cohesion presents a critical opportunity for building more resilient governance models, which currently covers only the participation part in the NEB Impact Model as outlined in Figure 9. The mobilisation of local knowledge, volunteer networks, and grassroots organisations in post-disaster recovery suggests a latent capacity for participatory environmental management. If institutional frameworks can better integrate these bottom-up efforts, the Savio River region could develop a more adaptive and inclusive approach to open space governance. The new system can bridge the current gap between strong socio-cultural capital and weak policy implementation. The undertaken analysis highlights significant limitations regarding governance and environmental performance, represented with the empty sectors in the diagram.

3.4 Practice 4: Civic Participation

The cultivation of social and economic capital is a fundamental criterion for sustainable urban development (European Commission 2010). This paradigm recognizes that resident participation in municipal and landscape management constitutes a critical determinant of environmental quality. Empirical evidence suggests that collaborative initiatives yield maximal impact, when originating from grassroots mobilisation rather than top-down implementation (Arnstein 1969). In Cesena, this phenomenon manifests through ongoing bottom-up recreational programs spearheaded by civil society organizations and academic institutions. In this framework two types of organisations stand out. Legambiente, an environmental organisation, has emerged as a principal facilitator of community engagement, exemplified by its April 2024 hydrogeological walk along the Savio River (Corriere Cesenate 2024). This intervention combined environmental education with practical waste management strategies, embodying the organisation's commitment to sustainable place-making. The University of Cesena, an academic institution, has maintained a consistent research focus on regional landscapes since its founding in 1999. With Professor Gianni Braghieri's supervision, numerous river-related theses were compiled since the early 2000s. Additionally, the University of Bologna's 2024-2025 'Design for sustainability' course, led by Professor Beatrice Turillazzi, employs the Savio River as a living laboratory for sustainable infrastructure design (University of Bologna 2024). Also, cross-sector collaborations impact the local participation practices. The 'Laboratorio del Cammino' initiative (2024) represents an innovative transdisciplinary model, blending academic research with community participation through experiential "walking with the water" programs that foster hydro-cultural awareness. Municipal authorities have complemented these efforts through strategic interventions. The annual 'Savio Trail' festival (Redazione Blog 2023) has successfully promoted slow tourism, while local cultural heritage and urban gardening programs (Comune di Cesena 2024a) have addressed dual objectives of food security and civic environmental stewardship. All these initiatives strengthen social cohesion through shared environmental practices, while generating economic value via sustainable tourism and local production. The initiatives also enhance ecological literacy across demographic groups, while creating replicable models for community-based landscape management.

The Cesena case study suggests that effective participatory frameworks require institutional support for grassroots innovation, academic-civil society knowledge exchange, multiscale programming (from hyperlocal gardens to regional trail networks) and continuous evaluation of both social and environmental outcomes. This analysis aligns with contemporary understandings of social-ecological systems (Berkes and Folke 1998), demonstrating how cultural practices and governance structures co-evolve with biophysical environments. The Savio River initiatives particularly exemplify Ostrom's (1990) principles of common-pool resource management, where local knowledge and graduated sanctions maintain sustainable systems.

Historically, civic engagement has fostered certain patterns, an organic development of cultural practices along the Savio River, as demonstrated in Figure 10 with full socio-cultural performance and governance segments. While the grassroots initiatives have largely emerged independently of municipal direction, the organisations demonstrate a well-balanced integration of socio-environmental priorities, providing a substantive foundation for more structured institutional collaboration. However, critical gaps persist in addressing systemic challenges and are left blank in the graphic. The total societal cost of ownership of riverine spaces is not clarified. Material circularity within landscape management practices is not fortified and connected. Sustainable mobility infrastructure and programming is not established. These lacunae suggest the need for enhanced policy frameworks that build upon existing community capital, while introducing more comprehensive sustainability metrics. The municipality's future stewardship efforts would benefit from formally incorporating these dimensions into participatory planning processes, thereby bridging the current divide between emergent cultural practices and structured environmental governance.

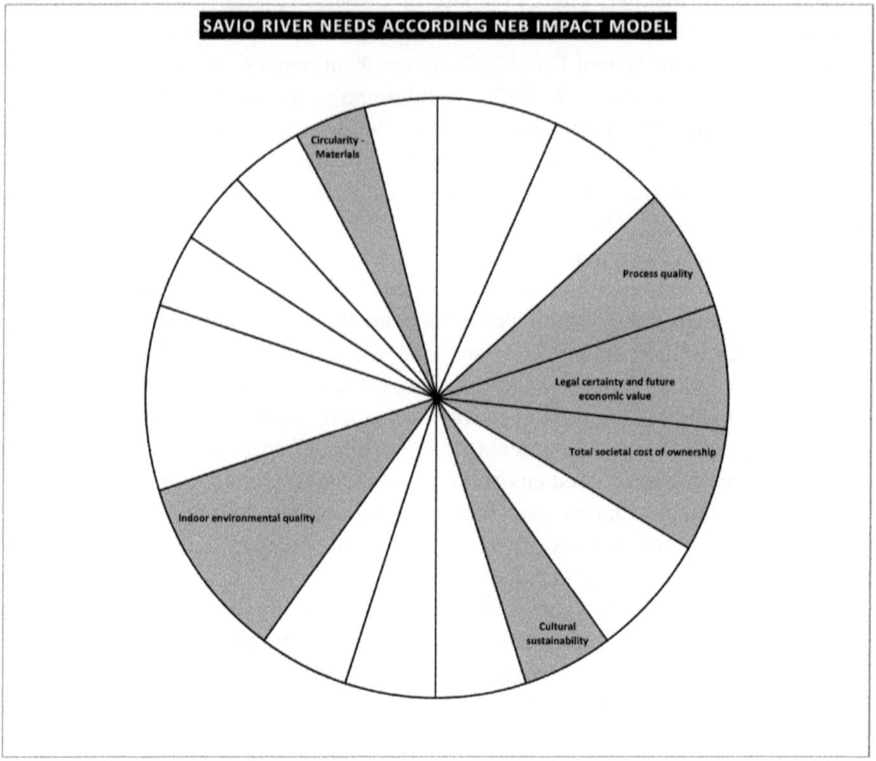

Figure 11: Needs in the Savio River ecosystem combined from all levels.

4. Discussion and Conclusions

Overall, the citizens and municipality have obtained significant experience in the work with and for the Savio River. The environmental, space quality, governance and socio-cultural aspects are often tackled with local actors. There is considerable knowledge and experience, which is a strength and continuing the work will be shaped by these practices. The collaborative efforts between citizens and municipal authorities have yielded substantial experience in managing the Savio River, particularly in addressing environmental quality, spatial organisation, governance frameworks, and socio-cultural dimensions. While this accumulated knowledge provides a strong foundation for future initiatives, significant challenges remain in aligning the diverse actors involved. The development of a cohesive vision and collective pathway is complicated by varying levels of responsibility, capacity, and collaborative experience among stakeholders.

Local initiatives demonstrate robust engagement in climate adaptation, resource management, and participatory governance, providing a solid basis for scalable solutions. However, several systemic weaknesses persist and are highlighted in grey in the Figure 11. Material circularity remains underdeveloped, limiting environmental efficacy. Cultural sustainability is not fully integrated into green space planning, despite its potential to unify stakeholders. Governance processes suffer from inconsistent quality, risking long-term trust in institutions. Economic and legal frameworks lack stability, jeopardizing the feasibility of future projects.

Practices around the Savio River are diverse in terms of climate, resources, outdoor quality, and participation. In most areas of sustainable development, local practices have rich experience or established structures to build impactful actions that can move the region towards a sustainable future. Many experiences can be enhanced into powerful practices and others can learn from each other or create good synergies. However, certain needs manifest in various aspects of the habitat surrounding the Savio River. The circularity of materials is not considered enough and weakens the environmental ambitions. The indoor environmental qualities were not included in the assessments, but these might gain greater significance with future actions. As a green or natural space, the Savio River area is not sufficiently linked to cultural sustainability but could be an umbrella term to integrate most of the local actors. In terms of governance, in most practices the process quality is stated as a weakness and this could impact the trust in governing bodies over time.

Next to these detail aspects, the biggest handicap is the economic challenge. On the one hand, the legal certainty, which guarantees future economic values is very weak. Additionally, the societal cost of ownership will significantly challenge future actions. The adopted vision and Water Management System has to address the economic shortcomings very soon, in order to be effective and impactful. The new management system has to introduce new parameters to bring the different experiences together. Stakeholder engagement and citizen participation is central both in the definition of the vision, but also in a long-term 'care' of the river. Future Action Plans should contain bottom-up practices and foresee roles for citizens. To encompass the duty/responsibility aspect, the municipality envisions to create a series of events focusing on the river and allow low threshold participation and sharing the holistic view on the Management System. The interventions have to go hand in hand with the framework for nature and hydraulic safety.

To achieve this, the Municipality of Cesena must take decisive steps towards better engagement of citizens and local stakeholders on environmental issues. Currently, the governance approach, particularly in areas like river maintenance and flood protection, is mainly top-down (Comune di Cesena 2024a). While this method provides a strong level of control, it often alienates the local community, preventing them from becoming active participants in the process. The review of the past activities and engagement practices, shows that there is a need for systemic citizen empowerment. Local communities can take on stronger roles in the future scenarios, given the proven and reliable partnership in crisis situations. This partnership and the public-private collaboration are particularly important for the success of the river management. Different stakeholders have different relations and expectations, which can be a factor hindering, delaying, or preventing work on the healthy Savio River ecosystems. This would not only affect the city and the citizens of Cesena, but the whole valley and the plain of Emilia-Romagna.

With reference to the research questions in this study, following observations about the practices using these quick assessments can be made.

4.1 What is the vision for the River and in Which Framework will it be Implemented?

One problem is the strive for consensus among different stakeholders and citizens, as there are very different entities and institutional stakeholders, who are responsible for the management of the Savio River. For this reason, it is very important to reflect on past experiences and emotions related to the issue.

4.2 How did and does the City Currently Work on Sustainability?

The rich local experiences and citizen activities need to be better integrated in the actions implemented by the municipality and it is essential to develop a culture to co-create a consensus which works for different parties related to the river. In an environment where there is no consensus, cooperation between people and public institutions is not possible. Currently, there are competing desires, perspectives and demands. For example, some are advocating for the use of Nature-Based-Solutions (NBS) to improve the resilience in and around the river, some are only concerned about the hydraulic safety and others are only focused on naturalizing the river area. A vision and plan to coordinate different interests and demands is needed.

4.3 What Roles and Opportunities are Created for the Stakeholders?

Local sustainability practices will always face various challenges along the way, and the biggest risk is that projects, public funding, and the trust of stakeholders will erode with every failure and discontinuation of these pilot practices. Therefore, the administration has to be very careful in dealing with people's patience and trust. Roles and opportunities for stakeholders and citizens have not been created yet.

Cesena's aspiration to become a sustainable river-city necessitates an innovative management system that balances anthropogenic and ecological needs. The city is trying to turn the tide, with capitalising on the partnerships and collaborations in the region, nation and internationally (Table 1). These partnerships not only allow networking with and learning from other cities and communities, but also allows the city to participate in projects and create additional capacities. Key priorities include mitigating flood risks, advancing sustainable energy solutions, and preserving ecosystem health. A holistic approach incorporating Nature-Based Solutions (NBS), sustainable agricultural practices, and Urban Drainage Systems (SUDS), is critical to achieving these goals while reinforcing nature's role in long-term hydraulic security.

The resilience topic as described and demonstrated with Figure 5, continues to shift the focus towards collaboration. The current top-down governance model, particularly in flood protection and river maintenance, has proven insufficient in fostering community ownership (Comune di Cesena 2024a). A shift toward inclusive decision-making is imperative, leveraging the demonstrated reliability of local actors during crises. The RIWET project (2023–2027) exemplifies this approach, facilitating cross-sectoral collaboration among European regions to address shared challenges in river and wetland restoration. Also, the climate neutrality objective (see figure 5) is localised with the Vigne-Railway Station district and is tackled in manifold ways. Next to the collaboration with different municipalities such is in the PED-EU-NET network (Table 1), new coalitions are formed to bring external expertise such as in the POSEIDON project and its creation of local Urban Living Labs to co-create with the residents and other stakeholders.

In summary, the inclusion of stakeholders and citizens is the most urgent and pressing topic that needs to be explored and strengthened. In this sense, within the projects RIWET and POSEIDON, a participatory process aimed at raising awareness and sense of care towards the Savio River has recently begun. The process will ensure finding and refining a shared vision, one which valorises the river and considers both the Savio River's potential as an ecological connector for the whole city, as well as the vulnerabilities, such as flooding risks. The vision is informed and dependent on the advancing climate change, which will undoubtfully make the future even more unpredictable. Therefore, the vision has to be built on a multiple stakeholder approach, which involves institutions, academia, local associations, schools, as well as private citizens. The new vision is strongly related to the elaboration of a collaboration framework, which will help the city turn the tide towards a more resilient, sustainable, and beautifully built environment.

References

Ambiente. 2021. Piano di gestione del rischio di alluvioni - secondo ciclo. Ambiente. https://ambiente.regione.emilia-romagna.it/it/suolo-bacino/sezioni/piano-di-gestione-del-rischio-alluvioni/piano-gestione-rischio-alluvioni-2021.

Armaroli, C., Ciavola, P., Perini, L., Calabrese, L., Lorito, S., Valentini, A. et al. 2012. Critical storm thresholds for significant morphological changes and damage along the Emilia-Romagna coastline, Italy. Geomorphology, 143-144: 34–51. https://doi.org/10.1016/j.geomorph.2011.09.006.

Balsamo, F. 2008. Un paese nei secoli: Savio, TipoLitografia Scaletta. Ravenna.
Brighi, G. 1997. Le acque devono correre: le centuriazioni fra Rimini, Cervia e Cesena, Il Ponte Vecchio, Cesena.
Brorström, S. 2015. Strategizing sustainability: The case of River City, Gothenburg, Cities, Volume 42: Part A, 25–30. https://doi.org/10.1016/j.cities.2014.08.005.
Casalini, M., and Zanotti, G. 2023. Nel fango. L'alluvione del 2023. I burdél int e' paciug. Il Ponte Vecchio. https://www.mondadoristore.it/Nel-fango-L-alluvione-del-2023-I-burdel-int-e-paciug-na/eai979125978229/.
Cederna A., Gallerani G., and Tovoli C. 1998. In nome del bel paese: scritti di Antonio Cederna sull'Emilia-Romagna (1954-1991). Grafiche Zanini, Bologna.
Comune di Cesena. 2024a. Regional Stakeholder Group Meeting. https://www.interregeurope.eu/sites/default/files/2024-11/RIWET%20Regional%20Stakeholder%20Group%20Meetings%20Cesena_1st%20semester.pdf.
Comune di Cesena. 2024b. Regolamento per la realizzazione e valutazione delle Dotazioni territoriali multiprestazionali e Ecologico Ambientali e delle Compensazioni Ambientali. Comune Di Cesena. https://www.comune.cesena.fc.it/documento_pubblico/regolamento-deam-ca/.
Comune di Cesena. n.d. Elaborati del PUG di Cesena-Montiano. Sititematici.comune.cesena.fc.it. Retrieved November 28, 2024, from https://sititematici.comune.cesena.fc.it/pug/elaborati.
Comune die Cesena. 2021. La strategia per la qualità urbana ed ecologico ambientale. https://sititematici.comune.cesena.fc.it/flex/cm/pages/ServeAttachment.php/L/IT/D/1%252Fb%252F1%252FD.c30e53b60a88caa88d01/P/BLOB%3AID%3D51906/E/pdf.
Corriere Cesenate. 2024. Legambiente, passeggiata idrogeologica lungo il fiume Savio. Corriere Cesenate. https://www.corrierecesenate.it/legambiente-passeggiata-idrogeologica-lungo-il-fiume-savio/.
CrAFt project. Horizon Europe. https://craft-cities.eu/about-the-project/.
Demetra. 2005. Legge regionale n.6 del 2005. Demetra – Normativa, atti e sedute della Regione Emilia-Romagna. https://demetra.regione.emilia-romagna.it/al/articolo?urn=er:assemblealegislativa:legge:2005;6.
Demetra. 2017. Legge regionale n.24 del 2017. Demetra – Normativa, atti e sedute della Regione Emilia-Romagna. (n.d.-b). https://demetra.regione.emilia-romagna.it/al/articolo?urn=er:assemblealegislativa:legge:2017;24.
EEA European Environment Agency. 2022. Exiting the Anthropocene? Exploring fundamental change in our relationship with nature. https://www.eea.europa.eu/publications/exiting-the-anthropocene/exiting-the-anthropocene-exploring-fundamental/pdfStatic.
European Commission. 2022. Green City Accord. Environment.ec.europa.eu. https://environment.ec.europa.eu/topics/urban-environment/green-city-accord_en.
European Commission 2021a. Climate Action. https://climate.ec.europa.eu/eu-action/adaptation-climate-change/eu-adaptation-strategy_en.
European Commission. 2021b. Climate-neutral and smart cities. Research-And-Innovation.ec.europa.eu. https://research-and-innovation.ec.europa.eu/funding/funding-opportunities/funding-programmes-and-open-calls/horizon-europe/eu-missions-horizon-europe/climate-neutral-and-smart-cities_en.
European Commission. n.d. Covenant of Mayors - Europe | Covenant of Mayors - Europe. Eu-Mayors.ec.europa.eu. https://eu-mayors.ec.europa.eu/en/home.
European Union. 2021. New European Bauhaus: beautiful, sustainable, together. New-European-Bauhaus.europa.eu. https://new-european-bauhaus.europa.eu/index_en.
Gambi, L. 2008. La cognizione del paesaggio: scritti di Lucio Gambi sull'Emilia Romagna e dintorni, Bononia University Press, Bologna.
Gibbs, D., and Krueger, R. 2012. Fractures in Meta-Narratives of Development: An Interpretive Institutionalist Account of Land Use Development in the Boston City-Region. International Journal of Urban and Regional Research, 36: 363–380.
Green City Network. 2021. Home - Green City Network. Green City Network. https://www.greencitynetwork.it/.
Grottoli, E., Cilli, S., Ciavola, P., and Armaroli, C., 2020. Sedimentation at river mouths bounded by coastal structures: A case study along the Emilia-Romagna coastline, Italy. *In:* Malvárez, G., and Navas, F. (eds.), Global Coastal Issues of 2020. Journal of Coastal Research, Special Issue No. 95, pp. 505–510.

I Percorsi del Savio. 2024. Sei nella Valle del Savio. Https://Ipercorsidelsavio.it/. https://ipercorsidelsavio.it/.

Jacobs, J. 1961.The Death and Life of Great American Cities. Random House.

Laboratorio del Cammino. 2024. Walking with Water | laboratorio. Laboratorio. https://www.laboratoriodelcammino.com/walkingwithwater.

Perini, L., Calabrese, L., Salerno, G., Ciavola, P., and Armaroli, C. 2016. Evaluation of coastal vulnerability to flooding: comparison of two different methodologies adopted by the Emilia-Romagna region (Italy), Nat. Hazards Earth Syst. Sci., 16: 181–194, https://doi.org/10.5194/nhess-16-181-2016.

Perini, L., Calabrese, L., Luciani, P., Olivieri, M., Galassi, G., and Spada, G. 2017. Sea-level rise along the Emilia-Romagna coast (Northern Italy) in 2100: scenarios and impacts, Nat. Hazards Earth Syst. Sci., 17: 2271–2287, https://doi.org/10.5194/nhess-17-2271-2017.

Piano di gestione del rischio di alluvioni - secondo ciclo. n.d. Difesa Del Suolo, Costa E Bonifica - Ambiente. https://ambiente.regione.emilia-romagna.it/it/suolo-bacino/sezioni/piano-di-gestione-del-rischio-alluvioni/piano-gestione-rischio-alluvioni-2021.

POSEIDON project. DUT. https://poseidon-project.eu.

Presidenza del Consiglio dei Ministri. 2024. Piano Speciale Preliminare. In https://commissari.gov.it/media/j2ie5eyc/piano-speciale-preliminare.pdf.

Redazione Blog. 2023. Savio Trail | I Percorsi del Savio. I Percorsi Del Savio. https://ipercorsidelsavio.it/2023/10/14/savio-trail/.

RIWET project. Interreg Europe. https://www.interregeurope.eu/riwet.

University of Bologna. 2024. 2024/2025 Design per la Sostenibilita'. Unibo.it. https://www.unibo.it/it/studiare/dottorati-master-specializzazioni-e-altra-formazione/insegnamenti/insegnamento/2024/434783.

Vandevyvere, H. 2011. How to cut across the catch-all? A philosophical-cultural framework for assessing sustainability. Int. J. of Innovation and Sustainable Development. 5. 403 - 424. 10.1504/IJISD.2011.043331.

Vandevyvere, H., and Wyckmans, A. 2022. NEB Impact Model. https://craft-cities.eu/wp-content/uploads/2023/07/D1.1-CrAFt-NEB-Impact-Model-updated.pdf.

Veggiani A. 1959. Note sul terremoto di S. Sofia del 1956, Studi Romagnoli n. X/1959, Fratelli Lega, Faenza, pp. 221-235.

Veggiani A. 1971. Le ultime vicende geologiche del territorio cervese e dell'area adriatica antistante, Studi Romagnoli n. XXII/1971 Fratelli Lega, Faenza 1971: 3–17.

Wirth, T., Fuenfschilling, L., Frantzeskaki, N., and Coenen, L. 2018. Impacts of urban living labs on sustainability transitions: mechanisms and strategies for systemic change through experimentation. European Planning Studies, 27(2): 229–257. https://doi.org/10.1080/09654313.2018.1504895.

13

The Game Inventa(rio) Cidades

Raising Multispecies and Environmental Dialogue for Urban Rivers

*Luciana Souza Bragança** and *Marcela Silviano Brandão*

1. Introduction

Architecture and urban planning and landscape architecture, historically based on anthropocentric approaches, neglect non-human agents in their work. Ontological questions have been focused on space and their human users, while nature has been treated as a backdrop or resource for human well-being. These agents, including fauna, flora, water and environmental processes, are often treated as obstacles to eliminating or resources to exploiting. However, in the constitution and maintenance of urban ecosystems their participation is fundamental. Such treatment highlights a reductionist and utilitarian approach to the relationship between humans and non-humans (Latour 2020). In Brazil, a resulting fragmentation between the various actors that shape the urban fabric reflects a systemic disconnection that impacts both the way we learn to plan and intervene in cities, as well as current urban planning practices.

From a socio-ecological point of view, this fragmented perspective poses a substantial challenge, since it ignores the complex interdependence between

Federal University of Minas Gerais, Architecture School, Belo Horizonte, Brazil.
Email: marcelasblarq@gmail.com, ORCID: 0000-0002-5248-5957
* Corresponding author: lubraganca@gmail.com, ORCID: 0000-0001-5707-624X

DOI: 10.1201/9781032711737-13

natural and social systems. By marginalizing non-human agents in urban projects, it perpetuates an approach that runs contradictory to the principles of urban sustainability and resilience. This jeopardizes the ability of cities to tackle contemporary environmental crises such as climate change. This attitude also leads to the loss of various species and the endangerment of urban waters, leading to the progressive loss of biodiversity. In Brazil, where biological, water and cultural diversity is vastly rich, this anthropocentric paradigm not only limits the potential for innovation in urban and landscape planning, but also it contributes to the degradation of ecosystems, exacerbating socio-environmental problems.

The aim of this chapter is to present the methodological tool developed by the Natureza Política Research Group at Architecture School of the Federal University of Minas Gerais (EA-UFMG): the *Inventa(rio) Cidades* game. The game's name is a wordplay in Portuguese, with the present tense of the verb "to invent" (inventa) plus the word "river" (rio) which together sound like inventory. The game is developed for the research projects Possible Gardens and Waters in Cities, in the context of research program Technical Assistance in Territories with Socio-environmental Conflicts. This game is based on the premise that urban projects must overcome anthropocentrism. It is based on the multispecies hypothesis (Bragança 2024, 2023a) and integrates a set of tools developed in the research project Possible Gardens (Bragança 2023b). Through fieldwork and thus backed by empirical data, the research group is establishing its conceptual foundations. The interaction between everyday spaces and theory allows for a deeper understanding of multispecies in urban spaces and the construction of a theoretical approach that reflects the analyzed reality and the developed research methodology. The research and extension group primarily operates in territories marked by socio-environmental conflicts. In this context, the development of the game aims to expand the dialogue with local communities through an approach grounded in empathy. It seeks to foster more inclusive and collaborative interactions in urban planning and intervention processes, guided by the premise of multispecies projects.

2. Anthropocentrism and Multispecies Projects

2.1 Anthropocentrism

In the Western world, the concept of an objective nature, external to the human being, and the idea that the human being does not belong to nature, began to consolidate in the medieval era, intensifying with the advance of industrialization. This dualism was essential to sustaining the notion of human superiority over other forms of life (Latour 2020), as well as over certain groups of human beings, who were considered part of nature from a modern perspective and were therefore treated as 'things' in the context of production. From the modern perspective, nature and culture are often understood as oppositions, with culture being positioned hierarchically above nature, controlling it (Latour 1994; 2020). This approach reinforces the anthropocentrism that permeates the social and territorial imaginary, especially regarding how cities are understood and designed. Scientific advances in the Renaissance reinforced a separation of humans and non-humans, consolidating the view of nature as a resource, in other words, as a subject to be exploited. At the same time, the intensification of global colonization, erased further ways of conceiving the world (Porto-Gonçalves 2016), in which the notion of nature as a category of exclusion made no sense.

This process of subjugation, objectification, and segregation of the "other" was fundamental to the colonial enterprise in the Americas, which was not restricted to the appropriation of territories and resources, but also involved the manipulation of local subjectivities (Mignolo 2003). This phenomenon resulted in the destruction of the common good and the imposition of an anthropocentric and rationalized view of life. In relationship to the nature, the colonization process in Brazil was expressed by perplexity of the Portuguese in the face of the dense forest. Regarded as hostile, this dense forest initially limited the colonizers' incursions, but they ended up by entering and dominating it. The "occupation" of the territory took mainly place through the exploitation of Brazil-wood and constant deforestation (Silva 2002). The confrontation between colonial conqueror and wild nature became a defining feature of the colonization process in Brazil (Silva 2002), in which landscape changes and the extraction of natural resources were central. Various economic cycles based on monoculture followed one another in the formation of the country, with direct implications for cities (Herzog 2013). In this context, the rise of cities ruled deeply against the natural landscape (Silva 2002). Urbanization and territorial expansion also implied the imposition of a European culture on local ontology and non-human elements, resulting in the degradation of original ecosystems. Brazilian urban landscapes are built against exuberant tropical nature, with its plants and animals being only accepted in hyper-domesticated gardens or scenic landscapes.

The urbanization process of modern Brazilian cities intensified the disregard for hydrologic and nature cycles. This phenomenon has consolidated a narrative that understands water either as a threat to be controlled or as a resource to be exploited. Water courses were progressively transformed into waste discharge sites, hidden under urban infrastructure and, consequently, marginalized and forgotten (Herzog 2013). In such dialectical relationships, for a long time the responses of nature were floods and the scarcity of fresh water. Brazil's modernization history shows also the exclusion of the agency for non-humans, as well as writing out of the non-hegemonic cultures. However, both continue to resist, as a non-confrontational resistance that is established through their pure existence. Such existence was evidenced by the research project Possible Gardens.

2.2 Multispecies Hypothesis

The research underpinning the hypothesis of multispecies (Bragança 2023a; 2023b; 2024) and the game *Inventa(rio) Cidades* show various agencies of non-human beings, including water. They point out how this agency can be integrated into the construction of urban spaces, where gardens provide the ground for agencies in a less hierarchical and conflicting way. This understanding was developed based on "everyday" gardens, which as open urban spaces, help shape a landscape where different beings build their common space. Referred as *possible gardens* - in contrast to hyper-domesticated gardens - they were implemented in river landscapes, as along the basin of the Arrudas stream in Belo Horizonte. The research applied multispecies cartography, participant observation and ethnographic interviews to shed light on the relationships between different species, multispecies landscape in water bodies and gardens, and their contribution to sustainability and management of water resources. Ideas about gardens, water bodies, social narratives, ecology and decolonization were mobilized to provide support for the research projects in overcoming anthropocentrism in urban practices. Diverse types of open spaces were analyzed, such as urban public and private gardens, where humans and non-humans interact directly, creating cosmopolitics. The interactions among various species were explored, underscoring the importance of space in fostering more balanced socio-ecological relationships and for a *possible* ecology. Various spaces and agencies were catalogued, along with their motivations and tangible ecological outcomes. These examples were analyzed as multispecies spaces where relationships unfold in a double impact (Stengers 2018), multiple agencies operating in a low-hierarchy structure minimize human dominance. Multispecies studies, or "multispecies anthropology," examine interdependent interactions between humans and other non-human species, including animals, plants, and other organisms. This approach goes beyond human-centered relationships, as it recognizes that cultural practices, knowledge systems, and technologies are shaped by these interactions. By challenging the traditional anthropocentric view, multispecies studies foster a broader understanding of social, cultural, and environmental dynamics (Latour 1994, 2020, Ingold 1995, Tsing 2019), opening new possibilities for projects and research.

Adopting a multispecies approach to exploring gardens along watercourses as territories where humans and non-humans coexist, raises the question about how this approach can influence architecture and urban planning practices. This will serve to reflect on the design of spaces to integrate such coexistence, and on the benefits of applying this approach in the production of space. Beyond disciplinary practices and backed by a comprehensive understanding of the world, architecture, urban planning and landscape design, have the potential to together shape multispecies spaces. This also requires a review of design methodologies. The multispecies hypothesis has been developed and tested in teaching, extension courses, and research projects by the Research Group Natureza Política, resulting in an experimental methodology. Experiences provide a critical approach to space, reasoned on the fact that animals, plants, and hydrologic cycles are active agents in the territory. Unlike traditional approaches, it recognizes the intrinsic agency of water, animals, and plants in the environment. Inspired by the multispecies perspective and insights gained in the field work, this methodological experiment broadens the concept of the right to the city, extending it not only to human beings, but also to other forms of life. Furthermore, the banks of urban rivers are territories of conflict and contention between humans and the environment (Andersen Cirera 2022, Tornieri 2024). The game discussed here was born in the context of multispecies, it is thus developed as a tool for promoting dialogue and generating empathy in territories of conflict. The game is grounded in a sociocultural and dialogic approach between the species' agents based on the centrality of water and urban rivers.

3. Games as dialogic tools

Games play an increasingly key role in various social contexts, offering users the opportunity to practice, explore, experiment, investigate and acquire knowledge in a safe and motivating environment. The act of playing can be understood as entering a 'magic circle' or even creating one as the game begins (Sperhacke and Bernardes 2017). Even after the game is over, it remains in the memory of the participants, turning it into a cultural element. A game provides as a break in everyday life, a temporary experience of fulfilment, described by Johan Huizinga (2000) as entering a 'magic circle'. The author defines this magic circle as a bounded space, either physical or metaphorical where the play takes place. The magic circle may have a material component, such as the game board, or be virtual. Within this magic circle, new ways of understanding or solving problems emerge, making play a significant part of everyday life, both for individual and social values.

Like an urban project, a game is a negotiation process that involves different actors, establishes a structure with defined objectives and helps participants manage complexity in a multifaceted way. *Inventa(rio) Cidades* is a dialogic tool that proposes a playful methodology aimed at engaging riverside communities. The inherent characteristics of games stand out for their great potential in solving real-world problems (Poplin 2012). The theory of gamification explores precisely this aspect by proposing the use of typical game elements and strategies - such as rules, objectives, rewards, penalties and challenges - to engage individuals, promote desirable behavior and tackle concrete problems in a creative way. As highlighted by Alves (2014), gamification can be understood as the process of applying thoughts, mechanisms and dynamics of games to urban contexts, with the aim of stimulating participation, facilitating learning, fostering motivation and ultimately providing solutions to problems. The latter is seen as one of the main potentials of a game (Consalvo 2009, Ribeiro and Falcão 2009, Lehdonvirta 2010, Salen and Zimmerman 2012a, 2012b), as it has an immaterial element at its core, performing a symbolic and meaningful function (Huizinga 2001). When designing spaces, it is possible to explore various correlations and proposals based on the magic circle principles. This concept delimits an environment where specific rules and interactions are temporarily accepted and shared between participants. This is widely used in game studies and experience design, as it allows for the creation of immersive and collaborative environments in which participants engage meaningfully and explore new forms of interaction and communication. In planning projects, the application of the magic circle opens new possibilities for designing spaces that prioritize negotiation and co-creation. This approach enables a dynamic and meaningful spatial experience in which functionality is integrated with a playful dimension, facilitating collaborative exchanges and decisions.

3.1 *Inventa(rio) Cidades* as an Empathy Game

The term "empathy games" refers to a category of games whose main objective is to promote the development of empathy and mutual understanding between the participants (Belman and Flanagan 2010, Diogo 2016). These games are designed to enable players to experience different perspectives and adopt the point of view of others, thus increasing the ability to connect emotionally with the experiences and feelings of others. The structure, objectives and dynamics in such a game, as the *Inventa(rio) Cidades*, mean that each player who gives his/her contribution brings his/her baggage of experiences. Empathetic games are designed to actively promote understanding of the situations presented (Belman and Flanagan 2010). To do this, four fundamental steps are proposed: (1) establish an explicit need for empathy; (2) provide feedback on the impact of the player's actions in the context of the game; (3) stimulate both emotional and cognitive empathy; and (4) emphasize the similarities between the player's reality and the shared experiences.

Empathy games are an effective pedagogical tool for raising awareness of complex social and cultural issues. They can also be used in situations of socio-environmental conflict, as suggested by *Inventa(rio) Cidades*. Prejudice, discrimination or injustice affects people's experiences, when players are exposed to narratives that transcend their own experiences, an empathy game can foster the development of a deeper and more critical understanding of these problems. In the case of *Inventa(rio) Cidades*, these situations also broaden the discussion by the introduction of non-humans as agents in their territories, as proposed in the multispecies projects. Such game immersion enables individuals not only to observe the marginalized groups' realities, but also to indirectly experience the impacts of spatial projects. As a result, empathy games not only promote the development of socio-emotional skills such as compassion and solidarity but also encourage players to reflect on their own social roles and be engaged as agents of change in their communities and spaces. In areas of socio-environmental conflicts and for the preservation of urban rivers, this is an essential action, as it makes their relevance explicit and strengthens the understanding of the consequences of urban projects on river basins.

3.2 Multispecies Indicators for Urban Rivers and the Need to Increase Dialogue

The game *Inventa(rio) Cidades* was developed as a strategy to enrich and ease the understanding of the Multispecies Indicators for Urban Rivers (Bragança 2024), especially in contexts where such indicators are crucial for project development and assessment. These indicators are developed to quantitatively assess the spaces around the riverbanks and guide projects in such sites. The indicators, based on those developed for Cuenca, Ecuador (Hermida et al. 2019), are divided into two main groups: Connectivity and Permanence. The Research Group Natureza Política expanded the number of indicators in these groups by including the multispecies perspective. New indicators include animals, plants, and water as relevant agents for urban design. The study area encompasses in Brazil the entire watershed, and not only the immediate riverbanks as in Cuenca. The new indicators, considering the interaction and coexistence of varied species - humans and non-humans - in the built environment, face communication and interpretation challenges that often hinder their effective use in urban design and planning processes. This is a complex process that, while understood to some degree, is still difficult to grasp fully due to the number and the intricacy of interconnected parts.

In this sense, the game takes on a dialogic role, acting as a mediation tool between different actors, such as architects, urban planners, biologists, ecologists, and the community members involved. It creates a space where these parties can discuss, in a playful and open way, the complexities associated with multispecies indicators, promoting an exchange of knowledge and the collective construction of solutions. Players take active roles in the production of space and can experience problems, for example, suffered by animals. By transforming complex data into interactive and collaborative dynamics, gamification allows the understanding of architecture, environment and sustainability to be expanded beyond technical metrics. It incorporates biodiversity and interspecies coexistence in a more engaging way. From an academic point of view, the use of games as a methodological tool has the potential to help overcome some of the limitations of traditional approaches, which are often technocratic and far from the realities experienced by the concerned communities.

By promoting active participation and co-creation, a game makes it possible to include diverse perspectives in the planning and evaluation process, which in turn can generate more shared results. In addition, playful interaction facilitates the communication of complex concepts, contributing to a greater democratization of technical and scientific knowledge. In this way, the game not only enriches the debate on multispecies but also broadens the horizon for quantified architecture, enabling more sensitive and contextualized approaches. In this regard, a game is also a means of strengthening the critical capacity of communities in relation to architectural projects that impact their territories, encouraging more active participation in decision-making and a broader vision of the role of architecture in balancing human and non-human needs. In the context of teaching architecture and urban planning, games also facilitate the students' understanding of the multispecies approach, as proposed by Research Group Natureza Política. With the aim of promoting empathy, the *Inventa(rio) Cidades* game invites players to experience the urban space as if they were other beings.

4. The game *Inventa(rio) Cidades*

4.1 Objectives and Structure

The *Inventa(rio) Cidades* is a board game, whose aim is to involve players in understanding the consequences of human actions on riverside spaces. Players, as agents, experience the impacts of urban and environmental interventions on riverbanks, making decisions that directly influence the fate of these ecosystems and their inhabitants. Players also experience the consequences suffered by other species. The game is a journey from upstream to downstream, following the urban riverbed. Players face various obstacles and opportunities along the way, including buildings, parks, unpaved streets and other elements that make the journey challenging and full of unexpected events. Each player must rely on luck and on own type attributes to progress; buildings and events can be a help or a trap, and the cards drawn can bring benefits or losses. In the end, the winner will be the one who first reaches the river mouth. The space built along the way reflects directly on the winning character, indicating for whom the city was built.

The players embody different types of agents that inhabit the urban space of Brazilian cities, representing the local ecosystem, these include flying animals like birds and insects, climbing animals like small mammals and reptiles, plants, water, wind and humans (Figure 1). The types of agents have been carefully selected and grouped together based on their specific interactions and actions in the urban space. The cards highlight the contribution each type of agent makes to ecological dynamics and the input they can provide for architecture and urban planning projects.

The game board has four different types of cards (Figure 1), which are placed in squares: Construction Square from where the player draws a card and builds a given space, Permanence, Connectivity Square from where the player draws a card and carries out the action described on it, and Empty Spaces where there is no need to perform any action. The board is not completely built; this happens throughout the game. This construction is like a magic circle for urban, landscaping and architectural projects. Each player, on his/her turn, rolls two dices and follows the instructions of the squares and cards. When landing on a construction square, the player has two options: if the corresponding space is already built, he must follow the environment's rules; if it is not yet built, he/she must build it. All players who pass through this house must follow the rules for the built environment described in the board legend. If the environmental rules provoke a stop at another building, the player must follow

Figure 1: Components of the *Inventa(rio) Cidades* game. Source: Natureza Política archive, 2024.

the rules again. When landing on connectivity or permanence square, the player must follow the instructions drawn on the corresponding card.

As the game progresses, players can either move forward or backward on the board, depending on the effects that buildings and cards have on the environment. By simulating the consequences of these actions, *Inventa(rio) Cidades* encourages a reflection on the impact of urbanization and construction, and the importance of environmental preservation. This improves the understanding that there are other lives in the urban environment besides humans. Once the participants are challenged to critically think about building projects in natural spaces, they develop greater sensitivity and empathy for environmental and social issues.

4.2 The Board as a City and a River

The board represents a water course and its banks to be built up. In a playful manner, the game creates a simulation where players can explore, in an intuitive and interactive way, the dynamics and impacts of urbanization on riverbanks, slopes, valley bottoms, watersheds and springs. By placing participants within a symbolic circle, the game creates an abstraction of environmental and urban issues. It allows players to experience, visualize, observe and reflect on the consequences of their actions in the riverine ecosystem. The game structure with different types of houses and elements to be met along the way, as well as empty spaces to be built according to the construction cards, makes the journey unpredictable. The play moves forward, backward or stands still according to the situation, making explicit the conflict between species within urban spaces. The concept of a magic circle plays a fundamental role in building a symbolic and cognitive representation of the interactions between the city and river environments. Within this magic circle, understood as a metaphorical space of experimentation, new ways of understanding the complex relationship between city and nature are established. By representing the watershed of an urban river, the game board becomes not only a means of entertainment, but a pedagogical tool that facilitates the analysis of real-world problems.

4.3 The Cards

According to the type of agent the player represents, the consequences of the squares and cards are different. The cards (Figure 1) allow players to experiment with different actions and observe their consequences. As previously mentioned, they are directly related to houses, and are construction cards, connection cards and permanence cards. The connectivity cards contain urban connectivity situations, such as planting trees, improving pedestrian access, building of a tall building in the path of prevailing winds. The permanence cards, in the same way, propose or prevent stay in space. They are directly related to the indicators and depict common situations in Brazilian cities. These cards provide positive or negative situations that affect the types of agents and show differences among them to create specific scenarios.

Using these cards, the basin area can be built as the game progresses. Depending on the luck on each player, these cards either help or hinder players in different ways. In this system, the final scenario of the game is built. The main aim is to highlight

Figure 2: The first test of the game. Source: Natureza Política archive, 2023.

the different experiences and perceptions of agents with different spatial elements. Thus, players can get involved with this construction and discuss sustainability, urban planning and environmental preservation in practice, transforming a fictional scenario into an experimental learning platform. In this way, the magic circle enables the creation of a temporary, controlled reality where the interactions between the city and the river can be explored with academic depth and complexity. The various agents, represented by the players, suffer the consequences of construction making it possible to create an atmosphere of empathy and negotiation. This negotiation takes place between different possible worlds such as those of urban plants and animals, pointing towards cosmopolitanism.

4.4 Gaming Experiences

During the development process, the game was tested several times by the research group (Figure 2) before it was applied in different research situations. Five different gameplay sessions provided a broad and diverse experience in terms of the profile and context of the participants. The first took place during the UFMG Exhibition (Figure 3), an event that brings together high school students interested in exploring the university's undergraduate course options. This event allows young people to explore their academic interests, and thus opened an opportunity to test the game, establishing a dialogue between young people's aspirations and the content explored. The main issues raised were related to the university entrance qualification. The

Figure 3: The first play outside the research group, with students. Source: Natureza Política archive, 2023.

Figure 4: The gameplay session with high school students. Source: Natureza Política archive, 2023.

competition between the different played agents created engagement among the students.

The second gameplay session was carried out with the support of the geography teacher with high school students of a private school with a constructivist-based pedagogy, which values student's autonomy and protagonism (Figure 4). In this environment, students were able to interact with the game in a reflective and independent way. They were encouraged to develop a critical perception of the issues

Figure 5: A gameplay session at the Cuenca University, Ecuador. Source: Natureza Política archive, 2023.

addressed. In this gameplay, discussions about empathy could be addressed in greater depth and complexity than in previous contexts, involving more detailed reflections, examples, and arguments, moving beyond a superficial understanding of this issue.

The third gameplay session involved students and researchers in Cuenca (Figure 5). Recalling that the indicators developed by the Cuenca researchers served as a reference for the development of multispecies indicators and the *Inventa(rio)*

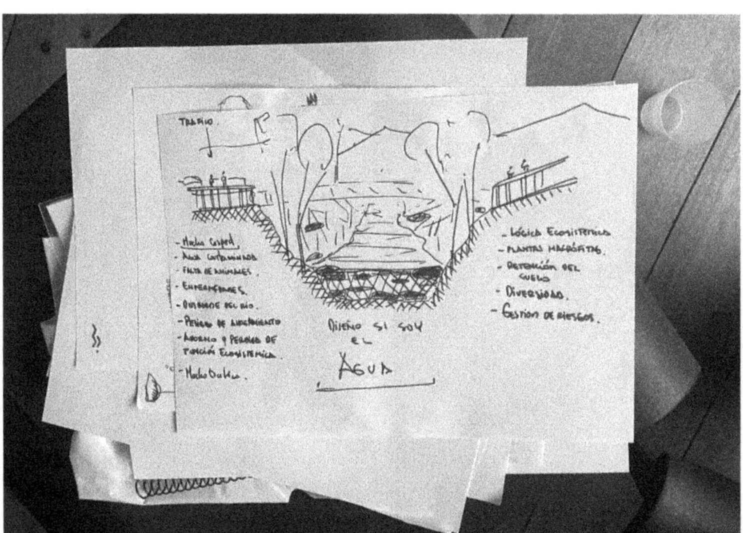

Figure 6: A proposal resulting from discussions in the gameplay session at the University of Cuenca. Source: Natureza Política archive, 2023.

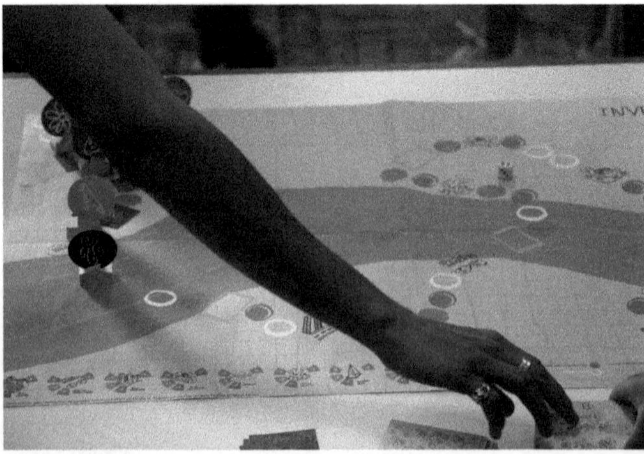

Figure 7: The gameplay at the Seminar on Political Nature. Source: Natureza Política archive, 2024.

Cidades game. This gameplay was especially enriching, as the participants had enough knowledge about the thematic issues being addressed, enabling them to offer informed and discerned perspectives to the game dynamics. Participants were also instigated build scenarios using the Multispecies Projects methodology (Figure 6), in which framework the game was also developed.

The fourth gameplay session was similar to the previous one in Cuenca. The game was played with participants of the Second International Seminar on Political Nature: Practical Cosmopolitics in Motion (Figure 7). The seminar brought together public managers, students, and social movements activists and was focused on socio-environmental conflicts (Figure 8).

Figure 8: Proposals resulting from discussions. Source: Natureza Política archive, 2024.

Figure 9: The gameplay session in the course "Multispecies Projects for Urban Rivers". Source: Natureza Política archive, 2024.

Figure 10: A student drawing a proposal based on the discussions in the course. Source: Natureza Política archive, 2024.

In the fifth gameplay session, the game was used in the course "Multispecies Projects for Urban Rivers" as a tool to support the projects developed by the students (Figure 9). In this academic environment, the game not only raised awareness of the complexity of interactions between species in the urban fabric but also provided a conceptual and methodological basis for developing projects for urban water ecosystems. Students were able to explore, in a practical way, topics such as cohabitation between human and non-human species and the preservation of water bodies. This playful and reflective interaction helped to understand the relevance of the multispecies project methodology, promoting innovative solutions to strengthen ecological and social relationships in territories experiencing socio-environmental conflicts (Figure 10).

In all sessions, the participants assumed simultaneously a playful and thought-provoking agent type. They were challenged to reflect on problems of contemporary cities, which are often designed and perceived as hostile for non-human species and water bodies. Through the game, this perception could be intensified, promoting collective awareness of the importance of a more integrated and respectful coexistence between humans and non-humans in the urban environment. Besides, explaining and reinforcing the role of architecture and urban planning in this process.

5. Conclusions: Raising a Multispecies Dialogue for Urban Rivers

The game *Inventa(rio) Cidades* proved to be a relevant tool for fostering ecological awareness and empathy. By simulating real-world situations with ecological and urbanization problems, the game allows participants to interact directly with the complexity of urban environments. The game seeks to integrate multiple perspectives, encouraging players to reflect on the consequences of their actions, especially from the point of view of non-human life, such as fauna, flora and ecosystems. This approach leads participants to experience and reflect on issues that may seem too abstract or distant in their daily life. In this way, the game provides a tool for discussion that is both playful and provocative, while highlighting the hostility of urban settlements for non-humans. This dynamic makes learning more immersive, promoting an experience that combines entertainment, the development of social and ecological skills, and above all to raise territorial awareness.

Integrating the concept of the 'magic circle' in empathy games, especially in conflicting contexts, involves reflecting on how such imaginative space facilitates the understanding and valuing diverse views, including the needs of non-human agents, which are often ignored in spatial conflict negotiations and urban agendas. Bringing non-human perspectives through empathy games opens up the urban spaces for a more complex and ethical representation of the environment and the living beings that inhabit it. Players are instigated to consider the needs of these agents, questioning the impact of human actions, while recognizing that these needs are legitimate and interdependent. This can contribute to a change of perspective in architecture, urban planning and landscape projects, leading to the development of more integrated solutions. In particular, when the cosmopolitical approach as applied in the multispecies project serves as guidance. The multispecies project within the game was developed, calls for considering not only human interests, but also multiple agents as active and critical parts of the process in decisions about space, environment and land use. Rather than seeing nature as a backdrop or a resource to be exploited, participants are encouraged to recognize the value and right of every non-human agent to contribute to decisions. The multispecies approach can influence architecture and urban planning practices by taking inspiration from gardens along watercourses, as they are territories where humans and non-humans coexist. The proposals developed with support of the game reflect this perspective, incorporating greenery, permeable surfaces, and food-providing vegetation to foster multispecies coexistence and ecological resilience.

As an empathy game, *Inventa(rio) Cidades* serves as both an educational tool and a tool for raising awareness in academic, community or public policy contexts. By focusing on environmental challenges, the game promotes a creative approach that encourages the search for innovative solutions to complex socio-environmental problems. In this context, architecture, urban and landscape planning, as disciplines that create spaces, have an important role to play, as they can drive the dialogue about and for urban rivers, broadening the socio-environmental perspective in these spaces.

Acknowledgements

Many people contributed to the research presented in this chapter. The authors acknowledge PRPQ/UFMG, FAPEMIG, CNPQ for funding and thank the students who engaged in the game development, and CYTED RUN research network. The authors would like also to thank several players: students, workshop participants, government officials, public councilors and the community members.

References

Alves, F. 2014. Gamefication: como criar experiências de aprendizagem engajadoras: um guia completo. São Paulo: DVS.

Andersen Cirera, K. (2022). Spatial equity in river access. Measuring the public space potential of urban riverbanks in Valdivia, Chile. International Journal of Sustainable Development and Planning, 17(1): 1–12. https://doi.org/10.18280/ijsdp.170101.

Belman, J., and Flanagan, M. 2010. Designing games to foster empathy. International Journal of Cognitive Technology 15.1 (2010): 11.

Bragança, L. S. 2024. Possible Gardens and the waters in the city: research and teaching. Cadernos De Pós-Graduação em Arquitetura E Urbanismo 24(1): 169–184. https://doi.org/10.5935/cadernospos.v24n1p185-199.

Bragança L. S. 2023. Possible Gardens: cosmopolitical worlds. Frontiers Environmental Sci. 11: 1234178. https://doi.org/10.3389/fenvs.2023.1234178.

Bragança, L. S. 2023a. Projeto multiespécies de paisagens e águas. Proc. QUAPÁ_CEL. Available at: https://www.researchgate.net/publication/384602926_PROJETO_MULTIESPECIES_DE_PAISAGENS_E_AGUAS_MULTISPECIES_LANDSCAPES_AND_WATERS_PROJECT_A_multispecies_approach_to_communities_in_urban_open_space_projects_based_on_teaching.

Bragança, L. S. 2023b. Jardins como possibilidade, Coleção Jardins Possíveis, vol. 1. Paisagens Híbridas, Rio de Janeiro.

Consalvo, M. 2009. There Is No Magic Circle. Games and Culture, 4(4): 408–417. http://dx.doi.org/10.1177/1555412009343575.

Diogo, G. 2016. Empathy Games: Life, Death, and Digital Narratives, Radboud University.

Herzog, C. P. 2013. Cidade para todos: reaprendendo a conviver com a natureza. Mauad, Rio de Janeiro.

Huizinga, J. 2000 [1938]. Homo Ludens: o jogo como elemento da cultura. São Paulo, Perspectiva.
Ingold, T. 1995. Humanidade e Animalidade. Revista Brasileira de Ciências Sociais, 28(10):1–15.
Latour, B. 1994. Jamais fomos modernos: Ensaio de antropologia simétrica, Rio de Janeiro: 34 Editora.
Latour, B. 2020. Diante de Gaia: Oito conferências sobre a natureza no antropoceno. Ubu: São Paulo/Rio de Janeiro.
Lehdonvirta, V. 2010. Virtual Worlds Don't Exist: Questioning the Dichotomous Approach in MMO Studies. Game Studies: the international journal of computer game research, 10(1). Disponível em: htt p://gamestudies.org/1001/articles/ lehdonvirta. Acesso em: 20/11/2014.
Mignolo, V. 2003. Histórias locais, projetos globais. Editora UFMG: Belo Horizonte.
Poplin, A. 2012. Playful public participation in urban planning: A case study for online serious games. Computers, Environment and Urban Systems, 36(3): 195–206. https://doi.org/10.1016/j.compenvurbsys.2011.10.003.
Porto-Gonçalves, C. W. 2016. Lucha por la tierra, lucha por la Tierra - ruptura metabólica y reapropiación social de la naturaleza. Revista de Pensamiento Cristiano - Iglesia Viva, vol. 2: 267.
Ribeiro, J., and Falcão, T. 2009. Mundos Virtuais e Identidade Social: Processos de Formação e Mediação através da Lógica do Jogo. Logos, 16(1): 84–96.
Salen, K., and Zimmerman, E. 2012a. Regras do Jogo: Fundamentos do Design de Jogos. São Paulo, Blucher, vol. 1.
Salen, K., and Zimmerman, E. 2012b. Regras do Jogo: Fundamentos do Design de Jogos. São Paulo, Blucher, vol. 3.
Silva, M. A. 2002. Histórias de paisagens: a natureza verde e o surgimento das vilas e cidades coloniais do Brasil. Revista Locus (UFJF) 13(1). https://periodicos.ufjf.br/index.php/locus/article/view/20554.
Sperhacke, S. L., and Bernardes, M. M. S. 2017. O processo de ludificação: como transformar métodos de design em jogo de tabuleiro?. *In:* Design em Pesquisa, 1st ed., 1: 270–294. Porto Alegre: Marcavisual.
Stengers, I. 2018. A proposição cosmopolítica. Revista do Instituto de Estudos Brasileiros, 69: 442–464.
Tsing A. 2019. Viver nas ruínas: paisagens multiespécies no Antropoceno. IEB Mil Folhas, Brasília.
Tornieri, S. (2024). Ordinary vs. Extraordinary: An Urban Comparison in the Delta Po Area. Urban Planning, 9, Article 8271. https://doi.org/10.17645/up.8271.
Vincent F. A., and Dan, Z. 2020. Basic empathy: Developing the concept of empathy from the ground up. International Journal of Nursing Studies (2020): 103695.

Index

N

O

P

R

S

T

U

V

W

Y

For Product Safety Concerns and Information please contact our EU
representative GPSR@taylorandfrancis.com
Taylor & Francis Verlag GmbH, Kaufingerstraße 24, 80331 München, Germany

www.ingramcontent.com/pod-product-compliance
Ingram Content Group UK Ltd.
Pitfield, Milton Keynes, MK11 3LW, UK
UKHW022210190926
475702UK00003B/289
* 9 7 8 1 0 3 2 6 8 8 6 9 5 *